Preparation and Application of Advanced Functional Membranes

Preparation and Application of Advanced Functional Membranes

Guest Editors

Annarosa Gugliuzza
Cristiana Boi

Basel • Beijing • Wuhan • Barcelona • Belgrade • Novi Sad • Cluj • Manchester

Guest Editors

Annarosa Gugliuzza	Cristiana Boi
National Research Council (CNR)	DICAM
Institute on Membrane Technology (ITM)	Alma Mater Studiorum– Università di Bologna
Rende	Bologna
Italy	Italy

Editorial Office
MDPI AG
Grosspeteranlage 5
4052 Basel, Switzerland

This is a reprint of the Special Issue, published open access by the journal *Membranes* (ISSN 2077-0375), freely accessible at: www.mdpi.com/journal/membranes/special_issues/SU3I891N24.

For citation purposes, cite each article independently as indicated on the article page online and using the guide below:

Lastname, A.A.; Lastname, B.B. Article Title. *Journal Name* **Year**, *Volume Number*, Page Range.

ISBN 978-3-7258-3260-6 (Hbk)
ISBN 978-3-7258-3259-0 (PDF)
https://doi.org/10.3390/books978-3-7258-3259-0

© 2025 by the authors. Articles in this book are Open Access and distributed under the Creative Commons Attribution (CC BY) license. The book as a whole is distributed by MDPI under the terms and conditions of the Creative Commons Attribution-NonCommercial-NoDerivs (CC BY-NC-ND) license (https://creativecommons.org/licenses/by-nc-nd/4.0/).

Contents

About the Editors vii

Preface ix

Annarosa Gugliuzza and Cristiana Boi
Editorial for the Special Issue "Preparation and Application of Advanced Functional Membranes"
Reprinted from: Membranes 2024, 14, 100, https://doi.org/10.3390/membranes14050100 1

Stefanos (Steve) Nitodas, Meredith Skehan, Henry Liu and Raj Shah
Current and Potential Applications of Green Membranes with Nanocellulose
Reprinted from: Membranes 2023, 13, 694, https://doi.org/10.3390/membranes13080694 11

Giovanni Ceccio, Jiri Vacik, Jakub Siegel, Antonino Cannavó, Andrey Choukourov and Pavel Pleskunov et al.
Etching and Doping of Pores in Polyethylene TerephthalateAnalyzed by Ion Transmission Spectroscopy and Nuclear Depth Profiling
Reprinted from: Membranes 2022, 12, 1061, https://doi.org/10.3390/membranes12111061 27

Anna Strzelewicz, Monika Krasowska and Michał Cieśla
Lévy Flights Diffusion with Drift in Heterogeneous Membranes
Reprinted from: Membranes 2023, 13, 417, https://doi.org/10.3390/membranes13040417 37

Giuseppe Di Luca, Guining Chen, Wanqin Jin and Annarosa Gugliuzza
Aliquots of MIL-140 and Graphene in Smart PNIPAM Mixed Hydrogels: A Nanoenvironment for a More Eco-Friendly Treatment of NaCl and Humic Acid Mixtures by Membrane Distillation
Reprinted from: Membranes 2023, 13, 437, https://doi.org/10.3390/membranes13040437 48

Dmitrii Butylskii, Vasiliy Troitskiy, Daria Chuprynina, Lasâad Dammak, Christian Larchet and Victor Nikonenko
Application of Hybrid Electrobaromembrane Process for Selective Recovery of Lithium from Cobalt- and Nickel-Containing Leaching Solutions
Reprinted from: Membranes 2023, 13, 509, https://doi.org/10.3390/membranes13050509 66

Niki Joosten, Weronika Wyrebak, Albert Schenning, Kitty Nijmeijer and Zandrie Borneman
On the Performance of a Ready-to-Use Electrospun Sulfonated Poly(Ether Ether Ketone) Membrane Adsorber
Reprinted from: Membranes 2023, 13, 543, https://doi.org/10.3390/membranes13060543 80

Ana Sofia Figueiredo, Ana Maria Ferraria, Ana Maria Botelho do Rego, Silvia Monteiro, Ricardo Santos and Miguel Minhalma et al.
Bactericide Activity of Cellulose Acetate/Silver Nanoparticles Asymmetric Membranes: Surfaces and Porous Structures Role
Reprinted from: Membranes 2022, 13, 4, https://doi.org/10.3390/membranes13010004 97

Luís Pereira, Frederico Castelo Ferreira, Filipa Pires and Carla A. M. Portugal
Magnetic-Responsive Liposomal Hydrogel Membranes for Controlled Release of Small Bioactive Molecules—An Insight into the Release Kinetics
Reprinted from: Membranes 2023, 13, 674, https://doi.org/10.3390/membranes13070674 119

Quan Liu, Zhonglian Yang, Gongping Liu, Longlong Sun, Rong Xu and Jing Zhong
Functionalized GO Membranes for Efficient Separation of Acid Gases from Natural Gas: A Computational Mechanistic Understanding
Reprinted from: Membranes 2022, 12, 1155, https://doi.org/10.3390/membranes12111155 147

Daniel Polak and Maciej Szwast
Analysis of the Influence of Process Parameters on the Properties of Homogeneous and Heterogeneous Membranes for Gas Separation
Reprinted from: *Membranes* **2022**, *12*, 1016, https://doi.org/10.3390/membranes12101016 . . . . **160**

Ziqi Cheng, Shen Li, Elena Tocci, Giacomo Saielli, Annarosa Gugliuzza and Yanting Wang
Pathway for Water Transport through Breathable Nanocomposite Membranes of PEBAX with Ionic Liquid [$C_{12}C_1$im]Cl
Reprinted from: *Membranes* **2023**, *13*, 749, https://doi.org/10.3390/membranes13090749 . . . . . **180**

Geani Teodor Man, Paul Constantin Albu, Aurelia Cristina Nechifor, Alexandra Raluca Grosu, Szidonia-Katalin Tanczos and Vlad-Alexandru Grosu et al.
Thorium Removal, Recovery and Recycling: A Membrane Challenge for Urban Mining
Reprinted from: *Membranes* **2023**, *13*, 765, https://doi.org/10.3390/membranes13090765 . . . . . **194**

About the Editors

Annarosa Gugliuzza

Annarosa Gugliuzza is a senior researcher at the National Research Council of Italy. She holds a degree with honors and a Ph.D. in Chemistry. She was a visiting researcher at the University of Ann Arbor, MI, USA, and POLYMAT, San Sebastián, Spain. Qualified as a full professor in Principles of Chem for Appl Technol (Academic field 03/B2), she is a lecturer and member of the Doctoral School Committee for the STFCM PhD program at the University of Calabria, Italy. Skilled in nanomaterials and nano-assembly technologies, her current research interests are mainly in membrane design dedicated to water desalination and reuse, agrofoods, environmental remediation, gas separation, and smart surfaces and interfaces. She has been the coordinator/PI of scientific projects on water desalination and breathable textiles. She has also been involved in numerous international and national projects focused on purification and separation practices for natural resources, and energy and environment management. She is a member of the scientific advisory committee for the CHross LAB of Chemical Sciences and Materials Technology, CNR-DSCTM, for cultural assets. She has been chair of workshops and co-chair within sessions of various international conferences on membranes and membrane processes. As an expert evaluator, she serves funding agencies such as ERC GRANT-EU and EWI-Singapore. Engaged in many editorial activities, she serves as an associate editor and editorial board member of various scientific journals and is an editor of four books on membranes and related technologies.

Cristiana Boi

Cristiana Boi is a research associate professor of chemical engineering at Alma Mater Studiorum, Università di Bologna, Italy, and a research associate professor in the Department of Chemical and Biomolecular Engineering at North Carolina State University in Raleigh, NC, USA. She holds the Italian Habilitation as a full professor for the academic field 09/D2—Systems, methods, and technologies of chemical and process engineering.

She graduated in chemical engineering from the University of Bologna in 1991 and obtained her Ph.D. in chemical and environmental engineering from the University of Bologna in 1996, with a joint project with Rensselaer Polytechnic Institute, Troy, NY, USA. Her research interests are in downstream processing in biotechnology with a focus on membranes and chromatographic processes for highly selective separations. In particular, she is interested in the preparation, modification, and characterization of functional materials for bioseparations, biomedical applications, water purification, and the production of biopharmaceuticals.

Cristiana Boi has published 50 articles in high-impact peer-reviewed journals, 3 book chapters, and presented more than 130 papers at international conferences. She is an editor for *Separation and Purification Technology* and *Membranes*.

She has supervised/co-supervised 9 Ph.D. students and 74 M.Sc. students and is currently supervising 5 Ph.D. students and 8 M.Sc. students. She has been the president of the European Membrane Society for 2017 and 2018, the funding secretary of the World Association of Membrane Societies for 2017–2020, and is the current chair of Area 2G, Bioseparations, of the AIChE.

Preface

This reprint is a collection of research studies on advances in membrane design and process, which fit into a sustainability framework where theoretical and experimental findings are targeted to the optimization of natural resource management, environmental remediation, recovery, and recycling of value-added products. Reading this compendium, one discovers how membranes can also be fabricated from bio-based and biodegradable materials, whose combination leads to enhanced process engineering designed to make purification and separation practices much more competitive for eco-friendliness, efficiency, and productivity. Case studies on water purification and reuse, chemical and biological pollutant removal, drug delivery, hydrogen recovery, and CO_2 capture are proposed through the synergic interplay of modelling, experimental activity, and discussion beyond the state of the art.

Authors from Europe, the USA, and Asia gave a strong contribution to raising interdisciplinary knowledge and environmental outreach through efforts connected with the proposal of innovative and ecofriendly solutions. This reprint is an excellent reference for researchers, investigators, students, and stakeholders who want to tackle scientific challenges through the use of the membrane science discipline.

We wish to express our gratitude to the Italian Ministry of Foreign Affairs and International Cooperation for funding the Great Relevance International 2DMEMPUR Project (MAE00694722021-05-20) within the Italy (MAECI)–China (NSFC) 2018–2020 Joint Collaboration on New Materials, with particular reference to two-dimensional systems and graphene, and also thank the University of Bologna for the project RFO_2023_BOI.

Annarosa Gugliuzza and Cristiana Boi
Guest Editors

Editorial

Editorial for the Special Issue "Preparation and Application of Advanced Functional Membranes"

Annarosa Gugliuzza [1,*] and Cristiana Boi [2,3,*]

1. Institute on Membrane Technology-National Research Council, CNR-ITM, Via Pietro Bucci 17C, 87036 Rende, Italy
2. Department of Civil, Chemical, Environmental and Materials Engineering, Alma Mater Studiorum, University of Bologna, Via Terracini 28, 40131 Bologna, Italy
3. Department of Chemical and Biomolecular Engineering, North Carolina State University, Raleigh, NC 27695-7905, USA
* Correspondence: a.gugliuzza@itm.cnr.it (A.G.); cristiana.boi@unibo.it (C.B.)

1. Introduction

Membrane science is a discipline that cuts across almost all fields of research and experimentation [1]. It is well suited to the demand for best practices that provide sustainable solutions for natural resource management [2,3], pollution remediation [4,5], the recovery and recycling of value-added products from waste [6], and energy production and storage [7]. Advanced technologies are also being proposed for sustainable agriculture and industrial practices [8], textiles [9], biomedicine [10], packaging [11], and cultural assets [12,13]. In each case, membrane science provides multidisciplinary approaches to understand and control phenomena at different length scales, so the design and development of membrane devices and technologies requires the interplay of complementary disciplines, including modeling and simulation, membrane fabrication and characterization, process implementation, the build up of prototyping, and technology transfer. Membrane technology also has the advantage of being easy to integrate with other more conventional technologies, being remotely controllable and adaptable to different volumetric spaces [14]. Additionally, its sustainability offers the opportunity to be powered by green energy as an alternative fuel source [15]. Membrane technology is built upon both challenging and basic concepts such as the development of interfaces enabling the selective passage of bioactive molecules, ions, liquids (such as water), and gases according to well-established transport mechanisms. The latter can be regarded as the result of the intricate channels, pathways, and pores generated inside the membranes [1], as well as the accessibility to chemical moieties working as temporary adsorption sites or donor–bridge–acceptor systems [16,17].

Combining sustainable membrane design and technology is a major challenge in membrane science [18,19]. The aim is to provide solutions and best practices to manage and recover resources according to the principles of the circular green economy and durability [20,21]. Accordingly, a multidisciplinary approach is often proposed to implement the desired membrane operation on an industrial scale through process optimization, which involves a step-by-step experimental investigation and theoretical analysis, along with feasibility and cost-effectiveness assessments. First of all, the development of membranes with specific morphological and chemical features allows for the manipulation of desired or undesired events conceivable at the nano and micro scale so that the corresponding effects can be amplified at the macro scale. Once the type of operation is identified, the choice of the fabrication process that will be used to obtain suitable structural and chemical features in a predefined volumetric space becomes crucial. In fact, a jigsaw puzzle of functions can generate specific relationships for cooperative mechanisms that allow one to obtain the targets of a specific separation process. In recent years, advanced manufacturing processes, along with diverse families of materials, have been proposed to shift the perspective of membranes from traditional physical barriers [1] to interactive [22] and, where required, smart selective interfaces [23]. Thus, traditional phase inversion [1], track-etching [24], stretching [25], and interfacial polymerization [26] have been optimized or combined with more advanced fabrication methods, including phase separation micromolding [27], electrospinning [28], copolymer self-assembly [29], breath figure self-assembly [30], lithography [31], layer-by-layer assembly [32], and many others. On the other hand, different classes of materials, including organic fillers [33,34], nanotubes [35,36], nanoparticles [37,38], 2D materials [39,40], and quantum dots [41], have recently been proposed to promote (i) the assisted transport of gas/vapors [42,43], liquids [44,45], and ions [46,47] through membranes; (ii) improved selectivity [48,49] and high recovery factors [50,51]; (iii) energy saving [52,53]; and (iv) long-term durability [54,55].

In the last two decades, the number of papers focusing on membrane fabrication has increased exponentially (Figure 1a), confirming the growing awareness of the need to develop new functional membranes for effective operations in different subject areas (Figure 1b).

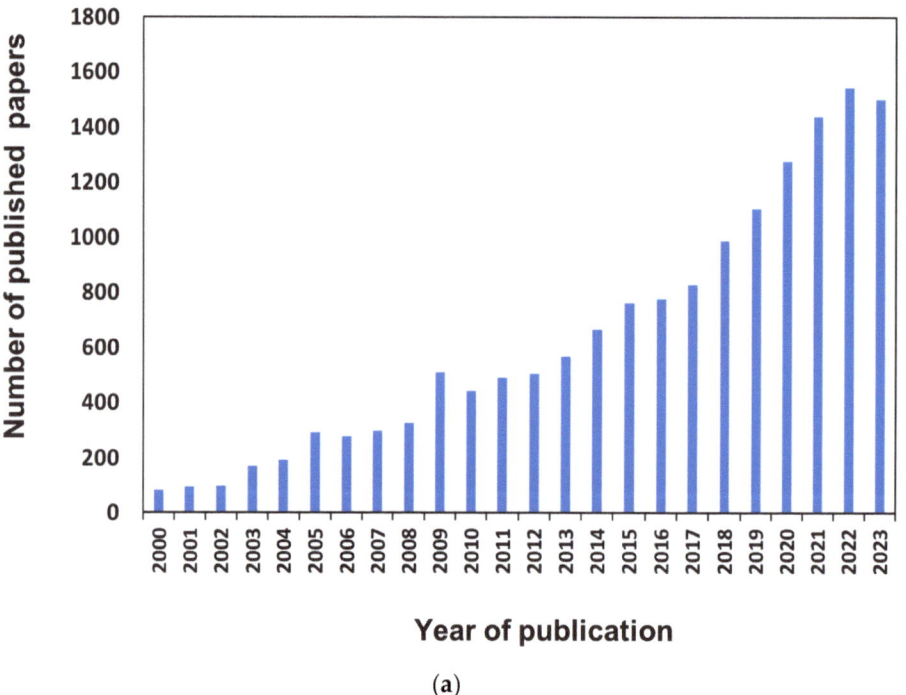

Figure 1. Cont.

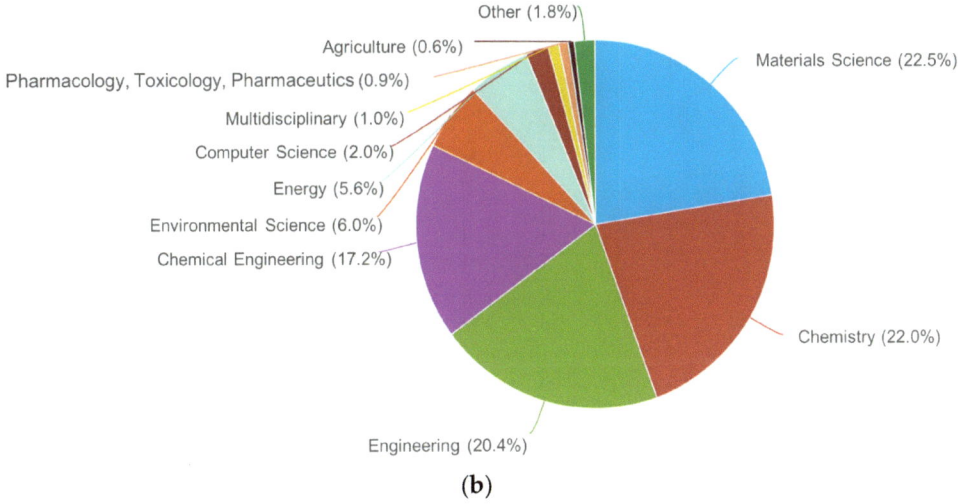

(**b**)

Figure 1. Annual publications on the subject of membrane fabrication (**a**) and subject areas' contributions to the literature (**b**) over the past 20 years (2000–2023). Source: Scopus.

At the same time, the use of membrane technology has expanded, as it is seen as a practical way to manage natural resources and provide tangible solutions for environmental remediation, including water purification, CO_2 capture, and pollution control. For example, the number of articles on water purification in the last 20 years (Figure 2a) has shown an increasing trend, reaching a maximum of almost 1464 in 2022. The rise in the number of publications dedicated to membranes for energy applications has been exponential, reaching almost 3000 articles in 2023 (Figure 2b). This is because membranes can play a fundamental role in the more efficient use of natural resources for energy applications, with membranes suitable for application in batteries, hydrogen separation, and fuel cells serving as important examples.

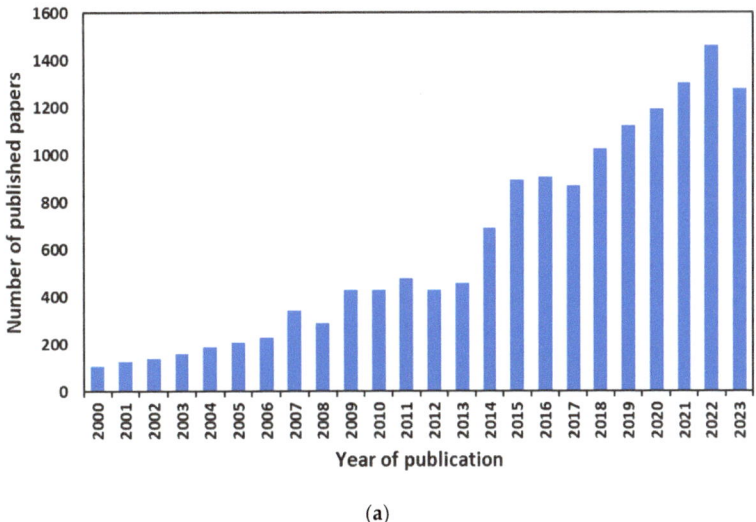

(**a**)

Figure 2. *Cont.*

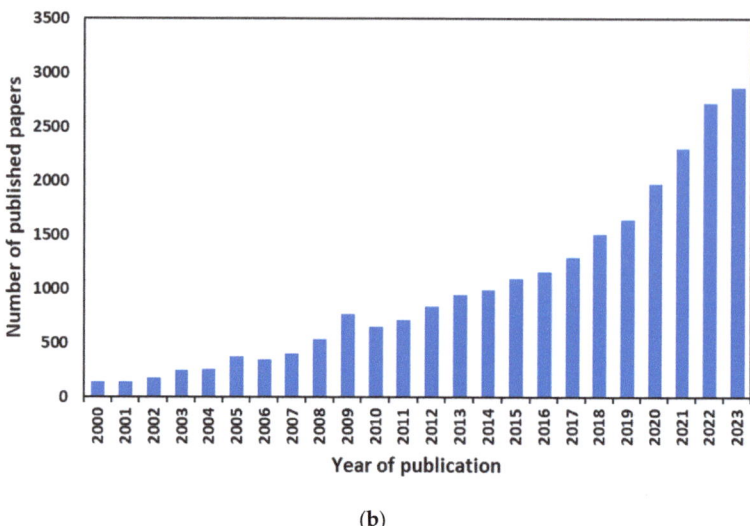

(**b**)

Figure 2. Annual publications on (**a**) membranes for water purification and (**b**) membranes for energy applications in the last 20 years (2000–2023). Source: Scopus.

This Special Issue, "*Preparation and Application of Advanced Functional Membranes*", offers some significant novel insights into the vast research dedicated to membrane manufacturing and applications, comprising ten interesting original articles and two reviews.

2. Overview of Published Articles

The ten original articles featured in this Special Issue deal with various issues related to water desalination (1), aqueous stream purification (2), antibacterial activity (1), controlled release (1), gas separation (2), the modeling of molecular transport through membranes (2), and nanohole formation in membranes (1). Two noteworthy reviews complete the Special Issue: the first focuses on nanocellulose green membranes and their related applications, and the second discusses the use of membranes for thorium removal, recovery, and recycling. Di Luca et al. [56] propose polyvinylidene fluoride (PVDF) membranes engineered with supramolecular complexes based on smart poly(N-isopropyl acrylamide) (PNIPAM) mixed hydrogels with aliquots of $ZrO(O_2C$-$C_{10}H_6$-$CO_2)$ (MIL-140) and graphene to make water desalination more sustainable and effective under relatively soft working conditions. Cooperative mechanisms take place at subnanometer scale to increase freshwater production, contrast fouling events, and facilitate in situ cleaning so that distinct effects can be induced by a simple switch of the density charge at the membrane interface without the need for additional chemicals or processing steps. Regarding the results, fluxes five times higher than pristine PVDF membrane were obtained after engineered membranes coming into contact with a mixture of NaCl (35 g L^{-1}) and humic acid (1 mg mL^{-1}); effective antifouling and safer and more sustainable self-cleaning actions helped to (i) limit the decline in the flux with time and (ii) allow for the recovery of up to 99% of the water permeation properties of the functional membranes.

Butylskii et al. [57] developed a new recycling method for lithium-ion batteries (LIBs) using a hybrid electrobaromembrane (EBM) approach. This technique selectively separates Li^+ and Co^{2+} ions using a special membrane with a pore diameter of 35 nm. The process combines an electric field and a pressure field to achieve highly efficient ion separation. The flux of lithium through the membrane is approximately 0.3 mol m^{-2} h^{-1}, and the presence of coexisting nickel ions does not interfere with lithium separation. Also, under conditions where the feeding and receiving solutions are identical, the flux of cobalt ions can be directed from the receiving to the feeding solution, leading to a result that is close to zero (−0.0025 mol m^{-2} h^{-1}) and an ion separation coefficient of Li$^+$ and Co^{2+} ions of −55. EBM conditions can be tailored to extract only lithium while leaving cobalt and nickel in the feed solution with a separation coefficient equal to infinity. This process holds promise for meeting the global demand for lithium and mitigating e-waste issues.

Joosten et al. [58] developed electrospun wire membrane adsorbers based on sulfonated poly(ether-ether-ketone) (sPEEK) for the recovery of valuable resources. These membrane adsorbers selectively bind lysozyme and have potential applications in the removal of heavy metals, dyes, and pharmaceutical components. Variations in fiber diameter have minimal effect on specific surface area and dynamic adsorption capacity. Different degrees of sulfonation of sPEEK do not proportionally affect the adsorption capacity. A dynamic lysozyme adsorption capacity of 59.3 mg/g can be achieved at 10% breakthrough regardless of flow rate. This study highlights the role of fiber diameter and functional group density in optimizing membrane adsorber performance. More specifically, variations in fiber diameter and functional group density do not significantly affect the binding capacity, meaning that a membrane absorber can be realized easily and used immediately to bind positively charged molecules without further morphological and chemical adjustments. This research article provides a valuable and prompt solution for efficient resource recovery and purification.

Composite ultrafiltration membranes have been prepared by Figuereido et al. [59] using cellulose acetate and silver nanoparticles. Varying the membrane structure results in different antibacterial effects against *E. coli*. The presence of silver nanoparticles significantly inhibits the growth of *E. coli*. More permeable membranes have higher silver contents and superior growth inhibition. When fully immersed in an *E. coli* suspension, the antibacterial activity of both silver-free and AgNP membranes are suitably correlated with surface chemical composition and silver accessibility, resulting in an efficient inhibition pattern for membranes with AgNPs. These results point towards potential applications in water purification and other areas related to antibacterial action.

The research article authored by Pereira et al. [60] focuses on magnetic-responsive hydrogels for liposomal drug delivery. They have developed systems with liposomes containing ferulic acid (FA) encapsulated in gelatin hydrogel membranes with iron oxide nanoparticles (MNPs). Their study compared these systems with conventional drug delivery methods. FA release from the liposomal gelatin followed the Korsmeyer–Peppas model, suggesting controlled diffusion in the absence of a magnetic field. However, under magnetic stimulation, due to the dispersed MNPs in the matrix, the release of FA from the liposomal gelatin membrane undergoes an increase in the constant rates; while keeping a diffusional controlled FA release mechanism, the FA release changes from Fickian diffusion to quasi-Fickian diffusion. Low-intensity magnetic fields stimulate FA release and shift the mechanism. The liposomal gelatin systems therefore offer a smoother and more controlled release of FA, with the potential for longer-term therapeutic use. This approach opens doors for novel magnetically controlled drug delivery approaches in biomedicine.

Liu et al.'s article [61] presents a computational mechanistic study that was conducted to understand how to remove acid gases from natural gas by using new functionalized GO membranes. They studied the adsorption and diffusion of several gases in 1,4-phenylenediamine-2-sulfonate (PDASA)-doped GO membrane channels, providing new insights into the solubility coefficient of CO_2 and H_2S. They estimated, for these two gases, a binding affinity that is higher than that computed for CH_4 and N_2. Their theoretical analysis suggested multilayer adsorption in functionalized GO membrane channels according to the Redlich–Peterson model, while the Langmuir model was used to describe weak adsorption for no polar gases. The result is an enhanced permeability and selectivity of acid gases such as CO_2 and H_2S—i.e., PCO_2 = 7265.5 Barrer, $\alpha CO_2/CH_4$ = 95.7; $P(H_2S + CO_2)$ = 42,075.1 Barrer, $\alpha H_2S/CH_4$ = 243.8—over CH_4, with a performance superior to that observed for more traditional GO membranes.

Polak et al. [62] provide an in depth-study of the transport properties of Pebax@2533 membranes filled with SiO_2, ZIF−8, and POSS-Ph over a broad range of temperatures and pressures, providing permeability, diffusivity, and solubility data for N_2, CH_4, and CO_2 gases. CO_2 selectivity was also examined as a function of temperature and pressure effects. The permeability increased with temperature, leading to a predominance of the diffusivity mechanism over the solubility one. Instead, the increase in pressure enhances CO_2 permeability, while greater benefits can be obtained for N_2 and CH_4 gases in terms of diffusivity.

Water vapor transport was studied through Pebax@2533 membranes filled with the ionic liquid [$C_{12}C_1$im]Cl (30 to 70 wt.%) by the authors of [63]. This theoretical study, which was inspired by prior experimental evidence, was conducted to establish how the addition of ionic liquid (IL) to the elastomeric membranes generates preferential pathways for water molecules, reaching values of water permeability of 85×10^{-3} g m^{-2} day^{-1} m at 318 K and with a content of [$C_{12}C_{1\mathrm{im}}$]Cl around 70 wt.%. Molecular dynamics simulations yield new insights into the role of the anions and cationic head groups of IL in directing water molecule diffusion. This pathway becomes wider with increases in temperature and IL concentration due to the larger water-accessible area. The use of smart organic nanofillers such as ionic liquids is herein demonstrated to make Pebax membranes breathable to a desired extent and in a reversible way. This type of engineered membrane is extremely interesting for the construction of environmental micro-regulation devices.

The study contributed by Strzelewicz et al. [64] explores particle diffusion in heterogeneous membrane-like structures by studying the interplay between membrane structure, external drift forces, and diffusion characteristics. They used Cauchy flight diffusion with drift and compared it to Gaussian random walk. The results show that strong drift can halt Gaussian diffusion while promoting superdiffusion with Cauchy flight. The membrane structures, which mimic real polymeric membranes with inorganic powders and designed obstacles, influence the transport behavior. More specifically, this paper proves that, under weak drift, the diffusion is controlled by the local environment, that is, the membrane. For stronger drift, superdiffusion is recognized, and in cases of excessively strong drift, the Brownian motion is almost stopped while tracers are pushed against obstacles. However, this pushing is overcome by random motion. Understanding these relationships is critical to improving the efficiency of processes that rely on membrane-based transport. This study underscores the role of structure and drift in shaping particle movement across membranes.

Ceccio et al. [65] propose the use of non-destructive ion transmission spectroscopy (ITS) and neutron depth profiling (NDP) to analyze, monitor, and quantify the evolution of etched micron-sized pores, the shape of which, in their study, transferred from latent tracks to a conical form (for one-sided etching) and, through a symmetrical (double-sided) etching process, to a well-developed cylindrical geometry. Using a traditional wet chemical etching process, they used dopants such as LiCl solution or boron to fill and generate defined pores and evaluated the effects on their shape and depth. As an example, the volume of the pore is demonstrated to increase linearly with both the etching temperature and time. Doping the etched tracks with 5 M LiCl, reductions of 24% and 11% can be obtained for the pore volume using etching times of 45 and 60 min, respectively. No destructive techniques allow for the investigation and quantification of every single effect of the working parameters on the final pore shape and size. This contribution falls into the list of studies dedicated to the fabrication and characterization of membranes with well controlled morphologies.

Nitodas et al. [66] focus on the sustainable use of nanocellulose to prepare membranes useful for a large range of applications, including water purification, desalination, antimicrobial applications, gas separation, and gas barrier applications. After an introduction to the synthesis and characterization of nanomaterials for nanocellulose membranes, case studies are illustrated through comparative analyses, highlighting the biodegradability, non-toxicity, low density, thermal stability, long-term reinforcement capabilities, and high mechanical strength of the proposed materials. Interestingly, this review provides further insights into the non-conventional uses of nanocellulose membranes in biomedical fields and other, more recreational practices, such as musical instruments.

Man et al. [67] discusses the use of membrane processes to address the removal, recovery, and recycling of thorium from industrial residues that often end up in municipal waste treatment facilities. They cover the biomedically relevant aspects of thorium, thorium detection techniques, classical extraction methods, and various membrane processes, such as electrodialysis and the use of liquid membranes. Special emphasis is placed on urban mining and the valorization of thorium. Moreover, this paper aims to provide insight into the safe handling of thorium and environmental protection through proposing practices and actions for the removal, recovery, and valorization of thorium.

3. Conclusions

The articles published in this Special Issue represent valuable contributions to the research dedicated to the development of new functional membranes for sustainable separation processes. Best practices are proposed to improve the performance of various separation processes, including water desalination, waste purification, CO_2 capture, antimicrobial applications, and smart drug delivery. Conceivable solutions are anticipated to make the operation of each membrane more ecofriendly through the selection of non-polluting, recyclable, and biodegradable materials, the use of low-environmental-impact manufacturing schemes, and the implementation of purification technologies for specific case studies. It is important to note that the output of each process is always related to the specific features of the membranes, including their morphological properties and chemical functionalities, which can be manipulated by external triggers. On the other hand, the understanding of transport mechanisms is always related to the intrinsic structural and chemical aspects of the membranes, as ascertained through the carrying out of theoretical analyses and experimental research.

As a final note, we would like to highlight that this Special Issue collects transversal and complementary contributions from China, the USA, and multiple European Countries, representing a shared and enhanced cross-technology research effort based upon issues regarding environmental and health protection.

Author Contributions: The authors equally contributed to the Editorial. All authors have read and agreed to the published version of the manuscript.

Funding: We acknowledge the financial grant we received from 'the Italian Ministry of Foreign Affairs and International Cooperation' within the framework of the Great Relevance International Project Italy (MAECI)-China (NSFC) 2018–2020—New Materials, with particular reference to Two-dimensional systems and Graphene (2DMEMPUR), MAE00694722021-05-20 and the University of Bologna for the project RFO_2023_BOI.

Acknowledgments: We wish to thank all of the authors for providing valuable contributions to this Special Issue, all of the reviewers for their valuable support and efforts made to improve the quality of the submitted papers, and our Assistant Editor for supporting the production of this Special Issue.

Conflicts of Interest: The authors declare no conflicts of interest.

References

1. Strathmann, H. *Introduction to Membrane Science and Technology*; Wiley-VCH Verlag GmbH: Weinheim, Germany, 2011; pp. 1–544.
2. Issaoui, M.; Jellali, S.; Zorpas, A.A.; Dutournie, P. Membrane technology for sustainable water resources management: Challenges and future projections. *Sustain. Chem. Pharm.* **2022**, *25*, 100590. [CrossRef]
3. Komaladewi, A.A.I.A.S.; Aryanti, P.T.P.; Subagia, I.D.G.A.; Wenten, I.G. Membrane technology in air pollution control: Prospect and challenge. *J. Phys. Conf. Ser.* **2019**, *1217*, 012046. [CrossRef]
4. Kang, Y.; Zhong, Z.; Xing, W. Functionalized membranes for multipollutants bearing air treatment. In *Hybrid and Combined Processes for Air Pollution Control*; Assadi, A., Amrane, A., Nguyen, T.A., Eds.; Elsevier: Amsterdam, The Netherlands, 2022; pp. 167–200. [CrossRef]
5. Cheng, Y.; Xia, C.; Garalleh, H.A.L.; Garaleh, M.; Lan Chi, N.T.; Brindhadevi, K. A review on optimistic development of polymeric nanocomposite membrane on environmental remediation. *Chemosphere* **2023**, *315*, 137706. [CrossRef] [PubMed]
6. Li, X.; Van der Bruggen, B.; Matsuyama, H.; Lin, Y.; Zheng, J. Resource Recovery and Recycling from Water Streams: Advanced Membrane Technologies and Case Studies. *ACS EST Water* **2023**, *3*, 1699–1701. [CrossRef]

7. Yaroslavtsev, A.B.; Stenina, I.A.; Golubenko, D.V. Membrane materials for energy production and storage. *Pure Appl. Chem.* **2020**, *92*, 1147–1157. [CrossRef]
8. Mzahma, S.; Duplay, J.; Souguir, D.; Ben Amar, R.; Ghazi, M.; Hachicha, M. Membrane Processes Treatment and Possibility of Agriculture Reuse of Textile Effluents: Study Case in Tunisia. *Water* **2023**, *15*, 1430. [CrossRef]
9. Van der Bruggen, B.B.; Curcio, E.; Drioli, E. Process intensification in the textile industry: The role of membrane technology. *J. Environ. Manag.* **2004**, *73*, 267–274. [CrossRef]
10. Hariharan, P.; Sundarrajan, S.; Arthanareeswaran, G.; Seshan, S.; Das, D.B.; Ismail, A.F. Advancements in modification of membrane materials over membrane separation for biomedical applications—Review. *Environ. Res. Part B* **2022**, *204*, 112045. [CrossRef]
11. Sommer, B. Membrane Packing Problems: A Short Review on Computational Membrane Modeling Methods and Tools. *Comput. Struct. Biotechnol. J.* **2013**, *5*, e201302014. [CrossRef]
12. Llorensa, J.; Zanelli, A. Structural membranes for refurbishment of the architectural heritage. *Procedia Eng.* **2016**, *155*, 18–27. [CrossRef]
13. Van, N.T.B.; Singyabuth, S. Cultural Sustainability of Hoi an Ancient Houses in the process of becoming the World Cultural Heritage City. *Int. J. Membr. Sci. Technol.* **2023**, *10*, 251–262. [CrossRef]
14. Basile, A.; Charcosset, C. *Integrated Membrane Systems and Processes*; Wiley: Chichester, UK, 2016; pp. 1–424.
15. Gugliuzza, A.; Basile, A. *Membranes for Clean and Renewable Power Applications*; Woodhead Publishing: Cambridge, UK, 2014; pp. 1–410.
16. Nghiem, L.D.; Schäfer, A.I.; Waite, T.D. Adsorptive Interactions between Membranes and Trace Contaminants. *Desalination* **2002**, *147*, 269–274. [CrossRef]
17. Siddiqui, S.A.; Abdullah, M.M. Molecular modeling and simulation of some efficient charge transfer materials using density functional theory. *Mater. Today Commun.* **2020**, *22*, 100788. [CrossRef]
18. Comitti, A.; Vijayakumaran, H.; Nejabatmeimandi, M.H.; Seixas, L.; Cabello, A.; Misseroni, D.; Penasa, M.; Paech, C.; Bessa, M.; Bown, A.C.; et al. Ultralight Membrane Structures Toward a Sustainable Environment. In *Sustainable Structures and Buildings*; Bahrami, A., Ed.; Springer: Cham, Switzerland, 2024; pp. 17–37. [CrossRef]
19. Dhume, S.; Chendake, Y. Membrane Technology for Green Engineering. In *Applied Biopolymer Technology and Bioplastics, Sustainable Development by Green Engineering Materials*; Rawat, N.K., Volova, T.G., Haghi, A.K., Eds.; Apple Academic Press: New York, NY, USA, 2021; pp. 1–27.
20. Landaburu-Aguirre, J.; Molina, S. Circular Economy in Membrane Technology. *Membranes* **2023**, *13*, 784. [CrossRef] [PubMed]
21. Ludovic, F.; Dumée, L.F.; Sadrzadeh, M.; Shirazi, M.M.A. *Green Membrane Technologies towards Environmental Sustainability*; Elsevier: Amsterdam, The Netherlands, 2023; pp. 1–646.
22. Gugliuzza, A.; Aceto, M.C.; Drioli, E. Interactive functional poly(vinylidene fluoride) membranes with modulated lysozyme affinity: A promising class of new interfaces for contactor crystallizers. *Polym. Int.* **2009**, *58*, 1452–1464. [CrossRef]
23. Pan, Y.; Liu, Y.; Yang, S.; Zhang, C.; Ullah, Z. Recent research progress on the stimuli-responsive smart membrane: A review. *Nanotechnol. Rev.* **2023**, *12*, 20220538. [CrossRef]
24. Apel, P. Track etching technique in membrane technology. *Radiat. Meas.* **2001**, *34*, 559–566. [CrossRef]
25. Li, K.; Zhang, Y.; Xu, L.; Zeng, F.; Hou, D.; Wang, J. Optimizing stretching conditions in fabrication of PTFE hollow fiber membrane for performance improvement in membrane distillation. *J. Membr. Sci.* **2018**, *550*, 126–135. [CrossRef]
26. Baig, U.; Waheed, A. Exploiting interfacial polymerization to fabricate hyper-cross-linked nanofiltration membrane with a constituent linear aliphatic amine for freshwater production. *npj Clean Water* **2022**, *5*, 46. [CrossRef]
27. Vogelaar, L.; Lammertink, R.G.H.; Barsema, J.N.; Nijdam, W.; Bolhuis-Versteeg, L.A.M.; van Rijn, C.J.M.; Wessling, M. Phase Separation Micromolding: A New Generic Approach for Microstructuring Various Materials. *Small* **2005**, *1*, 645–655. [CrossRef]
28. Ahmed, F.Z.; Lalia, B.S.; Hashaikeh, R. A review on electrospinning for membrane fabrication: Challenges and applications. *Desalination* **2015**, *356*, 15–30. [CrossRef]
29. Hamta, A.; Ashtiani, F.Z.; Karimi, M.; Moayedfard, S. Asymmetric block copolymer membrane fabrication mechanism through self-assembly and non-solvent induced phase separation (SNIPS) process. *Sci. Rep.* **2022**, *12*, 771. [CrossRef] [PubMed]
30. Perrotta, M.L.; Saielli, G.; Casella, G.; Macedonio, F.; Giorno, L.; Drioli, E.; Gugliuzza, A. An ultrathin suspended hydrophobic porous membrane for high-efficiency water desalination. *Appl. Mater. Today* **2017**, *9*, 1–9. [CrossRef]
31. Sabirova, A.; Florica, C.F.; Pisig, F.; Syed, A.; Buttner, U.; Li, X.; Nunes, S.P. Nanoporous membrane fabrication by nanoimprint lithography for nanoparticle sieving. *Nanoscale Adv.* **2022**, *4*, 1119–1124. [CrossRef] [PubMed]
32. Joseph, N.; Ahmadiannamini, P.; Hoogenboom, R.; Vankelecom, I.F.G. Layer-by-layer preparation of polyelectrolyte multilayer membranes for separation. *Polym. Chem.* **2014**, *5*, 1817–1831. [CrossRef]
33. Gugliuzza, A.; Drioli, E. Role of additives in the water vapor transport through block co-poly(amide/ether) membranes: Effects on surface and bulk polymer properties. *Eur. Polym. J.* **2004**, *40*, 2381–2389. [CrossRef]
34. Hao, L.; Li, P.; Chung, T.-S. PIM-1 as an organic filler to enhance the gas separation performance of Ultem polyetherimide. *J. Membr. Sci.* **2014**, *453*, 614–623. [CrossRef]
35. Wang, R.; Chen, D.; Wang, Q.; Ying, Y.; Gao, W.; Xie, L. Recent Advances in Applications of Carbon Nanotubes for Desalination: A Review. *Nanomaterials* **2020**, *10*, 1203. [CrossRef] [PubMed]

36. Amirkhani, F.; Mosadegh, M.; Asghari, M.; Parnian, M.J. The beneficial impacts of functional groups of CNT on structure and gas separation properties of PEBA mixed matrix membranes. *Polym. Test.* **2020**, *82*, 106285. [CrossRef]
37. Gnus, M.; Dudek, G.; Turczyn, R. The influence of filler type on the separation properties of mixed-matrix membranes. *Chem. Pap.* **2018**, *72*, 1095–1105. [CrossRef]
38. De Pascale, M.; De Angelis, M.G.; Boi, C. Mixed Matrix Membranes Adsorbers (MMMAs) for the Removal of Uremic Toxins from Dialysate. *Membranes* **2022**, *12*, 203. [CrossRef] [PubMed]
39. Suvigya, K.; Lalita, S.; Gopinadhan, K. Membranes for desalination and dye separation: Are 2D materials better than polymers? A critical comparison. *Sep. Purif. Technol.* **2023**, *325*, 124693. [CrossRef]
40. Cheng, L.; Liu, G.; Zhao, J.; Jin, W. Two-Dimensional-Material Membranes: Manipulating the Transport Pathway for Molecular Separation. *Acc. Mater. Res.* **2021**, *2*, 114–128. [CrossRef]
41. Kim, A.; Moon, S.J.; Kim, J.H.; Patel, R. Review on thin-film nanocomposite membranes with various quantum dots for water treatments. *J. Ind. Eng. Chem.* **2023**, *118*, 19–32. [CrossRef]
42. Frappa, M.; Del Rio Castillo, A.E.; Macedonio, F.; Pellegrini, V.; Drioli, E.; Gugliuzza, A. A few-layer graphene for advanced composite PVDF membranes dedicated to water desalination: A comparative study. *Nanoscale Adv.* **2020**, *2*, 4728–4739. [CrossRef] [PubMed]
43. Ferrari, M.C. Recent developments in 2D materials for gas separation membranes. *Curr. Opin. Chem. Eng.* **2023**, *40*, 100905. [CrossRef]
44. Zhang, H.; Zheng, Y.; Yu, S.; Chen, W.; Yang, J. A Review of Advancing Two-Dimensional Material Membranes for Ultrafast and Highly Selective Liquid Separation. *Nanomaterials* **2022**, *12*, 2103. [CrossRef] [PubMed]
45. Nasrollahi, N.; Aber, S.; Vatanpour, V.; Mahmoodi, N.M. Development of hydrophilic microporous PES ultrafiltration membrane containing CuO nanoparticles with improved antifouling and separation performance. *Mater. Chem. Phys.* **2019**, *222*, 338–350. [CrossRef]
46. Cao, T.-X.; Xie, R.; Ju, X.-J.; Wang, W.; Pan, D.-W.; Liu, Z.; Chu, L.-Y. Biomimetic Two-Dimensional Composited Membranes for Ion Separation and Desalination. *Ind. Eng. Chem. Res.* **2023**, *62*, 14772–14790. [CrossRef]
47. Kim, S.; Wang, H.; Moo Lee, Y.M. 2D Nanosheets and Their Composite Membranes for Water, Gas, and Ion Separation. *Ang. Chem. Int. Ed.* **2019**, *58*, 17483–17871. [CrossRef]
48. Galizia, M.; Chi, W.S.; Smith, Z.S.; Merkel, T.C.; Baker, R.W.; Freeman, B.D. 50th Anniversary Perspective: Polymers and Mixed Matrix Membranes for Gas and Vapor Separation: A Review and Prospective Opportunities. *Macromolecules* **2017**, *50*, 7809–7843. [CrossRef]
49. Zhang, Z.; Rahman, M.; Abetz, C.; AbetzJ, V. High-performance asymmetric isoporous nanocomposite membranes with chemically-tailored amphiphilic nanochannels. *Mater. Chem. A* **2020**, *8*, 9554–9566. [CrossRef]
50. Sardarabadi, H.; Kiani, S.; Karkhanechi, H.; Mousavi, S.M.; Saljoughi, E.; Matsuyama, H. Effect of Nanofillers on Properties and Pervaporation Performance of Nanocomposite Membranes: A Review. *Membranes* **2022**, *12*, 1232. [CrossRef] [PubMed]
51. Agboola, O.; Fayomi, O.S.I.; Ayodeji, A.; Ayeni, A.O.; Alagbe, E.E.; Sanni, S.E.; Okoro, E.E.; Moropeng, L.; Sadiku, R.; Kupolati, K.W.; et al. A Review on Polymer Nanocomposites and Their Effective Applications in Membranes and Adsorbents for Water Treatment and Gas Separation. *Membranes* **2021**, *11*, 139. [CrossRef] [PubMed]
52. Frappa, M.; Castillo, A.E.D.R.; Macedonio, F.; Di Luca, G.; Drioli, E.; Gugliuzza, A. Exfoliated Bi2Te3-enabled membranes for new concept water desalination: Freshwater production meets new routes. *Water Res.* **2021**, *203*, 117503. [CrossRef] [PubMed]
53. Zeng, H.; He, S.; Hosseini, S.S.; Zhu, B.; Shao, L. Emerging nanomaterial incorporated membranes for gas separation and pervaporation towards energetic-efficient applications. *Adv. Membr.* **2022**, *2*, 100015. [CrossRef]
54. Cay-Durgun, P.; McCloskey, C.; Konecny, J.; Khosravi, A.; Lind, M.L. Evaluation of thin film nanocomposite reverse osmosis membranes for long-term brackish water desalination performance. *Desalination* **2017**, *404*, 304–312. [CrossRef]
55. Janakiram, S.; Ahmadi, M.; Dai, Z.; Ansaloni, L.; Deng, L. Performance of Nanocomposite Membranes Containing 0D to 2D Nanofillers for CO_2 Separation: A Review. *Membranes* **2018**, *8*, 24. [CrossRef] [PubMed]
56. Di Luca, G.; Chen, G.; Jin, W.; Gugliuzza, A. Aliquots of MIL-140 and Graphene in Smart PNIPAM Mixed Hydrogels: A Nanoenvironment for a More Eco-Friendly Treatment of NaCl and Humic Acid Mixtures by Membrane Distillation. *Membranes* **2023**, *13*, 437. [CrossRef]
57. Butylskii, D.; Troitskiy, V.; Chuprynina, D.; Dammak, L.; Larchet, C.; Nikonenko, V. Application of Hybrid Electrobaromembrane Process for Selective Recovery of Lithium from Cobalt- and Nickel-Containing Leaching Solutions. *Membranes* **2023**, *13*, 509. [CrossRef]
58. Joosten, N.; Wyrębak, W.; Schenning, A.; Nijmeijer, K.; Borneman, Z. On the Performance of a Ready-to-Use Electrospun Sulfonated Poly(Ether Ether Ketone) Membrane Adsorber. *Membranes* **2023**, *13*, 543. [CrossRef] [PubMed]
59. Figueiredo, A.S.; Ferraria, A.M.; Botelho do Rego, A.M.; Monteiro, S.; Santos, R.; Minhalma, M.; Sánchez-Loredo, M.G.; Tovar-Tovar, R.L.; de Pinho, M.N. Bactericide Activity of Cellulose Acetate/Silver Nanoparticles Asymmetric Membranes: Surfaces and Porous Structures Role. *Membranes* **2023**, *13*, 4. [CrossRef] [PubMed]
60. Pereira, L.; Ferreira, F.C.; Pires, F.; Portugal, C.A.M. Magnetic-Responsive Liposomal Hydrogel Membranes for Controlled Release of Small Bioactive Molecules-An Insight into the Release Kinetics. *Membranes* **2023**, *13*, 674. [CrossRef] [PubMed]
61. Liu, Q.; Yang, Z.; Liu, G.; Sun, L.; Xu, R.; Zhong, J. Functionalized GO Membranes for Efficient Separation of Acid Gases from Natural Gas: A Computational Mechanistic Understanding. *Membranes* **2022**, *12*, 1155. [CrossRef] [PubMed]

62. Polak, D.; Szwast, M. Analysis of the Influence of Process Parameters on the Properties of Homogeneous and Heterogeneous Membranes for Gas Separation. *Membranes* **2022**, *12*, 1016. [CrossRef] [PubMed]
63. Cheng, Z.; Li, S.; Tocci, E.; Saielli, G.; Gugliuzza, A.; Wang, Y. Pathway for Water Transport through Breathable Nanocomposite Membranes of PEBAX with Ionic Liquid [C12C1im]Cl. *Membranes* **2023**, *13*, 749. [CrossRef]
64. Strzelewicz, A.; Krasowska, M.; Cieśla, M. Lévy Flights Diffusion with Drift in Heterogeneous Membranes. *Membranes* **2023**, *13*, 417. [CrossRef]
65. Ceccio, G.; Vacik, J.; Siegel, J.; Cannavó, A.; Choukourov, A.; Pleskunov, P.; Tosca, M.; Fink, D. Etching and Doping of Pores in Polyethylene Terephthalate Analyzed by Ion Transmission Spectroscopy and Nuclear Depth Profiling. *Membranes* **2022**, *12*, 1061. [CrossRef] [PubMed]
66. Nitodas, S.; Skehan, M.; Liu, H.; Shah, R. Current and Potential Applications of Green Membranes with Nanocellulose. *Membranes* **2023**, *13*, 694. [CrossRef]
67. Man, G.T.; Albu, P.C.; Nechifor, A.C.; Grosu, A.R.; Tanczos, S.-K.; Grosu, V.-A.; Ioan, M.-R.; Nechifor, G. Thorium Removal, Recovery and Recycling: A Membrane Challenge for Urban Mining. *Membranes* **2023**, *13*, 765. [CrossRef]

Disclaimer/Publisher's Note: The statements, opinions and data contained in all publications are solely those of the individual author(s) and contributor(s) and not of MDPI and/or the editor(s). MDPI and/or the editor(s) disclaim responsibility for any injury to people or property resulting from any ideas, methods, instructions or products referred to in the content.

Review

Current and Potential Applications of Green Membranes with Nanocellulose

Stefanos (Steve) Nitodas [1,*], Meredith Skehan [1,2], Henry Liu [1] and Raj Shah [2]

1. Department of Materials Science and Chemical Engineering, Stony Brook University, Stony Brook, NY 11794, USA; meredith.skehan@stonybrook.edu (M.S.); henry.liu@stonybrook.edu (H.L.)
2. Koehler Instrument Company Inc., Bohemia, NY 11794, USA; rshah@koehlerinstrument.com
* Correspondence: steve.nitodas@stonybrook.edu

Abstract: Large-scale applications of nanotechnology have been extensively studied within the last decade. By exploiting certain advantageous properties of nanomaterials, multifunctional products can be manufactured that can contribute to the improvement of everyday life. In recent years, one such material has been nanocellulose. Nanocellulose (NC) is a naturally occurring nanomaterial and a high-performance additive extracted from plant fibers. This sustainable material is characterized by a unique combination of exceptional properties, including high tensile strength, biocompatibility, and electrical conductivity. In recent studies, these unique properties of nanocellulose have been analyzed and applied to processes related to membrane technology. This article provides a review of recent synthesis methods and characterization of nanocellulose-based membranes, followed by a study of their applications on a larger scale. The article reviews successful case studies of the incorporation of nanocellulose in different types of membrane materials, as well as their utilization in water purification, desalination, gas separations/gas barriers, and antimicrobial applications, in an effort to provide an enhanced comprehension of their capabilities in commercial products.

Keywords: nanocellulose; membranes; water treatment; wastewater; gas separation; bacterial nanocellulose

1. Introduction

Research within the fields of material science and nanotechnology has been on a steady rise over the past several years. By isolating and modifying certain advantageous properties of newly developed materials, such as nanostructures, the latter can be applied to optimized processes to facilitate the improvement of several aspects of our lives. Given the unique size of nanomaterials, nanostructured materials can be modified at the scale of 1–100 nm, resulting in the improvement of a variety of properties within products used in everyday life. In recent years, one such material has been nanocellulose. Nanocellulose is a naturally occurring nanomaterial extracted from the cell walls of plant fibers, plant biomass, and algae [1,2]. This green nanomaterial can be obtained either as nanocrystalline/nanofiber cellulose via top-down biosynthesis by disintegration of plant materials or as bacterial nanocellulose through bottom-up biosynthesis [3–5].

Nanocellulose is characterized by chemical inertness, high tensile strength, and biocompatibility; in addition, it exhibits dimensional stability, a low coefficient of thermal expansion, and the ability to modify its surface chemistry [6–10]. In recent studies, these unique properties of nanocellulose have been enhanced and applied to processes in the areas of water purification [11], automotive [12], food [1,13], and membrane separations [14]. The industry prioritizes the low cost and efficient production involved in utilizing nanocellulose as an alternative to conventional cellulose. Compared to traditional cellulose, nanocellulose has a higher surface area, aspect ratio, and Young's modulus, allowing its application to adsorption, photocatalysis, flocculation, and membranes [15]. In addition, the research interest in nanocellulose-based materials for environmental applications is rapidly growing due to increasing environmental problems, water contamination, and the severe risk of oil pollution [16,17].

Regarding water purification, it has been shown that modified variants of nanocellulose reduce the concentration of contaminants in wastewater. For instance, carboxymethyl nanocellulose stabilized nano zero-valent iron has been proven to be effective in reducing the presence of hexavalent chromium within wastewater [11]. Nanocellulose has been the subject of a wide spectrum of research efforts for its utilization in membranes for multifunctional wastewater treatment and adsorption, with a primary focus on improving the permeability of nanocellulose and extracting purified liquid [18–21].

Nanocellulose has also been applied to gas separation processes such as regulating membranes [22,23]. These nanocellulose membranes allow for the capture and storage of certain gases, including CO_2, which is prevalent in many industrial processes involving power generation [24,25]. This action decreases carbon emissions, leading to lower greenhouse gas pollution. Nanocellulose membranes have also been applied to biomedical devices. This provides greater precision in chemical regulation given its dynamic permeability, which is optimal for medical environments [26]. The current study constitutes a comprehensive review of the aforementioned important applications of nanocellulose-based membranes, with emphasis on the environmental sector (water purification, gas separations).

2. Types of Nanocellulose

Over the past few years, nanocellulose has emerged as one of the most promising materials for a variety of applications, largely due to its biodegradable nature. A highlighted advantage of nanocellulose is also its diverse size/dimensions which enable its versatility across multiple fields. With pollution on the rise and environmental issues greater than ever before, this material may pose solutions to issues in this specific area. The market for nanocellulose has been on a steady increase, with estimates of it being worth $660 million by the end of 2023 [27] and $783 million by 2025 [28]. Paper and pulp products represent the majority of the nanocellulose market; nanocellulose has been used as an additive in papermaking to produce lighter paper that exhibits enhanced properties, such as higher printing quality, improved mechanical strength, and less transparency [28,29]. Similar trends have been observed with patents associated with cellulose nano-objects; from 2010 to 2017, about 4500 patents referring to nanocellulose were published [30].

Currently, nanocellulose can be produced in many different forms, including one-dimensional (nanofibers/microparticles), two-dimensional (films), and three-dimensional (hydrogels/aerogels) variants. There are three main types of nanocellulose: cellulose nanocrystals (CNC), cellulose fibers (CNF), and bacterial nanocellulose (BNC) [31]. Cellulose nanocrystals (CNC) are a crystalline derivative of nanocellulose extracted through strong acid hydrolysis at high temperatures [32]. CNCs possess high thresholds in aspect ratio, surface area, and mechanical strength, which renders them ideal for applications on surfaces that require reinforcement.

Cellulose fibers (CNF) are microfibrils separated from nanocellulose and obtained by breaking down complex nanocellulose structures through chemicals and mechanical means. CNFs possess high plasticity regarding the dimensions of the material. Depending on the plant source of the nanocellulose, CNFs can have varying width and diameter ranges, rendering them ideal for applications that require flexibility in the size of the applied nanomaterials.

Bacterial nanocellulose (BNC or BC) is a promising natural biopolymer that can be produced by specific bacterial species, such as an exopolysaccharide of β-D glucopyranose [33]. It can be obtained through cultivation in a bacterial environment saturated with glucose, phosphate, and oxygen [34]. The versatile nature of this form of nanocellulose has generated greater research related to biomedical applications. BNC exhibits very good mechanical properties, whereas its nanostructured morphology and water-holding capacity render bacterial nanocellulose an ideal material for cellular immobilization and adhesion [35,36]. The properties of the three types of nanocellulose are compared in Table 1, where it can be seen that both CNC and BNC are characterized by high crystallinity, while CNC and CNF exhibit high Young's Modules values [31].

Table 1. Comparison of structure parameters of different types of nanocellulose. (Adapted from Huo et al. [31]).

Nanocellulose Type	Structural Properties			Mechanical Properties	
	Length (nm)	Diameter (nm)	Crystallinity (%)	Young's Modulus (GPa)	Tensile Strength (GPa)
CNCs	100–500	3–50	~90	50–140	8–10
CNFs	$\geq 10^3$	3–60	50–90	50–160	0.8–1
BC	$\geq 10^3$	20–100	84–89	78	0.2–2

3. Applications of Nanocellulose-Based Membranes

3.1. Desalination and Wastewater Treatment

About 4 billion people are experiencing water scarcity at least one month per year, whereas 1.8 billion people are facing an absolute water shortage [37]. It has also been reported that more than 63 million Americans have been exposed to more water contamination in recent decades [38]. Membrane-based desalination offers one solution to the water scarcity issue, as it supplements the natural hydrological cycle with freshwater obtained from seawater and other non-potable sources [39]. Additionally, membrane-based desalination is more cost-effective than traditional chemical treatments. Nanocellulose in particular has potential in membrane-based desalination because its unique properties lend themselves to a more efficient and eco-friendly membrane [39]. Nanocellulose exhibits high surface area and high tensile strength, and because it is naturally derived, it is non-toxic to humans and the environment [40].

Membrane-based desalination can be accomplished through various methods, including nanofiltration, reverse osmosis, pervaporation, ultrafiltration, and distillation. Nanocellulose has been successfully employed in all aforementioned methods, with different forms of nanocellulose displaying varying efficiency in each method [41–43]. Cellulose nanocrystals (CNCs) have been found to have applications in nanofiltration, reverse osmosis, and pervaporation. Yang et al. found that polydopamine-modified CNCs have significant applications in nanofiltration. When deposited on a thin-film nanocomposite membrane, the addition of the modified CNC produced pure water permeability of 128.4 L m^{-2} h^{-1} (LMH)/bar, Congo red rejection of 99.91%, and salt permeation of 99.33% [44]. Comparatively, cellulose nanofibrils (CNFs) have been found to have applications in ultrafiltration as well as nanofiltration and reverse osmosis.

Mohammed et al. carried out a study to test the performance of reduced graphene oxide/cellulose nanofibrils (CNFs) in membranes used for nanofiltration [45]. The study found that reduced graphene (rGO) membranes with CNF exhibited a pure water permeance of 37.2 LMH/bar, whereas rGO membranes without CNF had a reduced pure water permeance of 0.33 L m^{-1} h^{-1} bar^{-1}. This result can be seen in Figure 1. The figure shows the varying pure water permeance of three rGO membranes for five different organic dyes. The membranes vary in their rGO loading but have a fixed 1:1 ratio of rGO to CNF. Additionally, all three membranes received three minutes of oxygen plasma etching treatment in order to increase the number of nanopores on the surface of the membrane. The dyes tested include methyl orange (MO), rhodamine B (RhB), acid fuchsin (AF), brilliant blue (BB), and rose bengal (RB). As shown in the figure, the membrane with a 23.87 mg m^{-2} loading had the highest pure water permeance but only had more than 90% rejection for 2/5 of the dyes. The membrane with a 39.79 mg m^{-2} loading displayed improved rejection but suffered from a lower pure water permeance. In addition, it was still not able to obtain 90% rejection for MO and RhB. Consequently, the membrane with a 31.83 mg m^{-2} loading was determined to be the most well-rounded membrane due to its ability to reject more than 90% of the AF, RB, and BB dyes while maintaining an acceptable pure water permeance [45].

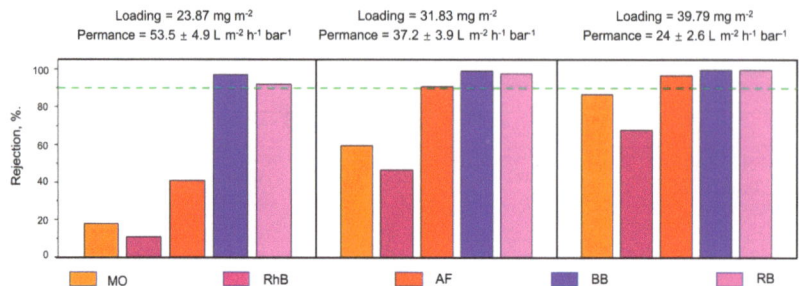

Figure 1. Comparison of pure water permeance and rejection of rGO:CNF$_{(1:1)}$ membrane for various loading after 3 min plasma treatment (Adapted from Mohammed et al. [45]).

Other research studies with nanocellulose-based materials have focused on the removal of heavy metals from wastewater. Heavy metals can stem from the effluents of several industries, including electroplating, metallurgical processes, and mining [46]. These activities constitute hazards to the environment and humans. For effective heavy metal removal by adsorption processes, the surface chemistry of the nanocellulose-based membrane must be tailored for the removal of specific metal species [47]. As an example, negatively charged carboxylated CNF coupled with trimethylolpropane-tris-(2-methyl-1-aziridine) propionate and graphene oxide was found to be an excellent adsorbent of numerous cations of heavy metals, including Pb^{2+}, Cd^{2+}, and Cu^{2+} [48]. Several surface modifications of nanocellulose have been investigated, with carboxylation being the most studied method for enhancing the sorption capacity of nanocellulose [49]. Tetramethylpiperidine-1-oxyl (TEMPO)-oxidized nanocellulose adsorbents present outstanding adsorption capabilities for divalent cations [50]. The carboxyl group provides a strong negative charge to nanocellulose, which permits the adsorption of even radioactive species [47].

In addition to the surface modification of the nanocellulose surface, the pH of the solution is critical for the selectivity of nanocellulose over heavy metal ions. Sharma et al. reported that the adsorption of Cd^{2+} from cadmium(II) nitrate solution using CNF modified by the nitro-oxidation method was optimum at pH 7 and decreased in acidic and basic conditions, as can be seen in Figure 2 [51]. At low pH values, some carboxylic acid groups became neutral, resulting in weaker electrostatic interactions overall. At high pH values, on the other hand, some CNF nanofibers were denatured, and consequently, the Cd^{2+} adsorption capacity of the system decreased.

Bacterial nanocellulose can also be employed for treating water contamination [52,53]. This material covers a smaller range of applications than CNC and CNF, with one of them being membrane distillation. Wu et al. produced a membrane for photothermal membrane distillation composed of polydopamine (PDA) particles and bacterial nanocellulose arranged in a bilateral composition and exposed to (tridecafluoro-1,1,2,2-tetrahydrooctyl)-trichlorosilane (FTCS) vapor [54]. They found that the membrane exhibited a salt rejection greater than 99.9%, a solar energy-to-collected water efficiency of 68%, and a 1.0 kg m^{-2} h^{-1} permeate flux. Additionally, the membrane allowed for ease of cleaning due to its ability to partially disinfect itself when exposed to sunlight.

Figure 3 shows how effective light irradiation has been at eliminating bacteria close to the FTCS-PDA/BNC membrane surface [54]. Fluorescent staining was used to color-tag dead and live cells red and green, respectively. Scanning electron microscope (SEM) imaging was used to observe the membrane surface, with red and green arrows used to indicate dead and live cells. In Figure 3a, which represents the control experiment, a solution of more than 324 live *E. coli* cells/mL was used to simulate bacteria-contaminated water. Under dark conditions and after 1 h, the fluorescence imaging shows the presence of live *E. coli* cells as well as the absence of dead cells. In Figure 3b, light irradiation (1 kW m^{-2}) was applied in addition to the *E. coli* solution to simulate in situ conditions. After 1 h, the presence of live and dead cells can be observed in Figure 3(b1–b3). Figure 3c depicts the membrane surface after the feed water from Figure 3b was drained and the membrane was exposed to light irradiation (1 kW m^{-2}) for a duration of 10 min. After this second dose of light irradiation, the membrane only exhibited dead cells, which suggests the rise in membrane temperature (~78 °C) in addition to the removal of the bulk top water was successful in killing the *E. coli*. Finally, Figure 3d depicts the membrane surface after being washed in DI water for 5 min. As seen in Figure 3(d1–d3), no *E. coli* cells were detected. This is a significant observation because it suggests that the membrane can be effectively cleaned without having to depend on invasive methods that can change the chemical composition or integrity of the membrane.

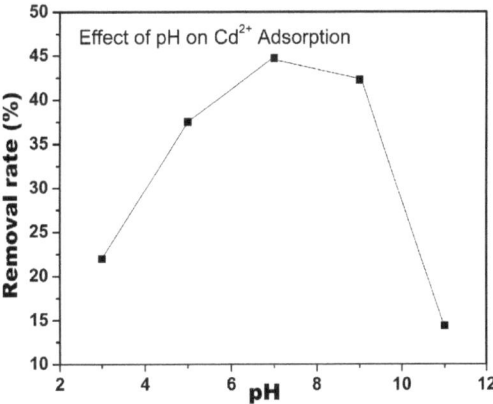

Figure 2. Effect of pH value on the Cd^{2+} adsorption efficiency of CNF (Adapted from Sharma et al. [51]).

Another variant of nanocellulose, nanocellulose acetate (NCA), also displayed promising desalination applications and anti-biofouling properties. In a study conducted by Morsy et al., NCA was prepared from rice straw waste by acidic hydrolysis, and the NCA was incorporated into reverse osmosis membranes via phase inversion [55]. The NCA membranes displayed decreased relative protein adsorption compared to the pristine membranes and increased water flux and salt rejection [55]. The aforementioned studies illustrate how the use of CNC, CNF, or BNC can differ according to which desalination method is preferred. However, as all of the prior materials are derived from nanocellulose, the latter can be considered a widely applicable material for desalination processes.

Another application of nanocellulose that has received attention in the last few years is in solar evaporators for producing freshwater from seawater. Solar evaporation is an attractive technology that combines water and solar energy. It has enabled an array of emerging applications, including contaminated water purification, seawater desalination, electric generation, steam sterilization, and fuel production, especially in resource-limited regions and countries [56]. Compared to many synthetic polymer-based evaporators, nanocellulose-based evaporators are expected to benefit from the NC's abundant reserves and renewable features [57]. In the study of Wu et al., it is shown that the microporous network of cellulose composite-based evaporators results in performance improvements such as high evaporation rates and salt resistance [57].

In another work by Jian et al., a flexible, scalable, and biodegradable photothermal bilayered evaporator for highly efficient solar steam generation was demonstrated [58]. The bilayered evaporator consisted of BNC loaded with a high concentration of polydopamine (PDA) particles during its growth. The size of the PDA particles was tailored to achieve light absorption properties matching the solar spectrum. This hybrid biodegradable material introduced in the evaporator exhibited good photothermal conversion and heat localization, leading to a high solar steam generation efficiency of 78%, thus showing a promising approach to tackle the global water crisis.

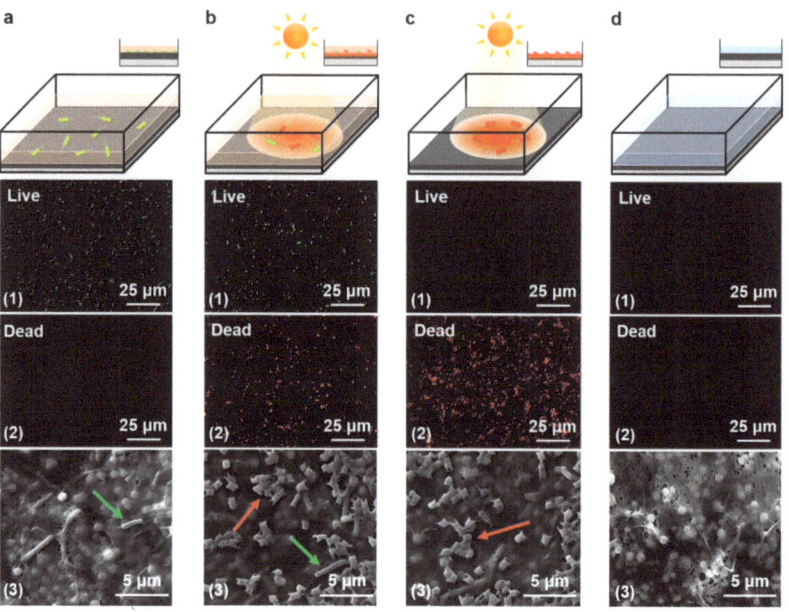

Figure 3. Activity measurements of photothermal disinfection. Schematic (1), fluorescence (2), and SEM images (3) of (**a**) FTCS-PDA/BNC membrane after exposure to water contaminated with *E. coli* for 1 h, (**b**) FTCS-PDA/BNC membrane after in situ PMD operation for 1 h with water contaminated with *E. coli*, (**c**) FTCS-PDA/BNC membrane after the water contaminated with *E. coli* was drained from the top surface, and exposure of the membrane to solar light (1 kW m^{-2}) for a duration of 10 min, (**d**) FTCS-PDA/BNC membrane after exposure to light and washing using distilled water (Adapted from Wu et al. [54]).

Table 2 summarizes the findings of the research works studied in this section.

Table 2. Summary of the performance of nanocellulose-based materials tested for water treatment applications.

Nanocellulose-Based Membrane Material	Summary of Results	Ref.
Polydopamine-modified Cellulose nanocrystals (CNCs)	✓ Pure water permeability of 128.4 L m^{-2} h^{-1} (LMH)/bar ✓ Congo red rejection of 99.91% ✓ Salt permeation of 99.33%	[44]
Reduced graphene oxide (RGO)/cellulose nanofibrils (CNFs)	Varying rGO loading at a fixed 1:1 ratio of rGO to CNF and testing of MO, RhB, AF, BB and RB dyes: • Pure water permeance of 37.2 LMH/bar. • Only BB and RB exhibited more than 90% rejection in all studied membranes.	[45]
Negatively charged carboxylated CNF with trimethylolpropane-tris-(2-methyl-1-aziridine) propionate and graphene oxide	Excellent adsorbent of numerous cations of heavy metals, including Pb^{2+}, Cd^{2+} and Cu^{2+}	[48]

Table 2. Cont.

Nanocellulose-Based Membrane Material	Summary of Results	Ref.
Tetramethylpiperidine-1-oxyl (TEMPO)-oxidized nanocellulose	Adsorption capabilities for divalent cations	[50]
CNF modified by nitro-oxidation	Adsorption of Cd^{2+} was optimal at pH 7 and decreased in acidic and basic conditions	[51]
Polydopamine (PDA) particles and bacterial nanocellulose	• Salt rejection > 99.9% • 1.0 kg m^{-2} h^{-1} permeate flux	[54]
Nanocellulose acetate (NCA)	• Salt rejection of 97.4% • Water flux of 2.2 L/m^2 h	[55]
Microporous network of cellulose composite	High evaporation rate and salt resistance within evaporator	[57]
BNC loaded with a high concentration of polydopamine (PDA)	Solar steam generation efficiency of 78% within evaporator	[58]

3.2. Gas Separation

A prominent application of nanocellulose is in gas separation technologies, with particular application in carbon capture [59–62]. Currently, carbon dioxide (CO_2) accounts for about 76% of total greenhouse gas emissions, and this number is only expected to rise, according to the U.S. Energy Information Administration [63]. As such, carbon capture and storage may play an important role in tackling this issue. Carbon capture can be broken into three strategies, including pre-combustion capture, oxyfuel processes, and post-combustion capture [63]. Among these strategies, post-combustion techniques hold particular interest in most carbon capture and storage (CSS) projects because incorporating different CO_2 separation technologies will not disturb existing processes [63].

Basic CO_2 gas separation methods common in CSS are adsorption and membrane separation, which can be realized through gas separation membranes [63]. Materials used for these membranes must be abundant, low-cost, and sustainable, and the membranes themselves must exhibit high CO_2 permeability and selectivity to be successful. Nanocellulose meets all these requirements, as cellulose can be produced at more than 100 million tons per year, and nanocellulose itself was found to be biodegradable and cheap at $2 USD per kg in 2011 [63–65]. Lastly, nanocellulose has properties that lend themselves to stronger and more effective membranes and adsorbents, such as a surface area ranging from 100 to 200 g/m^2, a tensile strength ranging from 7.5 to 7.77 GPA, and a Young's Modulus of 110–220 GPa [66].

Nanocellulose membranes can be created using a variety of methods, including vacuum filtration, solvent casting, dip coating, and electrospinning [67]. Without any modification, nanocellulose membranes exhibit low gas permeability [68]. As a result, they can be employed as gas barrier materials instead of gas permeation materials. On the other hand, cellulose and nanocellulose can be easily modified due to their profusion of hydroxyl groups. Common strategies for producing cellulosic CO_2 adsorbents include chemically modifying nanocellulose, incorporating inorganic particles into nanocellulose, and modifying nanocellulose with the addition of polymers [63].

An example of chemically modified nanocellulose is nanocellulose aerogel, which is created from the crosslinking and drying of nanocellulose [69]. Nanocellulose aerogel has a high surface area but must be chemically modified to achieve high CO_2 selectivity. Liu et al. modified spherical cellulose nanofibril (CNF) hydrogel by introducing 3–5 wt% N-(2-aminoethyl)-3 aminopropylmethyldimethoxysilane in water at 80 °C or 90 °C for a duration of 10 h, followed by freeze drying [70]. The resulting hydrogel attained a CO_2 adsorption capacity of 1.28–1.78 mmol/g at the higher temperature. This result was supported by a similar study conducted a year later by Zhang et al., who produced an N-(2-aminoethyl)-3 aminopropylmethyldimethoxysilane-modified spherical cellulose nanocrystal (CNC) aerogel with a CO_2 adsorption capacity of 1.68 mmol/g [71]. Inorganic particles have also been employed to increase the CO_2 adsorption of nanocellulose adsorbents. More specifically, Valencia et al. used silicalite-1 zeolite to modify a hybrid CNF-gelatin foam [72]. The resulting composite was able to adsorb up to 1.2 mmol CO_2/g, which is comparable to the CO_2 adsorption of pure silicalite-1. When the foam was further modified by incorporating a zeolitic imidazolate metal-organic framework (ZIF), the CO_2 adsorption and selectivity over nitrogen were found to be improved, and this was attributed to the hierarchical porous structure of ZIF that can facilitate strong interaction with CO_2 in the micropores [73].

Blending nanocellulose with hydrophilic polymers has also been found to improve a membrane's CO_2 permeability [74]. Venturi et al. incorporated 30% nanofibrilated cellulose (NFC) into a polyvinylamine membrane to increase its CO_2 permeability and CO_2/N_2 selectivity [75]. The modified membrane had a CO_2 permeability of 187 Barrer, a CO_2/N_2 selectivity of 100%, and a CO_2/CH_4 selectivity of 22% at 80% relative humidity. In another study, Dai et al. found that hybrid nanocellulose-80%/polyvinyl alcohol (PVA) membranes exhibit higher CO_2 permeance if they are prepared with cellulose nanocrystals (CNC) rather than cellulose nanofibrils (CNF) [76]. As it can be seen in Figure 4, the 80 wt% CNC/PVA membrane had a 65% increase in CO_2 permeance compared to the neat PVA membrane, while the 80 wt% CNF/PVA membrane only had a 15% increase in CO_2 permeance with respect to the PVA membrane. Comparatively, the type of nanocellulose had a negligible effect on the hybrid membranes' CO_2/N_2 selectivity. The 80 wt% CNC/PVA membrane was also able to maintain its high CO_2 permeance and CO_2/N_2 selectivity for over a year.

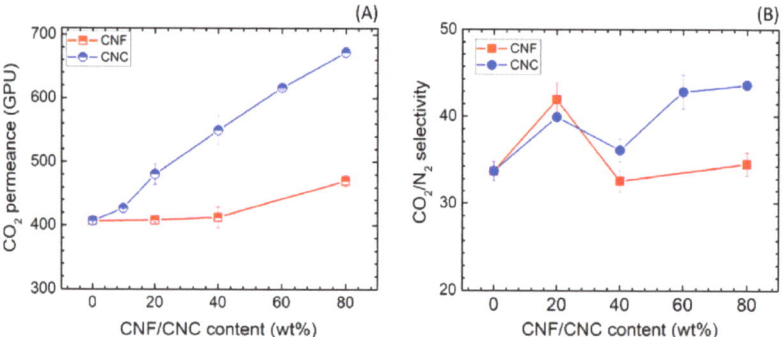

Figure 4. CO_2 permeance (**A**) and CO_2/N_2 selectivity (**B**) of the hybrid nanocellulose/PVA membranes as a function of the nanocellulose content (Adapted from Dai et al. [76]).

In a more recent study, Dai et al. blended cellulose nanocrystals with Polyether-block-amide (Pebax 1657) to produce Pebax/CNC hybrid membranes [74]. The membranes were characterized by mixed-gas permeation tests under dry (relative humidity (RH) = 0%) and humid (RH = 100%) conditions. The results can be summarized in Figure 5. Under dry conditions, it was observed that 5 wt% CNC loading enhanced CO_2 permeability by 29% (104.0 Barrer) compared to membranes with 0 wt% CNC loading. However, the CO_2/N_2 selectivity was not found to improve with the addition of CNC to the membrane. Under humid conditions, it was found that membranes with 5 wt% CNC loading increased CO_2 permeability by 42% (305.7 Barrer) and CO_2/N_2 selectivity by 18% (41.6 separation factor) compared to membranes with 0 wt% CNC loading. Increasing humidity led to higher CO_2 permeability and CO_2/N_2 selectivity, as can be seen in Figure 5. However, a common conclusion for both dry and humid conditions was that further increases in the CNC content above 5 wt% resulted in reduced CO_2 permeability and CO_2/N_2 selectivity. This was attributed to the fact that as the CNC content increases, CNC tends to form a more oriented alignment, resulting in highly packed and complicated hierarchical structures that limit the gas diffusion and, thus, the transport through the CNC/Pebax membranes [74].

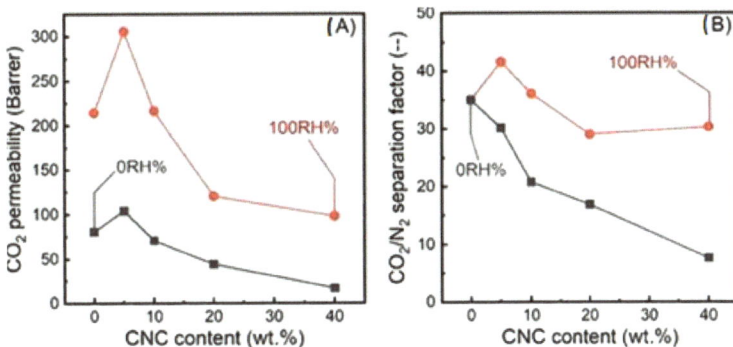

Figure 5. (**A**) CO_2 permeability and (**B**) CO_2/N_2 selectivity of CNC/Pebax membranes under dry (0 RH%) and fully humid (100 RH%) conditions, respectively (Adapted from Dai et al. [74]).

Table 3 summarizes the findings of the research works studied in this section.

Table 3. Summary of the performance of nanocellulose-based materials tested for water treatment applications.

Nanocellulose-Based Membrane Material	Results	Ref.
Modified spherical cellulose nanofibril (CNF) hydrogel	CO_2 adsorption capacity of 1.28–1.78 mmol/g at high temperature	[70]
N-(2-aminoethyl)-3 aminopropylmethyldimethoxysilane modified spherical cellulose nanocrystal (CNC) aerogel	CO_2 adsorption capacity of 1.68 mmol/g	[71]
Silicalite-1 zeolite modified hybrid CNF-gelatin foam	Adsorption up to 1.2 mmol CO_2/g	[72]
Hybrid CNF-gelatin foam with a zeolitic imidazolate metal-organic framework (ZIF)	Greater CO_2 adsorption and selectivity over nitrogen	[73]
Pebax/CNC hybrid membranes	Under dry conditions, 5 wt% CNC loading resulted in: • Enhancement of CO_2 permeability by 29% (104.0 Barrer) • No improvement in CO_2/N_2 selectivity compared to membranes with 0 wt% CNC loading.	[74]
Pebax/CNC hybrid membranes	Under humid conditions, 5 wt% CNC loading led to improvement of: • CO_2 permeability by 42% (305.7 Barrer) • CO_2/N_2 selectivity by 18% (41.6 separation factor) compared to membranes with 0 wt% CNC loading. Increasing humidity resulted in higher CO_2 permeability and CO_2/N_2 selectivity.	[74]
30% nano-fibrilated cellulose (NFC) into a polyvinylamine membrane	CO_2 permeability of 187 Barrer, CO_2/N_2 selectivity of 100%, and CO_2/CH_4 selectivity of 22% at 80% relative humidity	[75]
Hybrid nanocellulose (80%)/polyvinyl alcohol (PVA) membranes	80 wt% CNC/PVA membranes exhibited a 65% increase in CO_2 permeance compared to the pure PVA membrane. (80 wt% CNF/PVA membrane had a 15% increase in CO_2 permeance with respect to the pure PVA membrane.)	[76]

3.3. Biomedical Applications

Another area of interest for nanocellulose is biomedicine. Many unique properties of nanocellulose, including crystallinity, high specific surface area, good rheological properties, ease of alignment, barrier properties, surface chemical reactivity, biocompatibility, and most notably, a lack of toxicity, render this material ideal for use in various biomedical applications, such as immobilization of enzymes, prevention of microbial growth, drug delivery, and virus removal [77–83]. More specifically, nanocellulose hydrogels produced from bacterial or plant cellulose nanofibrils were found to promote cell regeneration and can be applied to tissue engineering scaffolds [84]. Tissue scaffolds assist with wound dressing and cartilage repair that require low cytotoxicity and biocompatibility with the extracellular matrix, which is what nanocellulose hydrogels can provide [85,86]. In addition to its low toxicity, nanocellulose is favored for biosensors due to its biodegradable nature [67]. Biosensors are devices that measure and monitor diagnostic, environmental, safety, and security parameters [33,87].

Another biomedical application of nanocellulose membranes is in Surgicel®, which is a bio-absorbable hemostatic material employed in the prevention of surgery-derived adhesions [88]. Bacterial nanocellulose membranes were produced through electrochemical oxidation with the Tetramethylpiperidine-1-oxyl (TEMPO) radical to be applied to Surgicel® for further improving its hemostatic performance [89]. This improvement was attributed to the enhanced oxidation degree, which increased from 4% to up to 15%. The in vivo biodegradability and biocompatibility of the resulting oxidized nanocellulose-based membranes were assessed through subcutaneous implantation of the membranes in rats and showed a highly biocompatible behavior, triggering only a mild inflammation process [89].

Figure 6 summarizes important biomedical applications of BNC, including biosensors and tissue engineering [33]. The realization of these applications depends on the production feasibility of BNC and nanocellulose in general. Therefore, Sharma et al. [33] address the BNC synthesis strategies in their study, suggesting the utilization of cost-effective substrates that may overcome the barriers associated with BNC production at large scale. These substrates can include agricultural wastes or wastewater rich in sugars from industrial effluents [33]. The challenges for the successful implementation of the nanocellulose biomedical applications are also made important in this study, including the tailoring of the cost of the substrates and the necessary legislation for product approval.

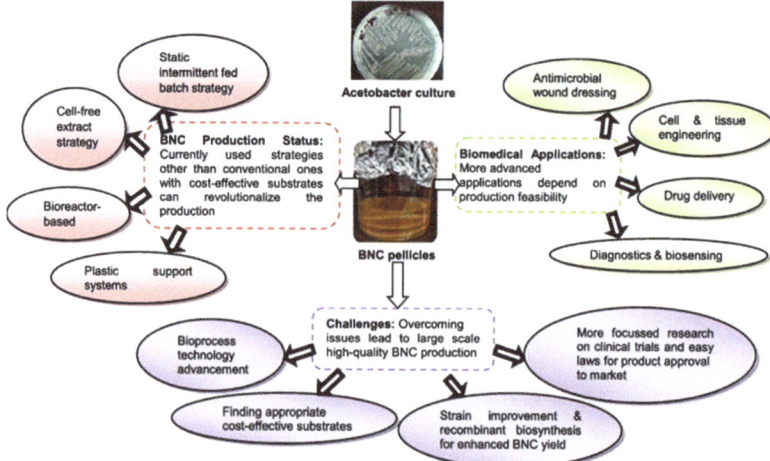

Figure 6. BNC production strategies, biomedical applications, and trends for overcoming challenges (Adapted from Sharma et al. [33]).

Nanocellulose hydrogels have also been employed as a medium for a gel-based blood type test [80]. Curvello et al. showed that traditional gel-based blood typing tests rely on microbeads, which do not always provide well-defined results [90]. This study found that gel columns containing at least 0.3 wt% 2,2,6,6-tetramethylpiperidine-1-oxyl (TEMPO)-oxidized CNFs were able to identify agglutinated and individual red blood cells (RBC) for forward blood typing [80]. TEMPO was chosen because TEMPO-oxidized cellulose can form nanofibers that have hydrophilic carboxylate groups [91]. For reverse blood typing, CNF can be crosslinked with hexamethylenediamine (HMDA) to achieve red blood cell agglutination and separation [90]. Compared to traditional materials used for blood typing tests, nanocellulose is inexpensive and sustainable. In addition to blood typing, nanocellulose hydrogels have shown promise as 3D cell cultures because they are able to accurately portray the extracellular matrix [92].

4. Future of Nanocellulose

Nanocellulose holds unique potential for a variety of applications that have already been addressed in this paper, including desalination, gas separation, automotive applications, and packaging. As such, new research should attempt to find novel uses for nanocellulose beyond pre-established applications. In addition to being investigated as a potential solution to water scarcity and global warming, this green nanomaterial may also pose a solution to the energy crisis; it can constitute an alternative option to fossil fuels by being employed as a sustainable and environmentally friendly material for renewable electronics [93]. Yang et al. posited that nanocellulose membranes can be used for osmotic energy harvesting [94]. Traditionally, osmotic energy harvesting technology has suffered from the required nanofluidic materials being too expensive to justify their practical application. However, Yang et al. found that inexpensive, yet effective membranes can be prepared by cross-linking CNF with 1,2,3,4-butanetetracarboxylic acid (BTCA) [94]. More research should be conducted to fully evaluate nanocellulose's potential in renewable energy harvesting.

Numerous research studies related to food packaging have also been carried out. With plastics being the prominent material in modern packaging, an abundance of plastic pollution along coastlines has resulted from their use. Recently, nanocellulose has emerged as a potential solution for packaging materials. Since nanocellulose is harvested from plant fibers, it can naturally decompose over time, eliminating the issue of waste generation. Representative research works conducted in this area include the isolation of high crystalline nanocellulose from Mimosa pudica plant fibers for packaging applications [1], as well as applying nanocellulose as a starch-based packaging material for food [13]. While at its current stage of development, nanocellulose may not replace plastics entirely as the primary packaging option due to the profit-prioritizing business model of industries; however, it does help promote improvements to address modern issues in our society, such as improved gas barrier properties of the packaging materials. The potential of nanocellulose is seemingly endless. From acoustic materials [95] and cosmetics [96] to complex battery matrices [97] and optical materials [98], a wide range of applications can be implemented with this unique material.

5. Conclusions

Recent studies have shown that nanocellulose has a promising future as a nanomaterial in applications in numerous industrial fields. Advantageous properties, including biodegradability, non-toxicity, low density, thermal stability, long-lasting reinforcing capabilities, and high mechanical strength, have garnered interest in areas such as desalination wastewater treatment, gas separation, and biotechnologies. However, despite all the aforementioned qualities of nanocellulose, further studies are required in order to optimize its production [99] and valorization [100]. Specifically, the high cost, long processing time, and low yield of standard production processes prove to be challenges for this nanomaterial. The elevated cost is mainly attributed to the high energy consumption of the process.

However, recent research has shown that these areas of concern can be addressed through the latest technologies. For instance, a drawback of nanocellulose has been its low thermal stability, but in a recent study by Chen et al., a highly thermally stable nanocellulose-based flexible material was developed that can be utilized in electronics [101].

In conclusion, nanocellulose holds limitless potential in both conventional and unconventional applications. Generally, researchers have focused on nanocellulose's uses in the preparation of chemicals or their handling, such as in separations, desalination, and packaging. But nanocellulose may also have uses in the natural and life sciences, including dermal care applications. Chantereau et al. have shown that bacterial nanocellulose membranes loaded with vitamin-based ionic liquids are ideal candidates for skin care applications due to their high thermal stability and increased solubility [102]. With further advancements, it may even have recreational uses, such as a material for musical instruments [95]. The future of nanocellulose should not be limited to one area of study but instead expanded to reach as many as possible.

Author Contributions: S.N.: Conceptualization, visualization, draft preparation, methodology, writing, review and editing, supervision, project administration, and formal analysis. M.S.: Draft preparation, investigation, writing, and resources. H.L.: Draft preparation, investigation, writing, and resources. R.S.: Draft preparation, visualization, resources, formal analysis, writing, review and editing, and supervision. All authors have read and agreed to the published version of the manuscript.

Funding: This research received no external funding.

Institutional Review Board Statement: Not applicable.

Informed Consent Statement: Not applicable.

Data Availability Statement: Not applicable.

Conflicts of Interest: Dr. Raj Shah is an employee of Koehler Instrument Company Inc., and Meredith Skehan has been an intern at Koehler Instrument Company Inc. The paper reflects the views of the scientists and not the company.

References

1. George, T.S.; Muhammadaly, S.A.; Kanoth, B.P.; Joseph, T.; Dominic, M.D.C.; George, N.; Balachandrakurupp, V.; John, H. Isolation of high crystalline nanocellulose from *Mimosa pudica* plant fibres with potential in packaging applications. *Packag. Technol. Sci.* **2021**, *35*, 163–174. [CrossRef]
2. Julkapli, N.M.; Bagheri, S. Nanocellulose as a green and sustainable emerging material in energy applications: A review. *Polym. Adv. Technol.* **2017**, *28*, 1583–1594. [CrossRef]
3. Pradhan, D.; Jaiswal, A.K.; Jaiswal, S. Emerging technologies for the production of nanocellulose from lignocellulosic biomass. *Carbohydr. Polym.* **2022**, *285*, 119258. [CrossRef] [PubMed]
4. Huang, P.; Wang, C.; Huang, Y.; Wu, M. Chapter 3: Surface Modification of Nanocellulose. *Nanocellulose* **2021**, 65–92. [CrossRef]
5. Nasir, M.; Hashim, R.; Sulaiman, O.; Asim, M. Nanocellulose: Preparation methods and applications. In *Cellulose-Reinforced Nanofibre Composites: Production, Properties and Applications*; Elsevier: Amsterdam, The Netherlands, 2017; pp. 261–276. [CrossRef]
6. Phanthong, P.; Reubroycharoen, P.; Hao, X.; Xu, G.; Abudula, A.; Guan, G. Nanocellulose: Extraction and application. *Carbon Resour. Convers.* **2018**, *1*, 32–43. [CrossRef]
7. Rajinipriya, M.; Nagalakshmaiah, M.; Robert, M.; Elkoun, S. Importance of Agricultural and Industrial Waste in the Field of Nanocellulose and Recent Industrial Developments of Wood Based Nanocellulose: A Review. *ACS Sustain. Chem. Eng.* **2018**, *6*, 2807–2828. [CrossRef]
8. Naz, S.; Ali, J.S.; Zia, M. Nanocellulose isolation characterization and applications: A journey from non-remedial to biomedical claims. *Bio Design Manuf.* **2019**, *2*, 187–212. [CrossRef]
9. Vineeth, S.K.; Gadhave, R.V.; Gadekar, P.T. Chemical Modification of Nanocellulose in Wood Adhesive. *Open J. Polym. Chem.* **2019**, *9*, 86–99. [CrossRef]
10. Köse, K.; Mavlan, M.; Youngblood, J.P. Applications and impact of nanocellulose based adsorbents. *Cellulose* **2020**, *27*, 2967–2990. [CrossRef]
11. Kumar, N.; Kardam, A.; Rajawat, D.S.; Jain, V.K. Suman Carboxymethyl nanocellulose stabilized nano zero-valent iron: An effective method for reduction of hexavalent chromium in wastewater. *Mater. Res. Express* **2019**, *6*, 1150f3. [CrossRef]
12. Rai, G.K.; Singh, V. Study of fabrication and analysis of nanocellulose reinforced polymer matrix composites. *Mater. Today Proc.* **2021**, *38*, 85–88. [CrossRef]

13. Mahardika, M.; Amelia, D.; Azril; Syafri, E. Applications of nanocellulose and its composites in bio packaging-based starch. *Mater. Today Proc.* **2023**, *74*, 415–418. [CrossRef]
14. Valentini, F.; Dorigato, A.; Rigotti, D.; Pegoretti, A. Polyhydroxyalkanoates/Fibrillated Nanocellulose Composites for Additive Manufacturing. *J. Polym. Environ.* **2019**, *27*, 1333–1341. [CrossRef]
15. Shak, K.P.Y.; Pang, Y.L.; Mah, S.K. Nanocellulose: Recent advances and its prospects in environmental remediation. *Beilstein J. Nanotechnol.* **2018**, *9*, 2479–2498. [CrossRef]
16. Tan, K.; Heo, S.; Foo, M.; Chew, I.M.; Yoo, C. An insight into nanocellulose as soft condensed matter: Challenge and future prospective toward environmental sustainability. *Sci. Total. Environ.* **2019**, *650*, 1309–1326. [CrossRef] [PubMed]
17. Lan, G.-X.; Liu, Y.; Zhou, N.; Guo, D.-Q.; Ma, M.-G. Multifunctional nanocellulose-based composites for potential environmental applications. *Cellulose* **2023**, *30*, 39–60. [CrossRef]
18. Zeng, H.; Hao, H.; Wang, X.; Shao, Z. Chitosan-based composite film adsorbents reinforced with nanocellulose for removal of Cu(II) ion from wastewater: Preparation, characterization, and adsorption mechanism. *Int. J. Biol. Macromol.* **2022**, *213*, 369–380. [CrossRef]
19. Yin, Z.; Li, M.; Li, Z.; Deng, Y.; Xue, M.; Chen, Y.; Ou, J.; Lei, S.; Luo, Y.; Xie, C. A harsh environment resistant robust Co(OH)$_2$@stearic acid nanocellulose-based membrane for oil-water separation and wastewater purification. *J. Environ. Manag.* **2023**, *342*, 118127. [CrossRef]
20. Pendergast, M.M.; Hoek, E.M. A review of water treatment membrane nanotechnologies. *Energy Environ. Sci.* **2011**, *4*, 1946–1971. [CrossRef]
21. Vineis, P.; Chan, Q.; Khan, A. Climate change impacts on water salinity and health. *J. Epidemiol. Glob. Health* **2011**, *1*, 5. [CrossRef]
22. Torstensen, J.Ø.; Helberg, R.M.; Deng, L.; Gregersen, O.W.; Syverud, K. PVA/nanocellulose nanocomposite membranes for CO$_2$ separation from flue gas. *Int. J. Greenh. Gas Control* **2019**, *81*, 93–102. [CrossRef]
23. Jaekel, E.E.; Kluge, S.; Tröger-Müller, S.; Tutuş, M.; Filonenko, S. Tunable Gas Permeation Behavior in Self-Standing Cellulose Nanocrystal-Based Membranes. *ACS Sustain. Chem. Eng.* **2022**, *10*, 12895–12905. [CrossRef]
24. Mithra, S.N.; Ahankari, S. Nanocellulose-based membranes for CO$_2$ separation from biogas through the facilitated transport mechanism: A review. *Mater. Today Sustain.* **2022**, *19*, 100191. [CrossRef]
25. Casadei, R.; Firouznia, E.; Baschetti, M.G. Effect of Mobile Carrier on the Performance of PVAm–Nanocellulose Facilitated Transport Membranes for CO$_2$ Capture. *Membranes* **2021**, *11*, 442. [CrossRef] [PubMed]
26. Patel, D.K.; Dutta, S.D.; Lim, K.-T. Nanocellulose-based polymer hybrids and their emerging applications in biomedical engineering and water purification. *RSC Adv.* **2019**, *9*, 19143–19162. [CrossRef] [PubMed]
27. Nguyen, L.H.; Naficy, S.; Chandrawati, R.; Dehghani, F. Nanocellulose for Sensing Applications. *Adv. Mater. Interfaces* **2019**, *6*, 1900424. [CrossRef]
28. Michelin, M.; Gomes, D.G.; Romaní, A.; Polizeli, M.d.L.T.M.; Teixeira, J.A. Nanocellulose Production: Exploring the Enzymatic Route and Residues of Pulp and Paper Industry. *Molecules* **2020**, *25*, 3411. [CrossRef]
29. Chen, M.; Ma, Q.; Zhu, J.Y.; Alonso, D.M.; Runge, T. GVL pulping facilitates nanocellulose production from woody biomass. *Green Chem.* **2019**, *21*, 5316–5325. [CrossRef]
30. Charreau, H.; Cavallo, E.; Foresti, M.L. Patents involving nanocellulose: Analysis of their evolution since 2010. *Carbohydr. Polym.* **2020**, *237*, 116039. [CrossRef]
31. Huo, Y.; Liu, Y.; Xia, M.; Du, H.; Lin, Z.; Li, B.; Liu, H. Nanocellulose-Based Composite Materials Used in Drug Delivery Systems. *Polymers* **2022**, *14*, 2648. [CrossRef]
32. Trache, D.; Tarchoun, A.F.; Derradji, M.; Hamidon, T.S.; Masruchin, N.; Brosse, N.; Hussin, M.H. Nanocellulose: From Fundamentals to Advanced Applications. *Front. Chem.* **2020**, *8*, 392. [CrossRef] [PubMed]
33. Sharma, C.; Bhardwaj, N.K. Bacterial nanocellulose: Present status, biomedical applications and future perspectives. *Mater. Sci. Eng. C* **2019**, *104*, 109963. [CrossRef] [PubMed]
34. Derami, H.G.; Gupta, P.; Gupta, R.; Rathi, P.; Morrissey, J.J.; Singamaneni, S. Palladium Nanoparticle-Decorated Mesoporous Polydopamine/Bacterial Nanocellulose as a Catalytically Active Universal Dye Removal Ultrafiltration Membrane. *ACS Appl. Nano Mater.* **2020**, *3*, 5437–5448. [CrossRef]
35. Numata, Y.; Sakata, T.; Furukawa, H.; Tajima, K. Bacterial cellulose gels with high mechanical strength. *Mater. Sci. Eng. C* **2015**, *47*, 57–62. [CrossRef]
36. El-Hoseny, S.M.; Basmaji, P.; de Olyveira, G.M.; Costa, L.M.M.; Alwahedi, A.M.; Oliveira, J.D.D.C.; Francozo, G.B. Natural ECM-Bacterial Cellulose Wound Healing—Dubai Study. *J. Biomater. Nanobiotechnol.* **2015**, *6*, 237–246. [CrossRef]
37. Mekonnen, M.M.; Hoekstra, A.Y. Four billion people facing severe water scarcity. *Sci. Adv.* **2016**, *2*, e1500323. [CrossRef]
38. Das, R.; Lindström, T.; Sharma, P.S.; Chi, K.; Hsiao, B.S. Nanocellulose for Sustainable Water Purification. *Chem. Rev.* **2022**, *122*, 8936–9031. [CrossRef]
39. Saud, A.; Saleem, H.; Zaidi, S.J. Progress and Prospects of Nanocellulose-Based Membranes for Desalination and Water Treatment. *Membranes* **2022**, *12*, 462. [CrossRef]
40. Reshmy, R.; Philip, E.; Paul, S.A.; Madhavan, A.; Sindhu, R.; Binod, P.; Pandey, A.; Sirohi, R. Nanocellulose-based products for sustainable applications-recent trends and possibilities. *Rev. Environ. Sci. Bio/Technol.* **2020**, *19*, 779–806. [CrossRef]
41. Chung, N.H.; Van Binh, N.; Dien, L.Q. Preparation of nanocellulose acetate from bleached hardwood pulp and its application for seawater desalination. *Vietnam J. Chem.* **2020**, *58*, 281–286. [CrossRef]

42. Rezaei-DashtArzhandi, M.; Sarrafzadeh, M.H.; Goh, P.S.; Lau, W.J.; Ismail, A.F.; Wong, K.C.; Mohamed, M.A. Enhancing the desalination performance of forward osmosis membrane through the incorporation of green nanocrystalline cellulose and halloysite dual nanofillers. *J. Chem. Technol. Biotechnol.* **2020**, *95*, 2359–2370. [CrossRef]
43. Sijabat, E.K.; Nuruddin, A.; Aditiawati, P.; Purwasasmita, B.S. Synthesis and Characterization of Bacterial Nanocellulose from Banana Peel for Water Filtration Membrane Application. *J. Physics Conf. Ser.* **2019**, *1230*, 012085. [CrossRef]
44. Yang, L.; Liu, X.; Zhang, X.; Chen, T.; Ye, Z.; Rahaman, S. High performance nanocomposite nanofiltration membranes with polydopamine-modified cellulose nanocrystals for efficient dye/salt separation. *Desalination* **2021**, *521*, 115385. [CrossRef]
45. Mohammed, S.; Hegab, H.M.; Ou, R.; Liu, S.; Ma, H.; Chen, X.; Sridhar, T.; Wang, H. Effect of oxygen plasma treatment on the nanofiltration performance of reduced graphene oxide/cellulose nanofiber composite membranes. *Green Chem. Eng.* **2020**, *2*, 122–131. [CrossRef]
46. Güzel, F.; Yakut, H.; Topal, G. Determination of kinetic and equilibrium parameters of the batch adsorption of Mn(II), Co(II), Ni(II) and Cu(II) from aqueous solution by black carrot (*Daucus carota* L.) residues. *J. Hazard. Mater.* **2008**, *153*, 1275–1287. [CrossRef]
47. Aoudi, B.; Boluk, Y.; El-Din, M.G. Recent advances and future perspective on nanocellulose-based materials in diverse water treatment applications. *Sci. Total. Environ.* **2022**, *843*, 156903. [CrossRef] [PubMed]
48. Mo, L.; Pang, H.; Lu, Y.; Li, Z.; Kang, H.; Wang, M.; Zhang, S.; Li, J. Wood-inspired nanocellulose aerogel adsorbents with excellent selective pollutants capture, superfast adsorption, and easy regeneration. *J. Hazard. Mater.* **2021**, *415*, 125612. [CrossRef]
49. Besbes, I.; Alila, S.; Boufi, S. Nanofibrillated cellulose from TEMPO-oxidized eucalyptus fibres: Effect of the carboxyl content. *Carbohydr. Polym.* **2011**, *84*, 975–983. [CrossRef]
50. Sehaqui, H.; De Larraya, U.P.; Liu, P.; Pfenninger, N.; Mathew, A.P.; Zimmermann, T.; Tingaut, P. Enhancing adsorption of heavy metal ions onto biobased nanofibers from waste pulp residues for application in wastewater treatment. *Cellulose* **2014**, *21*, 2831–2844. [CrossRef]
51. Sharma, P.R.; Chattopadhyay, A.; Sharma, S.K.; Geng, L.; Amiralian, N.; Martin, D.; Hsiao, B.S. Nanocellulose from Spinifex as an Effective Adsorbent to Remove Cadmium(II) from Water. *ACS Sustain. Chem. Eng.* **2018**, *6*, 3279–3290. [CrossRef]
52. Leitch, M.E.; Li, C.; Ikkala, O.; Mauter, M.S.; Lowry, G.V. Bacterial Nanocellulose Aerogel Membranes: Novel High-Porosity Materials for Membrane Distillation. *Environ. Sci. Technol. Lett.* **2016**, *3*, 85–91. [CrossRef]
53. Xu, T.; Jiang, Q.; Ghim, D.; Liu, K.-K.; Sun, H.; Derami, H.G.; Wang, Z.; Tadepalli, S.; Jun, Y.-S.; Zhang, Q.; et al. Catalytically Active Bacterial Nanocellulose-Based Ultrafiltration Membrane. *Small* **2018**, *14*, e1704006. [CrossRef] [PubMed]
54. Wu, X.; Cao, S.; Ghim, D.; Jiang, Q.; Singamaneni, S.; Jun, Y.-S. A thermally engineered polydopamine and bacterial nanocellulose bilayer membrane for photothermal membrane distillation with bactericidal capability. *Nano Energy* **2021**, *79*, 105353. [CrossRef]
55. Morsy, A.; Mahmoud, A.S.; Soliman, A.; Ibrahim, H.; Fadl, E. Improved anti-biofouling resistances using novel nanocelluloses/cellulose acetate extracted from rice straw based membranes for water desalination. *Sci. Rep.* **2022**, *12*, 4386. [CrossRef]
56. Chen, C.; Kuang, Y.; Hu, L. Challenges and Opportunities for Solar Evaporation. *Joule* **2019**, *3*, 683–718. [CrossRef]
57. Wu, W.; Xu, Y.; Ma, X.; Tian, Z.; Zhang, C.; Han, J.; Han, X.; He, S.; Duan, G.; Li, Y. Cellulose-based Interfacial Solar Evaporators: Structural Regulation and Performance Manipulation. *Adv. Funct. Mater.* **2023**, *33*, 2302351. [CrossRef]
58. Jiang, Q.; Derami, H.G.; Ghim, D.; Cao, S.; Jun, Y.-S.; Singamaneni, S. Polydopamine-filled bacterial nanocellulose as a biodegradable interfacial photothermal evaporator for highly efficient solar steam generation. *J. Mater. Chem. A* **2017**, *5*, 18397–18402. [CrossRef]
59. Janakiram, S.; Ansaloni, L.; Jin, S.-A.; Yu, X.; Dai, Z.; Spontak, R.J.; Deng, L. Humidity-responsive molecular gate-opening mechanism for gas separation in ultraselective nanocellulose/IL hybrid membranes. *Green Chem.* **2020**, *22*, 3546–3557. [CrossRef]
60. Janakiram, S.; Yu, X.; Ansaloni, L.; Dai, Z.; Deng, L. Manipulation of Fibril Surfaces in Nanocellulose-Based Facilitated Transport Membranes for Enhanced CO_2 Capture. *ACS Appl. Mater. Interfaces* **2019**, *11*, 33302–33313. [CrossRef]
61. Venturi, D.; Chrysanthou, A.; Dhuiège, B.; Missoum, K.; Baschetti, M.G. Arginine/Nanocellulose Membranes for Carbon Capture Applications. *Nanomaterials* **2019**, *9*, 877. [CrossRef]
62. Zhang, S.; Shen, Y.; Wang, L.; Chen, J.; Lu, Y. Phase change solvents for post-combustion CO_2 capture: Principle, advances, and challenges. *Appl. Energy* **2019**, *239*, 876–897. [CrossRef]
63. Ho, N.A.D.; Leo, C. A review on the emerging applications of cellulose, cellulose derivatives and nanocellulose in carbon capture. *Environ. Res.* **2021**, *197*, 111100. [CrossRef] [PubMed]
64. Panchal, P.; Ogunsona, E.; Mekonnen, T. Trends in Advanced Functional Material Applications of Nanocellulose. *Processes* **2018**, *7*, 10. [CrossRef]
65. Spence, K.L.; Venditti, R.A.; Rojas, O.J.; Habibi, Y.; Pawlak, J.J. A comparative study of energy consumption and physical properties of microfibrillated cellulose produced by different processing methods. *Cellulose* **2011**, *18*, 1097–1111. [CrossRef]
66. Ibrahim, H.; Sazali, N.; Salleh, W.N.W.; Abidin, M.N.Z. A short review on recent utilization of nanocellulose for wastewater remediation and gas separation. *Mater. Today Proc.* **2021**, *42*, 45–49. [CrossRef]
67. Dai, Z.; Ottesen, V.; Deng, J.; Helberg, R.M.L.; Deng, L. A Brief Review of Nanocellulose Based Hybrid Membranes for CO_2 Separation. *Fibers* **2019**, *7*, 40. [CrossRef]
68. Wu, C.-N.; Saito, T.; Fujisawa, S.; Fukuzumi, H.; Isogai, A. Ultrastrong and High Gas-Barrier Nanocellulose/Clay-Layered Composites. *Biomacromolecules* **2012**, *13*, 1927–1932. [CrossRef]
69. Sun, Y.; Chu, Y.; Wu, W.; Xiao, H. Nanocellulose-based lightweight porous materials: A review. *Carbohydr. Polym.* **2021**, *255*, 117489. [CrossRef]

70. Liu, S.; Zhang, Y.; Jiang, H.; Wang, X.; Zhang, T.; Yao, Y. High CO_2 adsorption by amino-modified bio-spherical cellulose nanofibres aerogels. *Environ. Chem. Lett.* **2018**, *16*, 605–614. [CrossRef]
71. Zhang, T.; Zhang, Y.; Jiang, H.; Wang, X. Aminosilane-grafted spherical cellulose nanocrystal aerogel with high CO_2 adsorption capacity. *Environ. Sci. Pollut. Res.* **2019**, *26*, 16716–16726. [CrossRef]
72. Valencia, L.; Rosas-Arbelaez, W.; Aguilar-Sanchez, A.; Mathew, A.P.; Palmqvist, A.E.C. Bio-based Micro-/Meso-/Macroporous Hybrid Foams with Ultrahigh Zeolite Loadings for Selective Capture of Carbon Dioxide. *ACS Appl. Mater. Interfaces* **2019**, *11*, 40424–40431. [CrossRef] [PubMed]
73. Valencia, L.; Abdelhamid, H.N. Nanocellulose leaf-like zeolitic imidazolate framework (ZIF-L) foams for selective capture of carbon dioxide. *Carbohydr. Polym.* **2019**, *213*, 338–345. [CrossRef]
74. Dai, Z.; Deng, J.; Ma, Y.; Guo, H.; Wei, J.; Wang, B.; Jiang, X.; Deng, L. Nanocellulose Crystal-Enhanced Hybrid Membrane for CO_2 Capture. *Ind. Eng. Chem. Res.* **2022**, *61*, 9067–9076. [CrossRef]
75. Venturi, D.; Grupkovic, D.; Sisti, L.; Baschetti, M.G. Effect of humidity and nanocellulose content on Polyvinylamine-nanocellulose hybrid membranes for CO_2 capture. *J. Membr. Sci.* **2018**, *548*, 263–274. [CrossRef]
76. Dai, Z.; Deng, J.; Yu, Q.; Helberg, R.M.L.; Janakiram, S.; Ansaloni, L.; Deng, L. Fabrication and Evaluation of Bio-Based Nanocomposite TFC Hollow Fiber Membranes for Enhanced CO_2 Capture. *ACS Appl. Mater. Interfaces* **2019**, *11*, 10874–10882. [CrossRef] [PubMed]
77. Lin, N.; Dufresne, A. Nanocellulose in biomedicine: Current status and future prospect. *Eur. Polym. J.* **2014**, *59*, 302–325. [CrossRef]
78. Gustafsson, S.; Manukyan, L.; Mihranyan, A. Protein–Nanocellulose Interactions in Paper Filters for Advanced Separation Applications. *Langmuir* **2017**, *33*, 4729–4736. [CrossRef]
79. dos Santos, C.A.; dos Santos, G.R.; Soeiro, V.S.; dos Santos, J.R.; Rebelo, M.D.A.; Chaud, M.V.; Gerenutti, M.; Grotto, D.; Pandit, R.; Rai, M.; et al. Bacterial nanocellulose membranes combined with nisin: A strategy to prevent microbial growth. *Cellulose* **2018**, *25*, 6681–6689. [CrossRef]
80. Sampaio, L.M.; Padrão, J.; Faria, J.; Silva, J.P.; Silva, C.J.; Dourado, F.; Zille, A. Laccase immobilization on bacterial nanocellulose membranes: Antimicrobial, kinetic and stability properties. *Carbohydr. Polym.* **2016**, *145*, 1–12. [CrossRef]
81. Norrrahim, M.N.F.; Kasim, N.A.M.; Knight, V.F.; Ong, K.K.; Noor, S.A.M.; Halim, N.A.; Shah, N.A.A.; Jamal, S.H.; Janudin, N.; Misenan, M.S.M. Emerging Developments Regarding Nanocellulose-Based Membrane Filtration Material against Microbes. *Polymers* **2021**, *13*, 3249. [CrossRef]
82. Abba, M.; Ibrahim, Z.; Chong, C.S.; Zawawi, N.A.; Kadir, M.R.A.; Yusof, A.H.M.; Razak, S.I.A. Transdermal Delivery of Crocin Using Bacterial Nanocellulose Membrane. *Fibers Polym.* **2019**, *20*, 2025–2031. [CrossRef]
83. Pachuau, L.S. A Mini Review on Plant-based Nanocellulose: Production, Sources, Modifications and Its Potential in Drug Delivery Applications. *Mini-Rev. Med. Chem.* **2015**, *15*, 543–552. [CrossRef] [PubMed]
84. Shi, Z.; Ullah, M.W.; Liang, X.; Yang, G. Chapter 5: Recent Developments in Synthesis, Properties, and Biomedical Applications of Cellulose-Based Hydrogels. *Nanocellulose* **2021**, 121–153. [CrossRef]
85. Zhang, M.; Guo, N.; Sun, Y.; Shao, J.; Liu, Q.; Zhuang, X.; Twebaze, C.B. Nanocellulose aerogels from banana pseudo-stem as a wound dressing. *Ind. Crop. Prod.* **2023**, *194*, 116383. [CrossRef]
86. Reshmy, R.; Philip, E.; Madhavan, A.; Arun, K.B.; Binod, P.; Pugazhendhi, A.; Awasthi, M.K.; Gnansounou, E.; Pandey, A.; Sindhu, R. Promising eco-friendly biomaterials for future biomedicine: Cleaner production and applications of Nanocellulose. *Environ. Technol. Innov.* **2021**, *24*, 101855. [CrossRef]
87. Eldhose, M.; Roy, R.; George, C.; Joseph, A. Sensing and Biosensing Applications of Nanocellulose. In *Handbook of Biopolymers*; Springer: Berlin/Heidelberg, Germany, 2023; pp. 1007–1032. [CrossRef]
88. Roshkovan, L.; Singhal, S.; Katz, I.S.; Galperin-Aizenberg, M. Multimodality imaging of Surgicel®, an important mimic of post-operative complication in the thorax. *BJR Open* **2021**, *3*, 20210031. [CrossRef]
89. Queirós, E.C.; Pinheiro, S.P.; Pereira, J.E.; Prada, J.; Pires, I.; Dourado, F.; Parpot, P.; Gama, M. Hemostatic Dressings Made of Oxidized Bacterial Nanocellulose Membranes. *Polysaccharides* **2021**, *2*, 80–99. [CrossRef]
90. Curvello, R.; Mendoza, L.; McLiesh, H.; Manolios, J.; Tabor, R.F.; Garnier, G. Nanocellulose Hydrogel for Blood Typing Tests. *ACS Appl. Bio Mater.* **2019**, *2*, 2355–2364. [CrossRef]
91. Saito, T.; Kimura, S.; Nishiyama, Y.; Isogai, A. Cellulose Nanofibers Prepared by TEMPO-Mediated Oxidation of Native Cellulose. *Biomacromolecules* **2007**, *8*, 2485–2491. [CrossRef]
92. Curvello, R.; Raghuwanshi, V.S.; Garnier, G. Engineering nanocellulose hydrogels for biomedical applications. *Adv. Colloid Interface Sci.* **2019**, *267*, 47–61. [CrossRef]
93. Lasrado, D.; Ahankari, S.; Kar, K. Nanocellulose-based polymer composites for energy applications—A review. *J. Appl. Polym. Sci.* **2020**, *137*, 48959. [CrossRef]
94. Yang, H.; Gueskine, V.; Berggren, M.; Engquist, I. Cross-Linked Nanocellulose Membranes for Nanofluidic Osmotic Energy Harvesting. *ACS Appl. Energy Mater.* **2022**, *5*, 15740–15748. [CrossRef]
95. Farid, M.; Purniawan, A.; Susanti, D.; Priyono, S.; Ardhyananta, H.; Rahmasita, M.E. Nanocellulose based polymer composite for acoustic materials. *AIP Conf. Proc.* **2018**, *1945*, 020025. [CrossRef]

96. De Amorim, J.D.P.; De Souza, K.C.; Duarte, C.R.; Da Silva Duarte, I.; de Assis Sales Ribeiro, F.; Silva, G.S.; De Farias, P.M.A.; Stingl, A.; Costa, A.F.S.; Vinhas, G.M.; et al. Plant and bacterial nanocellulose: Production, properties and applications in medicine, food, cosmetics, electronics and engineering. A review. *Environ. Chem. Lett.* **2020**, *18*, 851–869. [CrossRef]
97. Wang, Z.; Chen, Z.; Zheng, Z.; Liu, H.; Zhu, L.; Yang, M.; Chen, Y. Nanocellulose-based membranes for highly efficient molecular separation. *Chem. Eng. J.* **2023**, *451*, 138711. [CrossRef]
98. Qi, Y.; Guo, Y.; Liza, A.A.; Yang, G.; Sipponen, M.H.; Guo, J.; Li, H. Nanocellulose: A review on preparation routes and applications in functional materials. *Cellulose* **2023**, *30*, 4115–4147. [CrossRef]
99. Ribeiro, R.S.; Pohlmann, B.C.; Calado, V.; Bojorge, N.; Pereira, N., Jr. Production of nanocellulose by enzymatic hydrolysis: Trends and challenges. *Eng. Life Sci.* **2019**, *19*, 279–291. [CrossRef]
100. Pires, J.R.; Souza, V.G.; Fernando, A.L. Valorization of energy crops as a source for nanocellulose production—Current knowledge and future prospects. *Ind. Crop. Prod.* **2019**, *140*, 111642. [CrossRef]
101. Chen, Z.; Hu, Y.; Shi, G.; Zhuo, H.; Ali, M.A.; Jamróz, E.; Zhang, H.; Zhong, L.; Peng, X. Advanced Flexible Materials from Nanocellulose. *Adv. Funct. Mater.* **2023**, *33*, 2214245. [CrossRef]
102. Chantereau, G.; Sharma, M.; Abednejad, A.; Vilela, C.; Costa, E.; Veiga, M.; Antunes, F.; Pintado, M.; Sèbe, G.; Coma, V.; et al. Bacterial nanocellulose membranes loaded with vitamin B-based ionic liquids for dermal care applications. *J. Mol. Liq.* **2020**, *302*, 112547. [CrossRef]

Disclaimer/Publisher's Note: The statements, opinions and data contained in all publications are solely those of the individual author(s) and contributor(s) and not of MDPI and/or the editor(s). MDPI and/or the editor(s) disclaim responsibility for any injury to people or property resulting from any ideas, methods, instructions or products referred to in the content.

Article

Etching and Doping of Pores in Polyethylene Terephthalate Analyzed by Ion Transmission Spectroscopy and Nuclear Depth Profiling

Giovanni Ceccio [1,*], Jiri Vacik [1], Jakub Siegel [2], Antonino Cannavó [1], Andrey Choukourov [3], Pavel Pleskunov [3], Marco Tosca [3,4] and Dietmar Fink [1]

[1] Department of Neutron Physics, Nuclear Physics Institute (NPI) of the Czech Academy of Sciences (CAS), 250 68 Husinec, Czech Republic
[2] Department of Solid State Engineering, University of Chemistry and Technology Prague, 166 28 Prague, Czech Republic
[3] Department of Macromolecular Physics, Faculty of Mathematics and Physics, Charles University, V Holesovickach 2, 180 00 Prague, Czech Republic
[4] ELI —Beamlines Centre, Institute of Physics (FZU), Czech Academy of Sciences, 252 41 Dolni Brezany, Czech Republic
* Correspondence: ceccio@ujf.cas.cz

Citation: Ceccio, G.; Vacik, J.; Siegel, J.; Cannavó, A.; Choukourov, A.; Pleskunov, P.; Tosca, M.; Fink, D. Etching and Doping of Pores in Polyethylene Terephthalate Analyzed by Ion Transmission Spectroscopy and Nuclear Depth Profiling. *Membranes* **2022**, *12*, 1061. https://doi.org/10.3390/membranes12111061

Academic Editors: Annarosa Gugliuzza and Cristiana Boi

Received: 23 September 2022
Accepted: 25 October 2022
Published: 28 October 2022

Publisher's Note: MDPI stays neutral with regard to jurisdictional claims in published maps and institutional affiliations.

Copyright: © 2022 by the authors. Licensee MDPI, Basel, Switzerland. This article is an open access article distributed under the terms and conditions of the Creative Commons Attribution (CC BY) license (https:// creativecommons.org/licenses/by/ 4.0/).

Abstract: This work is devoted to the study of controlled preparation and filling of pores in polyethylene terephthalate (PET) membranes. A standard wet chemical etching with different protocols (isothermal and isochronous etching for different times and temperatures and etching from one or both sides of the films) was used to prepare the micrometric pores. The pores were filled with either a LiCl solution or boron deposited by magnetron sputtering. Subsequent control of the pore shape and dopant filling was performed using the nuclear methods of ion transmission spectroscopy (ITS) and neutron depth profiling (NDP). It turned out that wet chemical etching, monitored and quantified by ITS, was shown to enable the preparation of the desired simple pore geometry. Furthermore, the effect of dopant filling on the pore shape could be well observed and analyzed by ITS and, for relevant light elements, by NDP, which can determine their depth (and spatial) distribution. In addition, both non-destructive methods were proven to be suitable and effective tools for studying the preparation and filling of pores in thin films. Thus, they can be considered promising for research into nanostructure technologies of thin porous membranes.

Keywords: membrane; ion-track etching; nuclear methodologies; pores

1. Introduction

Polymeric membranes with etched ion tracks are known as polymer films with straight pore channels and have been investigated for decades and are still being studied today for their wide range of applications in applied and basic sciences [1–5]. A unique property of ion track membranes is that the pore density and pore diameter can be varied independently, which makes this polymeric membrane particularly interesting. An important aspect that must be thoroughly studied for this membrane is the full control of the membrane porous structure (i.e., the shape of the pores). Additional important properties that characterize the membrane are the transport properties and functionalization of pores by appropriate modification [6–9]. Different etching procedures allow for obtaining different pores geometries (cylinders, cones, double cones, cigar-like, and so on),; it has been shown that changes in the properties of membranes (e.g., diffusion and transport mechanisms, mechanical parameters, electric conductivity) can be achieved by variation of pores' geometry, whereby during fabrication of the pores (using an adequate etching protocol) or by post-etching processing (e.g., thermal annealing), or also by inserting (and grafting) of specific materials into the pores [10–12]. Specifically, this last aspect is important for modern applications, such as fabrication of efficient catalysts or separators in Li-ion batteries [13,14]. For this reason, it is particularly important to have a full control over the pore shape during the preparation, and this can be achieved with adequate methodologies for the study of such shapes during different steps of the etching protocol. The importance of pore diameter is crucial for some properties of the membranes. The wettability [5], ability of liquid to maintain the balance between the intermolecular interactions of adhesive and cohesive types on the solid, the diffusion and separations of fluids [15], such as gas or liquid, the ion transportation phenomena [16], or the sensing ability [4] of functionalized membranes are characteristics strictly related to the pore shape and diameter.

Various sophisticated methods are used for the determination of structural parameters of nuclear pores. They are either non-destructive (e.g., X-ray and neutron diffraction) or invasive, damaging locally or in large areas the investigated polymers (e.g., SEM cross-section imaging, FIB-based techniques) [17,18]. Is well known that SEM studies provide highly accurate measurements of pore geometry and surface morphology; unfortunately, the method is limited in surface area and highly costly due to the equipment needed. Another method to determine the pore diameters is conductometry [19], which allows for determining the effective pore diameter by analyzing the voltage characteristic both after and during the etching process, but the disadvantage lies in the time-consuming process and sophisticated calculations that are model dependent. Recently suggested was the investigation of ion track membranes by non-contact ultrasonic spectroscopy [20], which takes advantage of the two propagation paths through membranes prepared with the ion track technique for the measurements of pore diameter and density. However, the procedure is time consuming and dependent on the mathematical model. For the purpose of this study, it was necessary to frequently analyze geometrical parameters of the pores during the etching process (at selected etching time) and monitor the influence of dopants, and the degree of their filling, on the shape of the pores. Therefore, it was necessary to use appropriate methods that would not damage the samples during repeated measurements, i.e., that would keep the geometry of the pores unchanged and allow several measurements with direct results. For this reason, ion transmission spectroscopy (with a low ion fluence) was chosen, which was able to analyze the 3D geometry of the pores, degree of filling with selected dopants, and also their effect on the pore shape. The depth (spatial) distribution of the defined material inserted into the pores (light elements—lithium and boron) was determined by neutron depth profiling. Both nuclear analytical methods are available in the equipment portfolio of the CANAM infrastructure at NPI Řež. In particular, the ITS method results in low cost with immediate interpretation of the filling degree of the membranes, based only on stopping power of the ions in the polymer. The aim of this study was to find out whether it is possible to follow the formation of the pores (and their shape) at different etching steps using a wet chemical etching method, and whether it is possible to fill the pores (and to what extent) with a material suitable for specific purposes, such as Li-ion batteries or targets for proton–boron fusion. For this project, it was important to provide suitable diagnostic methods that would allow for obtaining relevant information about the shape of the pores and their filling with selected material at various stages of etching.

2. Materials and Methods

2.1. Membrane Preparation

Polymer membranes were prepared by wet chemical etching of ion tracks [21,22] in 19 μm thick polyethylene terephthalate (PET) films (provenance: Hostaphan® Mitsubishi Polyester Films, Wiesbaden, Germany) irradiated by Xe^{+26} ions with an energy of 1.2 MeV/u (the ion irradiation of PET films was carried out at JINR Dubna by Dr. P.Y. Apel). Several different etching protocols were applied to prepare membranes with different geometries of pores:

(i) Asymmetric etching (i.e., one-side etching protocol) of the irradiated foils, performed consecutively, was applied to study the gradual development of pores under both isochronal and isothermal conditions. Different pore shapes were obtained for different etching temperatures and exposure times. For the etching procedure, a 9M NaOH solution was used in a temperature range of 55–75 °C and etching times 0–60 min. For subsequent doping, 5M LiCl solution was selected as a dopant, and doping was carried out for 24 h at RT only from the side of etching at different stages of the pore development. After removing the sample from the dopant vessel, the sample surface was gently dried and cleaned from the excess dopant solution by wiping a smooth cloth over the sample surface.

(ii) In addition to one-side etching (performed in NPI Řež), a double-side etching protocol was also applied (JINR Dubna). Symmetric etching procedures made it possible to create membranes with cylindrical pores of several different diameters of 7 µm, 2.4 µm, and 0.53 µm (see Figure 1). Doping with boron was performed on only one side of the membrane (in addition to the PET films, a Si wafer was also used for the comparative analysis). Boron (99.9%, Kurt J. Lesker) was sputtered in Ar under a pressure of 5 Pa using a 3-inch planar magnetron powered by a radio frequency (13.56 MHz) power source. The power gradually increased (from 20 to 80 W) to avoid a thermal shock and cracking of the target. The total time of the deposition was 30 min, with a set thickness of 40 nm. Before deposition, the surface of the B target was pre-sputtered (cleaned) for 20 min.

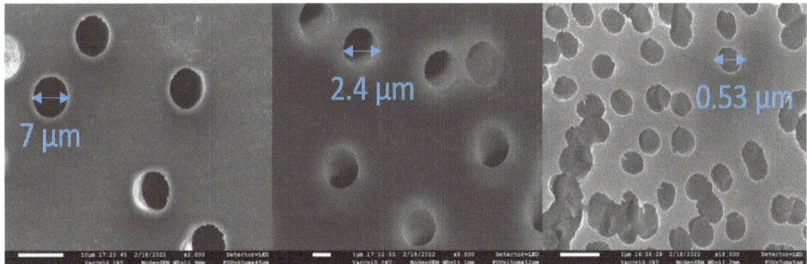

Figure 1. SEM micrographs of etched pores with diameters of 7.0 µm, 2.4 µm, and 0.53 µm.

2.2. Analytical Method

Specific nuclear analytical methods ITS and NDP, available in the NPI CANAM infrastructure [23], were applied to determine pore geometry and dopant incorporation in the PET membranes. Ion transmission spectroscopy (ITS) is a non-destructive nuclear technique based on the measurement of residual energies of MeV ions transmitted through thin films. It makes possible, by procession of the measured tomographic data, to determine the shape (3D geometry) of the micro-objects (e.g., etched pores or inclusions) in thin foils [24,25]. If ions pass through a porous membrane, they lose energy due to interactions with the polymeric material, and they are strongly affected by inhomogeneities (such as pores or dopants) along their path. Obviously, the transmission energy spectrum carries information about the average shape of the pores (and their areal density as well). If measured by a microbeam with a spatial resolution smaller than the pore size, the spectrum provides information on the shape of a single pore [26]. ITS sensitively reflects changes in the shape of pores, their areal density, and also their filling with other material. A thin, point-like ^{241}Am α-source with the main energy line of 5.486 MeV (82.7%) was used in this ITS study. It enabled the rapid and facile measurements of transmission spectra and evaluation of the instantaneous shape of the etched pores (Figure 2). The samples were measured in a small vacuum chamber, where the energy spectra were registered by a PIPS surface barrier detector, then stored in a TRUMP MCA and analyzed off-line using a Monte Carlo tomographic code (TOM), developed earlier by the author's team [27].

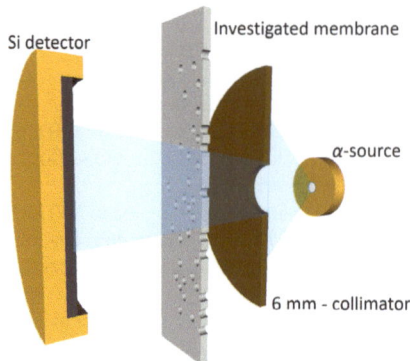

Figure 2. Ion transmission spectroscopy (ITS) setup assembled for the analysis of the PET membranes.

The filling of the pores with boron was investigated by neutron depth profiling (NDP) [28,29]. It is a non-destructive technique utilizing high intense reactions with emission of charged particles to determine depth (spatial) distribution of the NDP-relevant elements. The method is based on the simulation of residual energies of charged reaction products registered by the NDP spectrometer and evaluated off-line by the MC code (LiBor), also developed by the author's team [30]. In the case of boron, a nuclear reaction of $^{10}B(n_{th}, \alpha)^7Li$ was utilized with a high cross section of 3837 barns and relatively high reaction energy Q = 2.792 MeV. The NDP spectrometer is installed on the horizontal channel of the LWR15 research reactor (operated by the Research Centre Řež [31]); the NDP principle of operation is showed in Figure 3a. The samples were irradiated with thermal neutrons with a flux of $7 \times 10^7 n_{th}/cm^2$ provided by a supermirror neutron guide at a reactor power of 10 MW. The measurement was carried out in a large vacuum chamber with a background pressure of approx. 10^{-1} mbar. The samples were irradiated with a neutron beam at an angle of 15°; the solid angle (between the detector and the sample) was 0.382×10^{-2}. Figure 3b shows a SEM cross-section image of the boron film with a thickness of ~30 nm deposited on the Si wafer (JSM-7200F, Jeol Ltd., Tokyo, Japan).

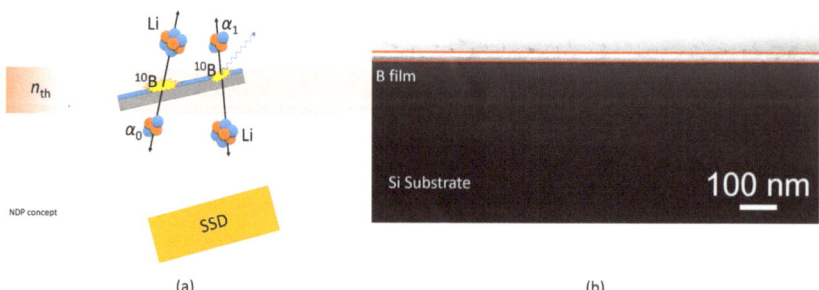

Figure 3. (a) Neutron depth profiling (NDP) setup, and (b) cross-section SEM of deposited B on the Si substrate.

3. Results

Figure 4 shows a series of ITS spectra recorded on the PET membranes etched isochronously in 9M NaOH, i.e., for 15 min at several selected etching temperatures (55 °C, 65 °C, and 75 °C).

As can be seen, the full energy peak (FEP, with a nominal alpha energy of 5.486 MeV) increases rapidly at higher etching temperatures, obviously because the etch rate of the latent tracks grows with the increasing temperature of the etchant and so does the size of the pores. In addition, the reduced energy peak (REP) shifts significantly to the higher energies (the reduced energy peak represents alpha particles passing through the unirradiated area of the membrane, i.e., with maximum energy loss). The shift is due to the reduction of the thickness of the membranes as a result of the etching of the non-irradiated area in PET. The ITS analysis of the transmission spectra in Figure 4 shows a sharp growth in the pore volume network (a spatial part of the membranes occupied by empty pores), which increased sharply (almost 25×) when the etching temperature changed from 55 °C to 75 °C. Pore volume growth is shown to be approximately linear in a given 15 min's isochronous etching (except for slightly slower growth at a lower temperature of 55 °C) with a steep gradient (slope) of ∼2.8 (°C^{-1}). Such a rapid change in the membranes' porosity is to be expected; the etching conditions are harsh, and so the latent tracks can be quickly dissolved and removed. However, it is important that the etched pore walls are fairly smooth without some rough morphology. A finer etching was also carried out (under milder conditions with a lower etchant molarity and a lower etching temperature), which however meant a longer etching time but better control of the pore shape as needed. Figure 5 shows ITS spectra for the isothermal pore etching. The PET film was exposed to 9M NaOH at a fixed etching temperature of 50 °C for different times of 15, 30, 45, and 60 min. In this part of the experiment, a lower etching temperature was chosen to allow better sensitivity in controlling the evolution of the pore shape.

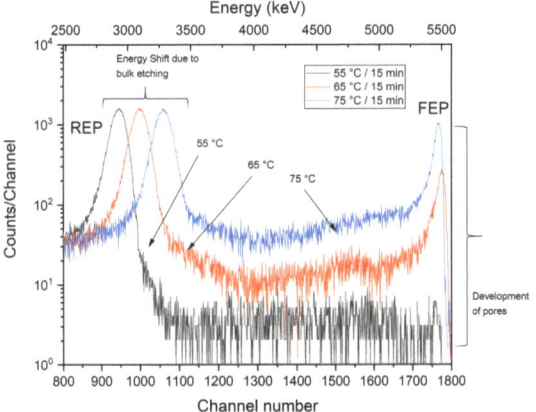

Figure 4. ITS results for porous membranes etched isochronously on one side of the PET foils for 15 min with different etching temperatures of 55 °C, 65 °C, and 75 °C.

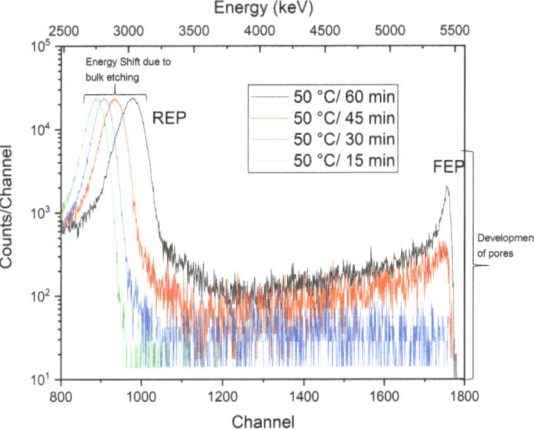

Figure 5. ITS results for porous membranes etched isothermally at 50 °C on one side of the PET foils with different etching times of 15, 30, 45, and 60 min.

As can be seen, the transmission spectra record a similar course as in Figure 5—that is, a sharp increase in the FEP peak, as well as the whole spectral area above REP, which means a dramatic evolution in the pore volume network. The analysis of the transmission spectra showed that, similar to the isochronous etching, the growth of pore volume is more or less linear with a very high slope of ~5.1 (min^{-1}) (except for the first 15 min etch). From the reconstruction of the spectra, evaluated by the TOM code, it was found that the breakthrough of the foil occurred just before 45 min of etching; the pores first acquired a conical shape, which gradually developed into a cylinder form after 60 min of etching. The two obtained profiles, calculated from the TOM code, are shown in Figure 6 (to be noted—the ITS spectra represent the averaged values of ion transmission data, so the TOM simulation gives an averaged form of individual pore shapes, where small irregularities are smoothed out). The obtained pore diameters are also in good agreement with literature value obtained with similar etching conditions and obtained with SEM microscopy [32,33]. These two pore shapes were then selected for the dopant incorporation and encapsulation tests.

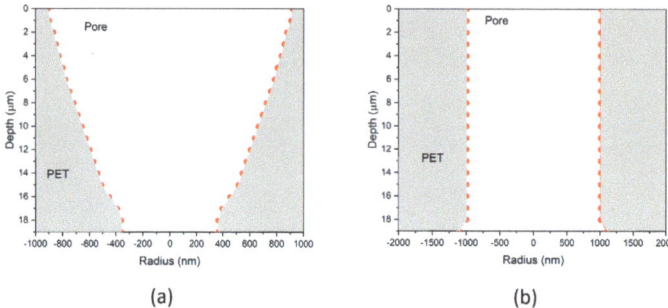

Figure 6. Pore shapes obtained by the TOM code simulation of the ITS spectra measured for the PET membranes etched isothermally at 50 °C for 45 min (**a**) and 60 min (**b**).

Figure 7 shows the effect of incorporating 5M LiCl dopant into both pore types performed under the same conditions (doping only from the etched-side of the membrane for 24 h at RT). As can be seen, LiCl was effectively encapsulated in both types of pores (either due to ^{7}Li capture by the electronegative ester groups -COOR exposed by etching on the pore walls, or simply in the form of loosely packed LiCl salt crystals trapped in the confined pore space), which completely closed their free passage of the probing alpha particles (the FEP area disappeared). On the other hand, a part of the spectra between REP and FEP increased, differently for both types of pores, which implies that the LiCl dopant is distributed in both pores in somewhat different ways.

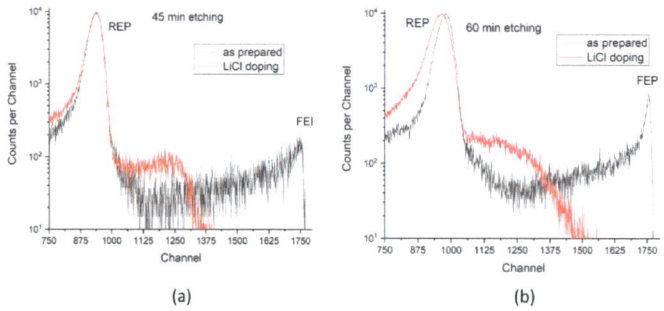

Figure 7. Comparison between the ITS spectra before LiCl doping (black line) and after LiCl doping (red line) for the PET membrane etched for 45 min (**a**) and 60 min (**b**).

From the ITS spectra, it can be concluded that the encapsulated LiCl substance does not form a compact structure (otherwise the spectrum above the REP would fully disappear), but acquires a somewhat dispersed form with a reduced average volume density, as compared with the textbook value of compact LiCl (2.07 g/cm^3). This reduction amounts to approximately 24% for the narrower pores etched for 45 min, passing from a calculated free volume of 59.66×10^9 nm^3 to a free volume of 53.09×10^9 nm^3, and to approximately 11% for the larger pores etched for 60 min, with a reduction of free volume from 25.45×10^9 nm^3 to 19.342×10^9 nm^3, possibly pointing at a reduced accessibility for dopants in narrower pores. With an increasing etched pore size, the maximum energy of the transmitted alpha particles increased, indicating that the surface cleaning procedure may have removed some of the initially present near-surface LiCl precipitates in the pores. The slight broadening of the REP peak of the doped sample in Figure 7b suggests that some traces of the dopant solution still survived the surface cleaning in this case, eventually due to stronger surface adhesion, by profiting from an enhanced surface roughness of this longer-etched sample.

For a more accurate quantification of the concentration and spatial distribution of the dopants (made up of some light elements), the method of neutron depth profiling can be used. In the final part of the study, the membranes (with pores of different diameters) were doped with boron sputtered on only one side of the double-etched membranes. The aim was to determine the efficiency of doping of pores with a different size. Three membranes with pore diameters of 7.0 µm, 2.4 µm, and 0.53 µm were prepared. The NDP analysis was carried out from the opposite side of the membranes to which boron was deposited (as shown in Figure 3a) to avoid measuring of boron from the membrane surface. Therefore, only boron from pores could be analyzed. Boron (^{10}B~19.8%, ^{11}B~80.2%) was chosen as a marker because of the high effective cross section (3837 b) of the reaction ^{10}B(n_{th}, α)^{7}Li with thermal neutrons. Figure 8 shows the NDP spectra obtained for all three boron-doped membranes with different pore sizes.

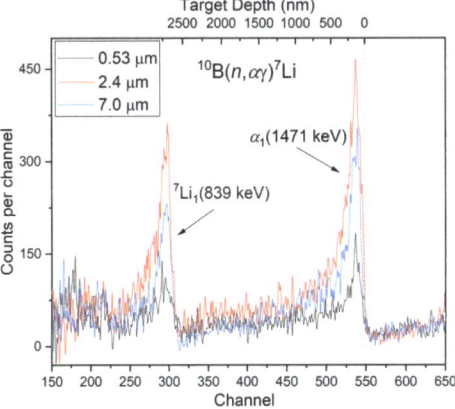

Figure 8. NDP spectra with boron sputtered on one side of a PET membrane with pores diameters of 7 µm, 2.4 µm, and 0.53 µm.

The spectra were normalized to the fluence of the neutron monitor (because of some variability of the LWR15 reactor power, it is needed to relate all NDP measurements to the fluence of the neutron flux, measured by a special spectrometer line). Monitoring the neutron fluence of the NDP measurements made possible an accurate analysis (concentration and depth distribution) of the boron dopant trapped in the pores. The peaks at channel 550 and 300 represent alpha and lithium particles, with full energy of 1.471 MeV and 839 keV, respectively, from the reactions generated in the pores. The spectra between the channels 550 and 300, utilizing only alpha particles, corresponds to an accessible depth of about 2.4 $\neq$ m for the NDP analysis. The following concentration of boron (^{10}B) in pores with different diameters was determined based on the comparison with the NIST standard:

4.39×10^{14} cm^{-2}, 5.96×10^{14} cm^{-2}, and 2.20×10^{14} cm^{-2} for pores with a diameter of 7 µm, 2.4 µm, and 0.53 µm, respectively. Measured values correspond to films of B attached to the pore walls (close to their openings) with an average thickness of several tens of nm: ca 5.1 nm for pores 2.4 µm, ca 3.8 nm for pores 7 µm, and 1.9 nm for pores 0.53 µm. Interestingly, boron is most easily trapped in pores (near the opening of the con-side of the membrane) with a 2.4 µm diameter, which points—as expected—at higher adhesion (and reduced surface mobility) of adsorbents to surfaces with larger curvature.

4. Conclusions

The ion transmission technique ITS with 5.486 MeV alpha particles was used to investigate the etching process in PET films irradiated with 1.2 MeV/u Xe^{+26} ions. Pore shape development, their filling with LiCl and B dopants, as well as the etching rate of the non-irradiated areas were studied for both isochronous and isothermal etching conditions. The ITS study showed that even in a simple arrangement with an α-source, it was possible to monitor and quantify the evolution of the etched micron-size pores: their shape transferred from latent tracks to a conical form (for one-side etching), and through a symmetrical (double-side) etching process to a well-developed cylindrical geometry; the pore volume network increased approximately linearly with both etching temperature (with a slope ∼2.8 (°C^{-1}) for etching time 15 min in a temperature range 55–75 °C) and etching time (with a slope ∼5.1 (°min^{-1}) for etching temperature 50 °C and exposure time range 15–60 min); bulk etched rate for isothermal etching (at 50 °C) was determined as ∼24 nm/min. Doping the etched tracks with 5M LiCl solution resulted in a reduction of the pore volume network by 24% and 11% for etching times of 45 and 60 min, respectively. Together with ITS, a method of neutron depth profiling was also applied (which is very suitable for in-depth profiling of several light elements that cannot be determined easily by other techniques). The NDP method enabled non-destructive analysis (concentration and depth distribution) of boron with which the PET membrane surface and etched circular pores were covered and filled, respectively. It turned out that boron deposited by magnetron sputtering penetrated easily into the pores and diffused to their opposite side, where it attached mostly near their openings. Interestingly, the most boron was found in pores with a diameter of 2.4 mm. This study showed that using traditional wet chemical etching, controlled by the ITS method, it is possible to develop pores with a defined structure and to fill them in a controlled manner with a relevant light material whose spatial distribution can be determined by the NDP method. Thus, both non-destructive nuclear techniques have proven to be suitable analytical tools that can be used for nanostructure technologies based on nuclear tracks in thin polymer membranes.

Author Contributions: Conceptualization, G.C.; methodology, G.C., P.P. and M.T.; validation, J.V. and D.F.; investigation, G.C. and J.V.; resources, J.V. and J.S.; data curation, P.P. and A.C. (Antonino Cannavò); writing—original draft preparation, G.C.; writing—review and editing, G.C., A.C. (Andrey Choukourov) and J.V.; funding acquisition, J.V. All authors have read and agreed to the published version of the manuscript.

Funding: The project was supported by the Grant Agency of the Czech Republic (Grantová Agentura České Republiky), project No. 22-17346S.

Institutional Review Board Statement: Not applicable.

Data Availability Statement: All the data presented in this study is already available in the manuscript itself.

Conflicts of Interest: The authors declare no conflict of interest.

References

1. Apel, P. Ion-Track Membranes and Their Use in Biological and Medical Applications. *AIP Conf. Proc.* **2007**, *912*, 488–494.
2. Kozlovskiy, A.; Borgekov, D.; Kenzhina, I.; Zdorovets, M.; Korolkov, I.; Kaniukov, E.; Kutuzau, M.; Shumskaya, A. PET ion-track membranes: Formation features and basic applications. In Proceedings of the International Conference on Nanotechnology and Nanomaterials, Kiev, Ukraine, 27–30 August 2018; Springer: Berlin/Heidelberg, Germany, 2018; pp. 461–479.

3. Liu, F.; Wang, M.; Wang, X.; Wang, P.; Shen, W.; Ding, S.; Wang, Y. Fabrication and application of nanoporous polymer ion-track membranes. *Nanotechnology* **2018**, *30*, 052001. [CrossRef]
4. Kaya, D.; Keçeci, K. Track-etched nanoporous polymer membranes as sensors: A review. *J. Electrochem. Soc.* **2020**, *167*, 037543. [CrossRef]
5. Vasi, S.; Ceccio, G.; Cannavò, A.; Pleskunov, P.; Vacík, J. Study of Wettability of Polyethylene Membranes for Food Packaging. *Sustainability* **2022**, *14*, 5863. [CrossRef]
6. Dutt, S.; Apel, P.; Lizunov, N.; Notthoff, C.; Wen, Q.; Trautmann, C.; Mota-Santiago, P.; Kirby, N.; Kluth, P. Shape of nanopores in track-etched polycarbonate membranes. *J. Membr. Sci.* **2021**, *638*, 119681. [CrossRef]
7. Khulbe, K.; Feng, C.; Matsuura, T. The art of surface modification of synthetic polymeric membranes. *J. Appl. Polym. Sci.* **2010**, *115*, 855–895. [CrossRef]
8. Korolkov, I.V.; Gorin, Y.G.; Yeszhanov, A.B.; Kozlovskiy, A.L.; Zdorovets, M.V. Preparation of PET track-etched membranes for membrane distillation by photo-induced graft polymerization. *Mater. Chem. Phys.* **2018**, *205*, 55–63. [CrossRef]
9. Vacik, J.; Ceccio, G.; Cannavò, A.; Lavrentiev, V. Effects of UV irradiation and thermal annealing on LiCl derivatives encapsulation in porous PET membranes coated with a thin Au film. *Radiat. Eff. Defects Solids* **2022**, *177*, 112–123. [CrossRef]
10. Mashentseva, A.A.; Korolkov, I.V.; Yeszhanov, A.B.; Zdorovets, M.V.; Russakova, A.V. The application of composite ion track membranes with embedded gold nanotubes in the reaction of aminomethylation of acetophenone. *Mater. Res. Express* **2019**, *6*, 115022. [CrossRef]
11. Kotál, V.; Švorčík, V.; Slepička, P.; Sajdl, P.; Bláhová, O.; Šutta, P.; Hnatowicz, V. Gold coating of poly(ethylene terephthalate) modified by argon plasma. *Plasma Process. Polym.* **2007**, *4*, 69–76. [CrossRef]
12. Agrawal, H.; Saraswat, V.K.; Awasthi, K. ZnO doping in PET matrix enhances conductivity of PET-ZnO nanocomposites. *Adv. Electrochem.* **2013**, *1*, 118–123. [CrossRef]
13. Miao, J.; Zhao, K.; Guo, F.; Xu, L.; Xie, Y.; Deng, T. Novel LIS-doped mixed matrix membrane absorbent with high structural stability for sustainable lithium recovery from geothermal water. *Desalination* **2022**, *527*, 115570. [CrossRef]
14. Liu, J.; Cao, D.; Yao, H.; Liu, D.; Zhang, X.; Zhang, Q.; Chen, L.; Wu, S.; Sun, Y.; He, D.; et al. Hexagonal Boron Nitride-Coated Polyimide Ion Track Etched Separator with Enhanced Thermal Conductivity and High-Temperature Stability for Lithium-Ion Batteries. *ACS Appl. Energy Mater.* **2022**, *5*, 8639–8649. [CrossRef]
15. Lee, P.L.J.; Thangavel, V.; Guery, C.; Trautmann, C.; Toimil-Molares, M.E.; Morcrette, M. Etched ion-track membranes as tailored separators in Li–S batteries. *Nanotechnology* **2021**, *32*, 365401. [CrossRef]
16. Siwy, Z.; Apel, P.; Dobrev, D.; Neumann, R.; Spohr, R.; Trautmann, C.; Voss, K. Ion transport through asymmetric nanopores prepared by ion track etching. *Nucl. Instrum. Methods Phys. Res. Sect. B Beam Interact. Mater. Atoms* **2003**, *208*, 143–148. [CrossRef]
17. Kaniukov, E.; Shumskaya, A.; Yakimchuk, D.; Kozlovskiy, A.; Ibrayeva, A.; Zdorovets, M. Characterization of pet track membrane parameters. In Proceedings of the International Conference on Nanotechnology and Nanomaterials, Lviv, Ukraine, 24–27 August 2016; Springer: Berlin/Heidelberg, Germany, 2016; pp. 79–91.
18. Apel, P.Y.; Ramirez, P.; Blonskaya, I.V.; Orelovitch, O.L.; Sartowska, B.A. Accurate characterization of single track-etched, conical nanopores. *Phys. Chem. Chem. Phys.* **2014**, *16*, 15214–15223. [CrossRef]
19. Chander, M.; Kumar, S. Estimation of the nano-pores diameter by conductometric measurements. *IOP Conf. Ser. Mater. Sci. Eng.* **2022**, *1221*, 012050. [CrossRef]
20. Álvarez-Arenas, T.G.; Apel, P.Y.; Orelovich, O. Characterization of ion-track membranes by non-contact ultrasonic magnitude and phase spectroscopy. *J. Membr. Sci.* **2007**, *301*, 210–220. [CrossRef]
21. Apel, P. Track etching technique in membrane technology. *Radiat. Meas.* **2001**, *34*, 559–566. [CrossRef]
22. George, J.; Irkens, M.; Neumann, S.; Scherer, U.; Srivastava, A.; Sinha, D.; Fink, D. Controlled ion track etching. *Radiat. Eff. Defects Solids* **2006**, *161*, 161–175. [CrossRef]
23. UJF. ústav Jaderné Fyziky. Available online: http://www.ujf.cas.cz/cs/ (accessed on 22 September 2022).
24. Vacik, J.; Hnatowicz, V.; Havranek, V.; Fink, D.; Apel, P.; Horak, P.; Ceccio, G.; Cannavo, A.; Torrisi, A. Ion track etching in polyethylene-terephthalate studied by charge particle transmission technique. *Radiat. Eff. Defects Solids* **2019**, *174*, 148–157. [CrossRef]
25. Ceccio, G.; Vacik, J.; Trusso, S.; Cannavò, A.; Horak, P.; Hnatowicz, V.; Apel, P. Ion transmission spectroscopy of pores filled with Au nanoparticles. *Nucl. Instrum. Methods Phys. Res. Sect. B Beam Interact. Mater. Atoms* **2021**, *491*, 29–33. [CrossRef]
26. Vacik, J.; Havranek, V.; Hnatowicz, V.; Horak, P.; Fink, D.; Apel, P. Study of ion tracks by micro-probe ion energy loss spectroscopy. *Nucl. Instrum. Methods Phys. Res. Sect. B Beam Interact. Mater. Atoms* **2014**, *332*, 308–311. [CrossRef]
27. Vacík, J.; Červená, J.; Hnatowicz, V.; Pošta, S.; Fink, D.; Klett, R.; Strauss, P. Simple technique for characterization of ion-modified polymeric foils. *Surf. Coat. Technol.* **2000**, *123*, 97–100. [CrossRef]
28. Park, B.; Sun, G. Analysis of depth profiles of 10B and 6Li in Si wafers and lithium ion battery electrodes using the KAERI-NDP system. *J. Radioanal. Nucl. Chem.* **2016**, *307*, 1749–1756. [CrossRef]
29. Chen-Mayer, H.H.; Lamaze, G.P. Depth distribution of boron determined by slow neutron induced lithium ion emission. *Nucl. Instrum. Methods Phys. Res. Sect. B Beam Interact. Mater. Atoms* **1998**, *135*, 407–412. [CrossRef]
30. Hnatowicz, V.; Vacik, J.; Fink, D. Deconvolution of charged particle spectra from neutron depth profiling using Simplex method. *Rev. Sci. Instrum.* **2010**, *81*, 073906. [CrossRef]
31. CVR. Centrum Výzkumu Řež. Available online: http://cvrez.cz/en/ (accessed on 22 September 2022).

32. Froehlich, K.; Scheuerlein, M.C.; Ali, M.; Nasir, S.; Ensinger, W. Enhancement of heavy ion track-etching in polyimide membranes with organic solvents. *Nanotechnology* **2021**, *33*, 045301. [CrossRef]
33. Yang, L.; Zhai, Q.; Li, G.; Jiang, H.; Han, L.; Wang, J.; Wang, E. A light transmission technique for pore size measurement in track-etched membranes. *Chem. Commun.* **2013**, *49*, 11415–11417. [CrossRef]

Lévy Flights Diffusion with Drift in Heterogeneous Membranes

Anna Strzelewicz [1,*], Monika Krasowska [1] and Michał Cieśla [2]

[1] Faculty of Chemistry, Silesian University of Technology, Strzody 9, 44-100 Gliwice, Poland; monika.krasowska@polsl.pl
[2] Institute of Theoretical Physics, Jagiellonian University, Łojasiewicza 11, 30-348 Kraków, Poland; michal.ciesla@uj.edu.pl
* Correspondence: anna.strzelewicz@polsl.pl

Abstract: The modelling of diffusion in membranes is essential to understanding transport processes through membranes, especially when it comes to improving process efficiency. The purpose of this study is to understand the relationship between membrane structures, external forces, and the characteristic features of diffusive transport. We investigate Cauchy flight diffusion with drift in heterogeneous membrane-like structures. The study focuses on numerical simulation of particle movement across different membrane structures with differently spaced obstacles. Four studied structures are similar to real polymeric membranes filled with inorganic powder, while the next three structures are designed to show which distribution of obstacles can cause changes in transport. The movement of particles driven by Cauchy flights is compared to a Gaussian random walk both with and without additional drift action. We show that effective diffusion in membranes with an external drift depends on the type of the internal mechanism that causes the movement of particles as well as on the properties of the environment. In general, when movement steps are provided by the long-tailed Cauchy distribution and the drift is sufficiently strong, superdiffusion is observed. On the other hand, strong drift can effectively stop Gaussian diffusion.

Keywords: diffusion; drift; heterogeneous membrane; structure; simulation; Lévy flights

1. Introduction

Simulations of diffusion in membranes are fundamental for understanding transport processes through membranes, especially when it comes to improving process performance. The diffusion type is determined by the microscopic dynamics and properties of an environment, such as the polymer structure and related internal changes resulting from the permeation of mixture components [1–4]. Typically, a particle passing through a medium constantly interacts with other particles and other components of the system. These collisions result in irregular observable movements, called Brownian motion. An example of an equation describing the dynamics of a single particle at the microscopic level is the Langevin equation. It is a stochastic differential equation, and its solutions, i.e., the trajectories corresponding to the same initial conditions, differ. Numerous collisions of the studied particle with other particles and the medium can be approximated by a random force, the properties of which depend on the chosen assumptions [5–7]. In the idealized model of such motion, the random force **R**(t) acting on a particle corresponds to a Gaussian process. This means that the collisions are independent and are described by a probability distribution with moments that are well defined [8,9]. This assumption leads to so-called normal diffusion; as a consequence of the assumption of collision independence, it has a Markovian character and linear scaling of the mean square displacement with time:

$$\langle \Delta r^2 \rangle = 2nDt \qquad (1)$$

This relation is called the Einstein relation, where D is the diffusion coefficient, n is the spacial dimension ($n = 1, 2, 3$), t is time, and $\langle \Delta r^2 \rangle$ is the mean square displacement (MSD) [10,11].

The above relation can be generalized to other processes where the mean square displacement depends on time, as follows:

$$\langle \Delta r^2 \rangle \sim t^\alpha \qquad (2)$$

In general, when $\alpha = 1$ we have normal diffusion (see Equation (1) and Figure 1). When $\alpha < 1$, the particle covers a smaller area than in normal diffusion at the same time. This movement is called subdiffusion. When $\alpha > 1$, the particle covers a larger area; such a movement is called superdiffusion. We call both cases anomalous diffusion [10,12]. Anomalous diffusion behavior (see Equation (2)) is closely related to the breakdown of the central limit theorem, and is caused by either broad distributions or long-range correlations. Alternatively, anomalous diffusion relies on the validity of the Lévy–Gnedenko generalized central limit theorem for situations in which not all moments of the underlying elementary transport events exist. Thus, wide spatial jumps or waiting time distributions lead to non-Gaussian propagators, and possibly non-Markovian time evolution of the system [13,14].

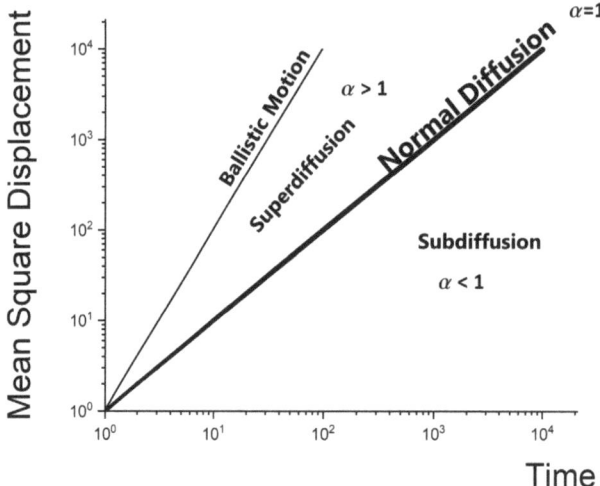

Figure 1. Anomalous diffusion regimes as characterized by Equation (2), i.e., power-law scaling of the mean squared displacement (MSD) with time.

Subdiffusion is usually caused by the trapping of a randomly moving object, while superdiffusion is the consequence of unlimited variance in the distance between successive object positions [9,15]. However, the following point remains worth considering. Using a continuous-time random walk (CTRW) description [16], it can be shown that if the jump length and the times between jumps are distributed via the power-law asymptotics $p(r) \propto |r|^{-1-\mu}$, $p(t) \propto t^{-1-\nu}$ (with $0 < \mu < 2, 0 < \nu < 1$), and $p(r,t) = t^{-\nu/\mu} p(rt^{-\nu/\mu}, 1)$, then the second moment of $p(r,t)$ and $t^{2\nu/\mu}$ scale similarly. It may happen that $2\nu = \mu$, in which case the ensemble average leads to a linear signature of time-dependent MSD even for jump lengths from heavy-tailed distribution. This is called paradoxical diffusion [15,17,18].

Lévy walks [19,20] and Lévy flights [21] are anomalous diffusive forms of random motion that are widely observed in natural systems, including movements of animals [22]. They are characterized by a grouping of short steps that are occasionally interspersed with longer movements; as such, they represent superdiffusive processes and can travel much further from the starting position than a Brownian walk of the same duration. The main difference between Lévy walks and Lévy flights is that the former has a finite propagation velocity and continuous trajectory, which makes it inertial and easier to apply to physical systems. From the numerical perspective, in both these types of movements the jump lengths are selected from the same distribution, while for Lévy walk a constant velocity is assumed, which means that the time of a single movement varies. In Lévy flights, the jump is applied instantaneously. It should be noted that Levy flights can be used as an approximation of Lévy walks [23].

In this study, we focus on one specific type of Lévy flight in which jump lengths are drawn independently from the Cauchy distribution:

$$p_C(x) = \frac{\sigma^2}{\pi(\sigma^2 + x^2)}, \tag{3}$$

where σ denotes the width of the distribution. Because the Cauchy distribution is α-stable, the tracer displacement after t independent jumps is the Cauchy distributed random variable, though now of a width σt [24]. Considering that we are interested in square displacement, let $y(t) = x^2(t)$. Thus, the probability distribution function of $y(t)$ is

$$p[y] = \frac{\sigma t}{\pi \sqrt{y}(\sigma^2 t^2 + y)} \tag{4}$$

However, the mean square displacement for the Cauchy distributed random variable is ill-defined; therefore, the median is used instead. In this case, the median of y is provided by the following equation:

$$\frac{1}{2} = \int_0^{M[y]} p[y']dy' = \frac{2\tan^{-1}\left(\frac{\sqrt{M[y]}}{\sigma t}\right)}{\pi}, \tag{5}$$

which leads directly to

$$M[y] = \sigma^2 t^2. \tag{6}$$

Figure 2 shows example trajectories of particle motion in four cases: Gaussian (Brownian) motion with and without drift and Lévy flight with and without drift. A motion analysis of a single particle is a standard tool to probe the local physical properties of complex systems because it delivers the complete, or at least projected, trajectory of an individual particle. In open space, the particle moves around its initial position and slowly away from it, with no preference for the direction of movement (Figure 2a). The addition of drift causes a distinct movement of the particle in the direction of the drift (Figure 2b). In the case of Lévy flights, the particle moves around a certain point while alternating with a 'long jump' along a straight line (Figure 2c). Introducing drift into this type of motion results in more jumps having a preferred direction (Figure 2d).

The main purpose of this study is to check how the above general picture of diffusion applies to transport properties in heterogenic membranes. In our previous studies [25,26], we modelled structures of heterogenic membranes which resemble real membrane structures, i.e., sodium alginate membranes filled with iron oxide nanoparticles [26]. Sub-diffusive motion is observed in a crowded environment regardless of whether particle movement is induced by a Gaussian or Lévy flight process [24,27]. On the other hand, the existence of an external force that causes constant drift can speed up the transport of Brownian particles through a crowded environment [26,28], while when the amount of the drift is too high the transport is practically stopped. In this study, we want to test the last possible situation, i.e., when the external drift is combined with Lévy flights. Because we compare the obtained results with previous studies on transport in heterogenic membranes, we use numerically modelled structures that resemble them [29].

The rest of this paper proceeds as follows: Materials and Methods, Results, Discussion, and Conclusions. Section 2 (Materials and Methods) discusses the membranes and the methodology used to study Lévy flight diffusion on these membranes with and without drift. Section 3 (Results and Discussion) provides a detailed description of the study results and graphs, analyzes the results, and provides an interpretation. Finally, the manuscript closes with our Conclusions (Section 4).

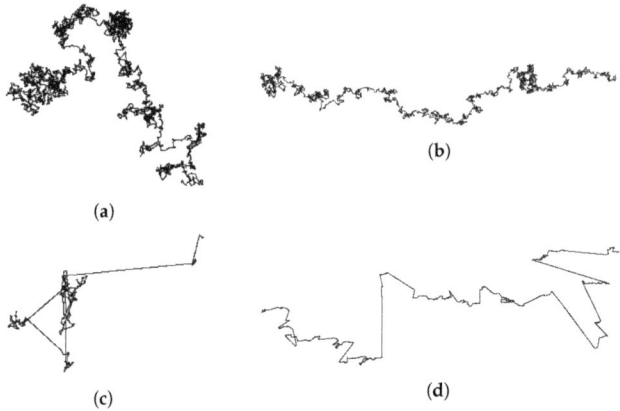

Figure 2. Comparison of Gaussian (Brownian) and Lévy flight in open space. The top panels show diffusion caused by the random Gaussian distribution of steps and the bottom ones show Cauchy distributed steps. In the left panels, there is no external drift, while in the right panels there is a drift along the horizontal axis. (**a**) The trajectory of Brownian motion without drift, (**b**) trajectory of Brownian motion with drift, (**c**) trajectory of Lévy flight without drift, and (**d**) trajectory of Lévy flight with drift.

2. Materials and Methods

2.1. The Model Membranes

We generated images of the structure of artificial membranes using the method described in [25,29]. Similarly, we generated four stuctures, named MS1, MS2, MS3, and MS4 (see Figure 3), with the following parameters: MS1 ($\rho = 0.85$, $d_f = 1.9693$, $\Delta D = 0.4259$); MS2 ($\rho = 0.85$, $d_f = 1.9664$, $\Delta D = 1.1809$); MS3 ($\rho = 0.9$, $d_f = 1.9797$, $\Delta D = 0.3667$); and M4 ($\rho = 0.9$, $d_f = 1.9783$, $\Delta D = 1.6377$).

When analyzing Lévy flight diffusion with drift on structures MS1–MS4 where obstacles are randomly distributed, we decided to prepare structures with deterministically distributed obstacles. We prepared three artificial membrane structures with differently positioned obstacles in order to show which distributions of obstacles can cause changes in transport. The first structure has three breaks (structure S1 in Figure 4 (left panel)). The second structure has only one break, at the central point of the membrane (structure S2 in Figure 4 (central panel)). The third structure has obstacles aligned along the main diagonal of the structure (structure S3 in Figure 4 (right panel)). The breaks between the obstacles are equally distributed.

The obstacle positioning we chose allowed us to determine how the placement of obstacles affects the movement of particles and whether effective diffusion in membranes with and without external drift depends on the type of internal mechanism that causes the movement of particles.

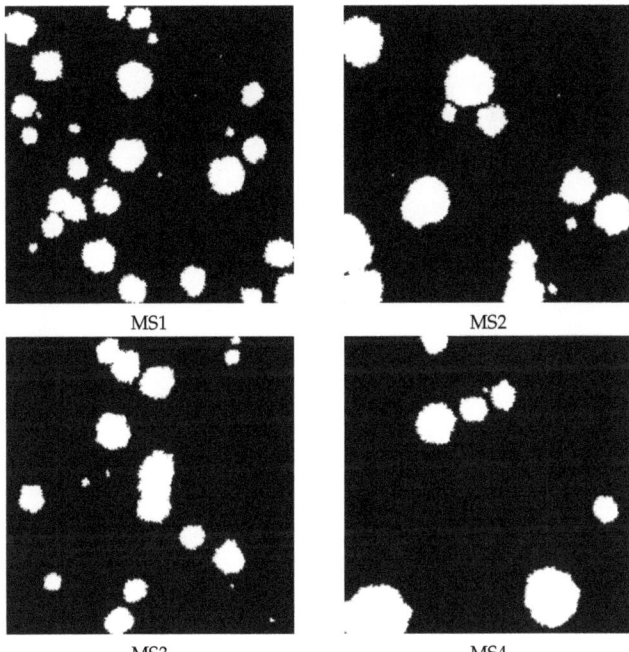

Figure 3. The artificial membrane structures with different structural parameters: MS1 ($\rho = 0.85$, $d_f = 1.9687$, $\Delta D = 0.7102$); MS2 ($\rho = 0.85$, $d_f = 1.9664$, $\Delta D = 1.1809$); MS3 ($\rho = 0.9$, $d_f = 1.9792$, $\Delta D = 0.8084$); and M4 ($\rho = 0.9$, $d_f = 1.9783$, $\Delta D = 1.6377$).

Figure 4. Artificial membrane structures with obstacles in different positions. These three structures are designed to show which distributions of obstacles can cause changes in transport. **Left panel:** The first structure has three breaks (structure S1). **Central panel:** The second structure has only one break, at the central point of the membrane (structure S2). **Right panel:** The third structure has obstacles aligned along the main diagonal of the structure (structure S3). The breaks between the obstacles are equally distributed.

2.2. Simulations of Gauss and Cauchy flights

We used numerical simulations to analyze the diffusion in the membrane systems mentioned above. A set of 100 randomly distributed independent tracers penetrated the available space, i.e., the black regions of the membranes shown in Figures 3 and 4, according to the conditions presented below.

A single movement of a tracer is

$$\vec{x}_i = [x_i \cos \phi, x_i \sin \phi], \tag{7}$$

where $|x_i|$ denotes the length of the step and ϕ is its direction.

The length of the step x_i is randomly selected from Gaussian

$$p_G(x) = \frac{1}{\sqrt{2\pi}\sigma} \exp\left[-\frac{x^2}{2\sigma^2}\right] \tag{8}$$

or Cauchy (3) distributions with parameter $\sigma = 1$. The direction ϕ is selected according to the uniform distribution on the interval $[0, 2\pi)$. To generate normally distributed random variables, we used the Box–Muller transform [30]. The Cauchy distribution can be obtained from the uniform one using $\tan[\pi(u - 1/2)]$, where u is the uniformly distributed random number on the interval $[0, 1)$. Having determined the step $\vec{x}_t$, we next checked whether or not a line between the present position $\vec{r}(t)$ and the target position $\vec{r}(t) + \vec{x}_{t+1}$ crossed any obstacle or ended on it, and whether the tracer changed its position ($\vec{r}(t+1) = \vec{r}(t) + \vec{x}_t$). Otherwise, it was assumed that the tracer was reflected and returned to its previous position ($\vec{r}(t+1) = \vec{r}(t)$). The simulation was stopped after 10^7 such iterations. During the numerical calculations, the set of 100 independent tracer trajectories was recorded. The time was measured using the number of iterations and the distance; if not explicitly stated otherwise, it was measured in pixels. To avoid saturation of square displacement due to the finite size of the membrane, periodic boundary conditions were used. The dependence of the median time for Cauchy flights and the mean value of the square displacement x^2 of the time for Gaussian processes determined the effective diffusion exponent α defined in Equation (2).

Additionally, the diffusion was analyzed in terms of the effective exponent α (see (2)); however, for Cauchy flights we used the median instead of the mean value:

$$M[x^2](t) \sim \langle x^2 \rangle \sim t^\alpha \text{ for large } t. \tag{9}$$

The effective diffusion exponents were estimated by fitting the power law (9) to the MSD data for $t > 10^5$. Thus, the effective diffusion exponent at long time limits was a slope of the line fitted to the data in the double logarithmic plot:

$$\log \langle x^2 \rangle = \alpha \log t. \tag{10}$$

3. Results and Discussion

In the previous section, we have described the basis of Gaussian and Cauchy flight simulations. Simulations of Cauchy flights with and without drift were performed first for the artificial membrane structures MS1, MS2, MS3, and MS4 presented in Figure 3. These artificial structures resemble the real membrane morphology of sodium alginate membranes filled with iron oxide nanoparticles. The distribution of obstacles in the membrane is random. For real membranes, the distribution of powder depends on the method of membrane preparation and the type of powder used. The type of powder affects the random distribution of powder and aggregate formation. The generated structures correspond to real structures; consequently, the distribution of obstacles is random. The dependence of the median of the square displacement x^2 on time driven by Cauchy flights moving in the structures MS1, MS2, MS3, and MS4 is shown in Figure 5.

For comparison, the mean square displacement for Gaussian random walk on the same membranes is shown in Figure 6.

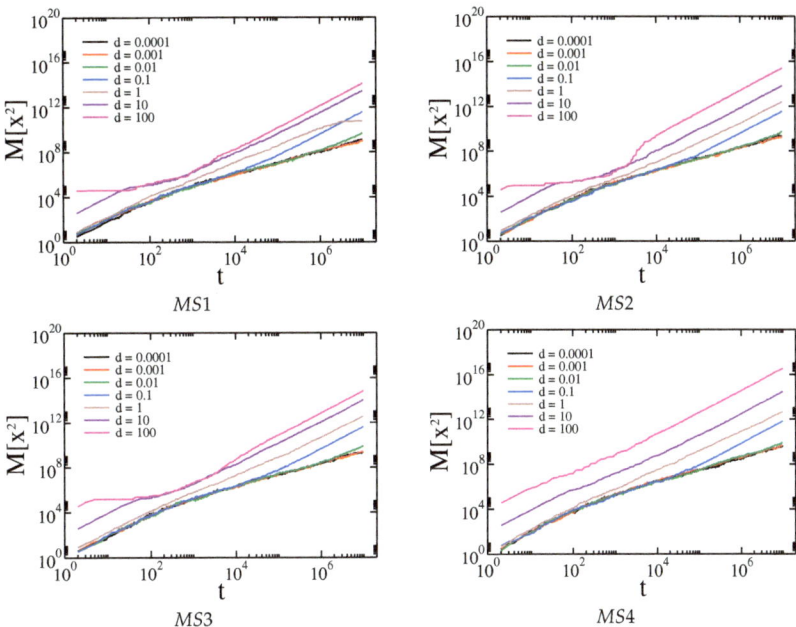

Figure 5. Dependence of the median of the square displacement x^2 on time driven by Cauchy flights moving in the structures MS1, MS2, MS3, and MS4, respectively.

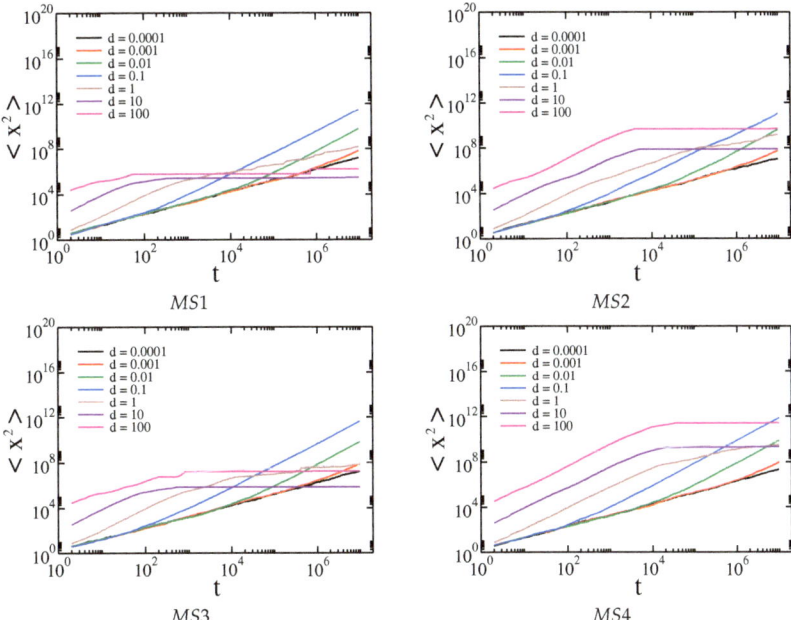

Figure 6. Dependence of the mean square displacement x^2 on time driven by Gaussian random walk in the structures MS1, MS2, MS3, and MS4, respectively.

For each membrane, a similar pattern is observed. In the beginning, the median of square displacement for Cauchy flights grows superlinearly. Then, for relatively small drift, the movement slows down, which confirms the previous results that the effective diffusion between obstacle types is generally determined by the properties of the environment and not by the internal process that causes the movement of the tracers [24–29,31]. For medium and high drifts, the movement is asymptotically superlinear; however, in the case of high drift, we observe a slowing down at medium times. This is because when a tracer hits an obstacle it requires time to avoid it. This time is longer when the drift is higher, because the tracer is pushed harder into the obstacle and the probability of drawing a step large enough to cancel the drift is relatively small. It is worth noting that for large drifts and Gaussian random walks the movement is entirely stopped, as the probability of drawing a large number decreases exponentially with its value. The above observations can be supported by the analysis of the effective diffusion exponent α, which is shown in Figure 7.

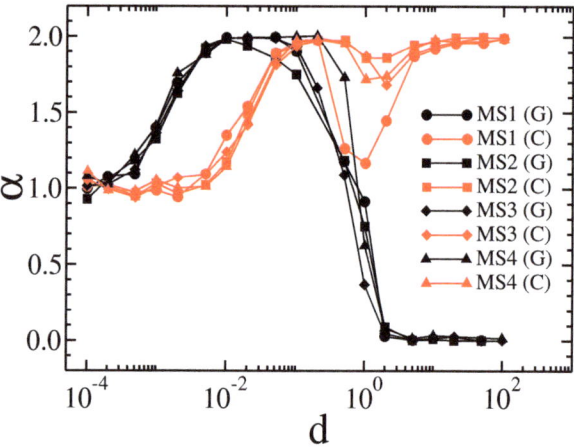

Figure 7. The dependence of the effective diffusion exponent α on the drift d for Gaussian random walks and Lévy flights on membranes MS1, MS2, MS3, and MS4.

Here, we notice that the effective exponent α for Gaussian random walks begins to grow with a smaller amount of drift than is the case for Cauchy flights. Additionally, for a larger drift that stops Gaussian diffusion, the local minimum of α is present for Cauchy flights.

In addition to the qualitatively same results for all membranes studied, there are a number of quantitative differences. For example, for Cauchy flights diffusion slowdown is not observed for the MS4 membrane, whereas it is present for the others. Therefore, we decided to generate specific sample structures of obstacles to determine the key factor in the type of observed results. Artificial membrane structures with differently placed obstacles are presented in Figure 4. The first structure has three breaks (structure S1). The second structure has only one break, at the central point of the membrane (structure S2). The third structure has obstacles aligned along the main diagonal of the structure (structure S3). Gaussian and Cauchy flight simulations with and without drift were performed on all these structures. The dependence of the median square displacement for Cauchy flight is shown in Figure 8, and the mean square displacement for the Brownian motion is shown in Figure 9.

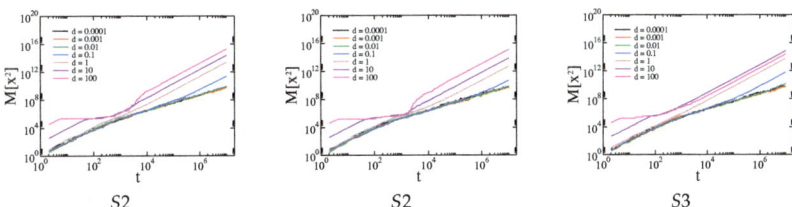

Figure 8. Dependence of the median of the square displacement x^2 on the time driven by Cauchy flights moving in the structures S1, S2, and S3, respectively.

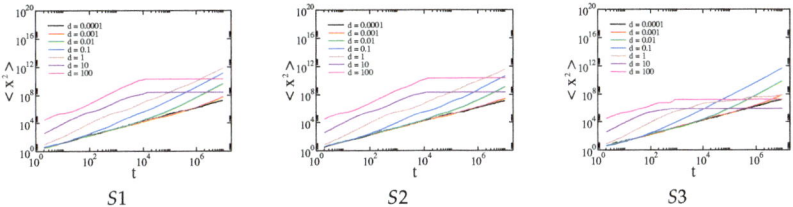

Figure 9. Dependence of the average of the square displacement x^2 on time driven by Gaussian random walk in the structures S1, S2, and S3, respectively.

The effective diffusion exponent for both types of transport is shown in Figure 10.

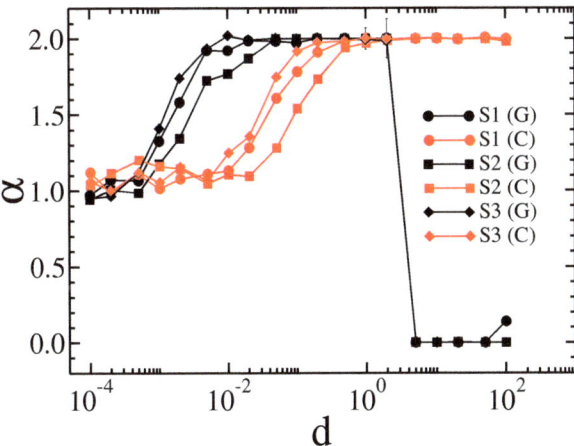

Figure 10. The dependence of the effective diffusion exponent α on the drift d for Gaussian random walks and Lévy flights in the membranes S1, S2, and S3.

In general, all the previously observed effects are present here. However, we can conclude that larger and more solid obstacles cause diffusion slowdown for intermediate times and Cauchy flights (see the results for structures S1 and S2 in comparison with those for S3). Because obstacles block most of the area and there are only small spaces in which to avoid them, the Gaussian random walk is efficiently timed by a sufficiently large drift. On the other hand, a small amount of drift $d \ll \sigma$ mainly speeds up Gaussian diffusion. This is due to the character of the Gaussian and Cauchy distributions. In the first case, most of the steps are comparable to σ. For the Cauchy distribution, there is generally a smaller step size, while from time to time a large leap occurs. However, these leaps are stopped by obstacles, while the small steps are not as efficient as in the case of the Gaussian distribution.

When $d \gg \sigma$ and the tracer hits an obstacle, it requires a relatively large step size to cancel the effect of the drift, which is quite probable for the Cauchy distribution and almost impossible for the Gaussian distribution. Therefore, a rapid decrease in effective transport is observed for Gaussian diffusion, while the movement induced by Cauchy flight is barely affected.

4. Conclusions

The present study was designed in order to understand the relationship between membrane structures and their diffusive transport characteristics in the presence of external drift. The study focused on numerical simulation of particle movement in different membrane structures with differently spaced obstacles. Four of the studied structures were similar to the structures of sodium alginate membranes filled with iron oxide nanoparticles. The other three structures were designed specially to show the crucial factors that affect most diffusional transport under the influence of drift. We investigated the movement of the particles as driven by Gaussian random walks and Cauchy flights, showing that effective diffusion in membranes with external drift depends on the amount of drift, the type of internal mechanism that causes the movement of the particles, and the distribution of the obstacles. In cases of weak drift, the effective diffusion is fully determined by the environment (i.e., the properties of the membranes), whereas the internal mechanism (i.e., Cauchy flight or Brownian motion) does not matter. For higher drift, superdiffusion is recognized; however, when the drift is too strong, Brownian motion is almost stopped, as the tracers are constantly pushed against the obstacles. Due to the heavy-tailed distribution of Cauchy flights, this pushing can be overcome by random motion; thus, transport can continue with the effective exponent $\alpha \approx 2$. In general, the observed relations do not depend qualitatively on the morphology of the studied structures.

Author Contributions: Conceptualization, A.S., M.K. and M.C.; methodology, A.S., M.K. and M.C.; software, M.C.; validation, A.S., M.K. and M.C.; formal analysis, A.S., M.K. and M.C.; investigation, A.S., M.K. and M.C.; and writing—original draft preparation, review and editing, A.S., M.K. and M.C. All authors have read and agreed to the published version of the manuscript.

Funding: This research received no external funding.

Institutional Review Board Statement: Not applicable.

Informed Consent Statement: Not applicable.

Data Availability Statement: The numerical data, analyzed images, and necessary software are available on reasonable request.

Conflicts of Interest: The authors declare no conflict of interest.

Abbreviations

The following abbreviations are used in this manuscript:

α	Effective diffusion exponent
d	Drift
d_f	Fractal dimension of polymer matrix
ΔD	Degree of multifractality
MSD	Mean square displacement
ρ	Amount of polymer matrix
σ	Width of the Cauchy distribution/standard deviation of the Gaussian distribution

References

1. Sokolov, I.M. Models of anomalous diffusion in crowded environments. *Soft Matter.* **2012**, *8*, 9043. [CrossRef]
2. Barrer, R.M.; Petropoulos, J.H. Diffusion in heterogeneous media: Lattices of parallelepipeds in a continuous phase. *Br. J. Appl. Phys.* **1961**, *12*, 691. [CrossRef]
3. Kabbej, M.; Guillard, V.; Angellier-Coussy, H.; Wolf, C.; Gontard, N.; Gaucel, S. 3D Modelling of Mass Transfer into Bio-Composite. *Polymers* **2021**, *13*, 2257. [CrossRef]
4. Kosztołowicz, T.; Dworecki, K.; Lewandowska, K.D. Subdiffusion in a system with thin membranes. *Phys. Rev. E* **2012**, *86*, 021123. [CrossRef] [PubMed]

5. Kosztołowicz, T.; Piwnik, M.; Lewandowska, K.D.; Klinkosz, T. The solution to subdiffusion-reaction equation for system with one mobile and one static raeactant. *Acta Phys. Pol. B* **2013**, *44*, 967–975. [CrossRef]
6. Ghosh, S.K.; Cherstvy, A.G.; Grebenkov, D.S.; Metzler, R. Anomalous, non-Gaussian tracer diffusion in crowded two-dimensional environments. *New J. Phys.* **2016**, *18*, 013027. [CrossRef]
7. Santamaria-Holek, I.; Rubi, J.M.; Gadomski, A. Thermokinetic Approach of Single Particles and Clusters Involving Anomalous Diffusion under Viscoelastic Response. *J. Phys. Chem. B* **2007**, *111*, 2293–2298. [CrossRef]
8. Sokolov, I.M.; Klafter, J. From diffusion to anomalous diffusion: A century after Einstein's Brownian motion. *Chaos* **2005**, *15*, 026103. [CrossRef]
9. Klafter, J.; Sokolov. I.M. *First Steps in Random Walks. From Tools to Applications*; Oxford University Press: New York, NY, USA, 2011. [CrossRef]
10. Oliveira, F.A.; Ferreira, R.M.S.; Lapas, L.C.; Vainstein, M.H. Anomalous diffusion: A basic mechanism for the evolution of inhomogeneous systems. *Front. Phys.* **2019**, *7*, 18. [CrossRef]
11. Vainstein, M.H.; Costa, I.V.L.; Oliveira, F.A. Mixing, Ergodicity and the Fluctuation-Dissipation Theorem in Complex Systems. In *Jamming, Yielding, and Irreversible Deformation in Condensed Matter*; Miguel, M.C., Rubi, M., Eds.; Springer: Berlin/Heidelberg, Germany, 2006; pp. 159–188. [CrossRef]
12. Palyulin, V.V.; Blackburn, G.; Lomholt, M.A.; Watkins, N.W.; Metzler, R.; Klages, R.; Chechkin, A.V. First passage and first hitting times of Lévy flights and Lévy walks. *New J. Phys.* **2019**, *21*, 103028. [CrossRef]
13. Metzler, R.; Klafter, J. The random walk's guide to anomalous diffusion: A fractional dynamics approach. *Phys. Rep.* **2000**, *339*, 1–77. [CrossRef]
14. Kuśmierz, Ł.; Chechkin, A.; Gudowska-Nowak, E.; Bier, M. Breaking microscopic reversibility with Lévy flights. *EPL* **2016**, *114*, 60009. [CrossRef]
15. Dybiec, B.; Gudowska-Nowak, E. Discriminating between normal and anomalous random walks. *Phys. Rev. E* **2009**, *80*, 061122. [CrossRef] [PubMed]
16. Montroll, E.W.; Weiss, G.H. Random Walks on Lattices. II. *J. Math. Phys.* **1965**, *6*, 167–181. [CrossRef]
17. Sokolov, I.M.; Mai, J.; Blumen, A. Paradoxal Diffusion in Chemical Space for Nearest-Neighbor Walks over Polymer Chains. *Phys. Rev. Lett.* **1997**, *79*, 857. [CrossRef]
18. Sokolov, I.M. Lévy flights from a continuous-time process. *Phys. Rev. E* **2000**, *63*, 011104. [CrossRef]
19. Klafter, J.; Shlesinger, M.F.; Zumofen, G. Beyond Brownian Motion. *Phys. Today* **1996**, *49*, 33. [CrossRef]
20. Zaburdaev, V.; Denisov, S.; Klafter, J. Lévy walks. *Rev. Mod. Phys.* **2015**, *87*, 483. [CrossRef]
21. Shlesinger, M.F.; Zaslavsky, G.M.; Frisch, J. (Eds.) *Lévy Flights and Related Topics in Physics*; Springer: Berlin/Heidelberg, Germany, 1995. [CrossRef]
22. Metzler, R.; Chechkin, A.V.; Gonchar, V.Y.; Klafter, J. Some fundamental aspects of Lévy flights. *Chaos Soliton. Fract.* **2007**, *34*, 129. [CrossRef]
23. Dybiec, B.; Gudowska-Nowak, E.; Barkai, E.; Dubkov, A.A. Lévy flights versus Lévy walks in bounded domains. *Phys. Rev. E* **2017**, *95*, 052102. [CrossRef]
24. Krasowska, M.; Strzelewicz, A.; Dudek, G.; Cieśla, M. Structure-diffusion relationship of polymer membranes with different texture. *Phys. Rev. E* **2017**, *95*, 012155. [CrossRef] [PubMed]
25. Strzelewicz, A.; Krasowska, M.; Dudek, G.; Cieśla, M. Optimal hybrid membrane structure based on experimental results and simulation analysis of diffusion process. *J. Mater. Sci.* **2022**, *57*, 11491. [CrossRef]
26. Krasowska, M.; Strzelewicz, A.; Dudek, G.; Cieśla, M. Numerical Study of Drift Influence on Diffusion Transport through the Hybrid Membrane. *Membranes* **2022**, *12*, 788. [CrossRef]
27. Cieśla, M.; Dybiec, B.; Sokolov, I.; Gudowska-Nowak, E. Taming Lévy flights in confined crowded geometries. *J. Chem. Phys.* **2015**, *142*, 164904. [CrossRef] [PubMed]
28. Kubala, P.; Cieśla, M.; Dybiec, B. Diffusion in crowded environments: Trapped by the drift. *Phys. Rev. E* **2021**, *104*, 044127. [CrossRef]
29. Strzelewicz, A.; Krasowska, M.; Dudek, G.; Cieśla, M. Design of polymer membrane morphology with prescribed structure and diffusion properties. *Chem. Phys.* **2020**, *531*, 110662. [CrossRef]
30. Box, G.E.P.; Muller, M.E. A note on the generation of random normal deviates. *Ann. Math. Stat.* **1958**, *29*, 610. [CrossRef]
31. Cieśla, M.; Capała, K.; Dybiec, B. Multimodal stationary states under Cauchy noise. *Phys. Rev. E* **2019**, *99*, 052118. [CrossRef]

Disclaimer/Publisher's Note: The statements, opinions and data contained in all publications are solely those of the individual author(s) and contributor(s) and not of MDPI and/or the editor(s). MDPI and/or the editor(s) disclaim responsibility for any injury to people or property resulting from any ideas, methods, instructions or products referred to in the content.

Article

Aliquots of MIL-140 and Graphene in Smart PNIPAM Mixed Hydrogels: A Nanoenvironment for a More Eco-Friendly Treatment of NaCl and Humic Acid Mixtures by Membrane Distillation

Giuseppe Di Luca [1], Guining Chen [2], Wanqin Jin [2] and Annarosa Gugliuzza [1,*]

[1] Institute on Membrane Technology, National Research Council (CNR-ITM), Via Pietro Bucci 17C, 87036 Rende, Italy; g.diluca@itm.cnr.it

[2] State Key Laboratory of Materials-Oriented Chemical Engineering, College of Chemical Engineering, Nanjing Tech University, 30 Puzhu Road, Nanjing 211816, China; gnchen@njtech.edu.cn (G.C.); wqjin@njtech.edu.cn (W.J.)

* Correspondence: a.gugliuzza@itm.cnr.it

Citation: Di Luca, G.; Chen, G.; Jin, W.; Gugliuzza, A. Aliquots of MIL-140 and Graphene in Smart PNIPAM Mixed Hydrogels: A Nanoenvironment for a More Eco-Friendly Treatment of NaCl and Humic Acid Mixtures by Membrane Distillation. *Membranes* 2023, *13*, 437. https://doi.org/10.3390/membranes13040437

Academic Editor: Wojciech Kujawski

Received: 22 March 2023
Accepted: 13 April 2023
Published: 17 April 2023

Copyright: © 2023 by the authors. Licensee MDPI, Basel, Switzerland. This article is an open access article distributed under the terms and conditions of the Creative Commons Attribution (CC BY) license (https://creativecommons.org/licenses/by/4.0/).

Abstract: The problem of water scarcity is already serious and risks becoming dramatic in terms of human health as well as environmental safety. Recovery of freshwater by means of eco-friendly technologies is an urgent matter. Membrane distillation (MD) is an accredited green operation for water purification, but a viable and sustainable solution to the problem needs to be concerned with every step of the process, including managed amounts of materials, membrane fabrication procedures, and cleaning practices. Once it is established that MD technology is sustainable, a good strategy would also be concerned with the choice of managing low amounts of functional materials for membrane manufacturing. These materials are to be rearranged in interfaces so as to generate nanoenvironments wherein local events, conceived to be crucial for the success and sustainability of the separation, can take place without endangering the ecosystem. In this work, discrete and random supramolecular complexes based on smart poly(N-isopropyl acrylamide) (PNIPAM) mixed hydrogels with aliquots of $ZrO(O_2C-C_{10}H_6-CO_2)$ (MIL-140) and graphene have been produced on a polyvinylidene fluoride (PVDF) sublayer and have been proven to enhance the performance of PVDF membranes for MD operations. Two-dimensional materials have been adhered to the membrane surface through combined wet solvent (WS) and layer-by-layer (LbL) spray deposition without requiring further subnanometer-scale size adjustment. The creation of a dual responsive nanoenvironment has enabled the cooperative events needed for water purification. According to the MD's rules, a permanent hydrophobic state of the hydrogels together with a great ability of 2D materials to assist water vapor diffusion through the membranes has been targeted. The chance to switch the density of charge at the membrane–aqueous solution interface has further allowed for the choice of greener and more efficient self-cleaning procedures with a full recovery of the permeation properties of the engineered membranes. The experimental evidence of this work confirms the suitability of the proposed approach to obtain distinct effects on a future production of reusable water from hypersaline streams under somewhat soft working conditions and in full respect to environmental sustainability.

Keywords: layer-by-layer; thermal and pH responsive membranes; responsive interfaces; water purification; membrane distillation

1. Introduction

Water pollution, extraction from groundwater basins, climate changes, and geochemical cycles are some of the major causes of risk for human health, biodiversity, and planet survival [1,2]. Water stresses as well as upward domestic, industrial, and livestock water consumption are seriously compromising the quality and quantity of usable water [3–5]. For such a reason, there is vigilant activity in identifying competitive and eco-sustainable technologies that support sophisticated and sustainable water management practices [6–8]. Membrane distillation (MD) is an example of green-powered technology, which allows freshwater to be recovered from wastewater and seawater through changes in the state of water [9,10]. In the most common thermally driven direct contact (DC) configuration, the MD technology uses hydrophobic porous membranes to separate two aqueous phases, one containing the stream to treat and the other containing pure water [11]. Applying a difference of temperature across the membrane, the water is evaporated from the hot side (feed = aqueous stream), is diffused as a vapor through the pores of the membrane and, is then collected at the cold side after condensation (permeate = pure water). A suitable combination of chemistry and morphology can make the membrane waterproof and permeable to water vapor at the same time [12,13]. In the logic of energy sustainability, MD can be also powered by solar, wind, and wave energy [14–16]. The competitiveness and sustainability of this technology is, however, strongly dependent on the choice of the membrane.

A membrane with hydrophobic properties is strongly desirable. Polymers such as polypropylene (PP), polytetrafluoroethylene (PTFE), and polyvinylidene fluoride (PVDF) are in fact the most used. PTFE is a highly crystalline polymer with excellent thermal stability and chemical resistance. However, the sintering method combined with the melt extrusion method is the unique practical route for the fabrication of hydrophobic membranes. PP is another crystalline polymer used to fabricate porous membranes by stretching and melt extrusion processes. However, PP membranes exhibit higher surface free tension and moderate thermal stability at high temperature. Unlike PTFE and PP, PVDF is marked by better solubility in common organic solvents and is easily adaptable when being moulded in distinctive and modifiable morphologies through the use of different and combined manufacturing techniques, including well-established phase separation. It is also a thermally stable and highly hydrophobic polymer with high resistance to most corrosive chemicals and organic materials. Over the last few years, PTFE, PP, and PVDF polymers and related copolymers have been used to fabricate commercial hydrophobic membranes, which have been initially applied in microfiltration and successively adapted to MD applications. While PTFE and PP membranes are often used in commercial and pilot MD systems, the potential of PVDF membranes in making MD operations much more competitive at scale is still under investigation. However, attempts at solutions in that direction have provided reasonable and encouraging results in the recent past [17]. Despite the fact that MD processes based on PVDF membranes are still a technology validated in a relevant environment, a lot of research has been carried out to explore the potential of this polymer in combination with other hydrophilic, organic, and inorganic materials. This is because improved surface and transport membrane properties are highly desirable for scaled MD processes [17]. For handling, PVDF is indeed one of the most used polymers for the fabrication of high-performing nanocomposite hydrophobic membranes, which are expected to catalyze the passage from traditional physical barriers to interactive chemical interfaces with amplified performance [18–29].

Among the various materials used, 2D materials have been demonstrated to be particularly attractive for improving the outputs of membrane separations [23,26,28]. However, less consideration has been given to managing their quantities with a regard for safety and respect for the environment. A desired target is to use the minimum quantity of nanofiller to enhance water yield. An option could be to confine small quantities of materials to the membrane surface considering that interfacial interactions between the membrane and surrounding environment can decide the final result [23,30–35]. In every membrane process, including MD operations, engineered membrane surfaces are in fact already expected to establish selective interactions with approaching solutions at the early stage [36–49]. Affinity and repulsive forces can be established at the interface so that permeation and/or refoulement can be well addressed.

This study provides experimental evidence about the effects of a dual responsive nanoenvironment, generated throughout the membrane surface, on the sustainability of the overall MD process. Aliquots of nanofiller are adhered to the membrane skin as a practical route for promoting sub-nanometer control over the surface properties so as to obtain (a) enhanced production of freshwater from hypersaline solutions containing NaCl and humic acid, (b) contrasted adhesion of foulants on the membrane surface, and (c) in situ self-cleaning action without the use of additional harsh chemical agents.

Specifically, a layer-by-layer (LbL) spray technique [50–60] is used for adsorbing randomly discrete complexes of a Zr-MOF compound and graphene and hydrogels on the surface of a PVDF membrane. While graphene together with other 2D materials has been successfully tested in MD operations [23,26,35,61–71], there is not yet a solid indication about the potential of metal–organic framework compounds (MOFs) in DCMD applications [72–74].

Large specific surface area, regular porous structure, and good thermal and chemical stability have made MOFs of significant interest for gas separation, catalysis, storage, and drug delivery [75–79], while their effective role in MD has yet to be proven. Cao et al. [67] proposed a study of molecular dynamic simulation through an ultrathin conductive MOF film, revealing the capability of the latter to permeate water to be three to six orders of magnitude higher than traditional membranes and one order of magnitude higher than single-layer nanoporous graphene or molybdenum disulfide (MoS_2). Yang et al. [72] proposed the fabrication via electrospinning of superhydrophobic poly(vinylidene fluoride) with Fe-MOF up to 5 wt.% for equipping DCMD devices. Fluxes up to 2.87 $Lm^{-2}h^{-1}$ have been obtained along with NaCl rejection of 99.99%. Cheng et al. [74] operated aluminum fumarate MOF/PVDF hollow fiber membranes in DCMD plants, yielding fluxes of 8.04 $Lm^{-2}h^{-1}$ at 50 °C and 15.64 $Lm^{-2}h^{-1}$ at 60 °C. A gain in flux was obtained by MOF-functionalized alumina tubes with values of 16.7 $Lm^{-2}h^{-1}$ at 50 °C and 32.3 $Lm^{-2}h^{-1}$ at 60 °C, though working in a more expensive vacuum configuration (VMD) [80]. In all these cases, MOFs have been used as physical spacers inside the polymer matrix to increase the intrinsic porosity of the membrane and open additional free gaps among the polymer chains. In this way, resistance to mass transfer was reduced and water flux was increased.

In the present work, we confine aliquots of $ZrO(O_2C-C_{10}H_6-CO_2)$ (MIL-140) and graphene to the surface of a PVDF membrane with the intent of triggering effective chemical interactions without affecting the original morphology and packing of the host polymer matrix. We use complexes of ionic thermo- and pH-responsive poly(N-isopropyl acrylamide) (PNIPAM) mixed hydrogels, i.e., acid- and amine-terminated hydrogels, to provide 2D materials with a chemical environment allowing (a) a random and electrostatically driven deposition of the nanofillers over the membrane skin (active surface), (b) hydrophobic thermal effects that prevent detachment or leaking of the adhered materials in aqueous media, and (c) modulated negative charge density for improved resistance to fouling and the stimulation of self-cleaning actions.

Random and discrete aliquots of MIL-140 and graphene have proven to be effective in enhancing water flux through cooperative interfacial forces established at the membrane–solution interface. The 'hydrogel–2D materials' complexes have also been demonstrated to contrast fouling events and remove foulants away from the surface through a simple switch of the ionic charge. Repulsion forces are further addressed at facilitating eco-friendly cleaning procedures, with a full recovery of the initial performance of the membranes bringing charged surfaces. These types of engineered membranes are promising and seem to bridge the distance between more traditional and new attractive families of responsive interfaces, which in the near future could make DCMD a more reliable, efficient, and environmentally friendly operation for the recovery of freshwater.

2. Experimental Section

2.1. Materials

PVDF (Solef®6020, Solvay Solexis: water adsorption <0.040% at 23 °C after 24 h; dp = 1.78 kg/m^3) was kindly supplied by Solvay Specialty Polymers (Milan, Italy). Graphene (G) flakes were purchased from Sigma Aldrich (carbon > 95 wt.%; oxygen < 2 wt.%, Milan, Italy). MIL-140 [ZrO(O$_2$C-C$_{10}$H$_6$-CO$_2$)] (M) was synthesized according to the procedure reported in [81]. This choice was due to the fact that Zr-based MOFs are among the most water and solvent stable and mechanically resistant due to the strong Zr-O bonds [68]. N-Methyl-2-pyrrolidinone (NMP, Riedel de Häem: max 0.05% in water, d. 1.03 kg m^{-3}) and 2-propanol (IPA, WWR PROLAB: d. 0.78 kg m^{-3}, Milan, Italy) were used as the respective solvent and nonsolvent for the preparation of microporous PVDF membranes. N-ethyl-o,p-toluenesulphonamide (Sigma Aldrich, water solubility < 0.01 g/100 mL at 18 °C, Milan, Italy) was used as a WS. Carboxylic acid-terminated poly(N-isopropylacrylamide) (A, M$_n$ 5000) and amine-terminated poly(N-isopropylacrylamide) (N, M$_n$ 5000) were purchased from Sigma Aldrich. Fluorinert (FC-40, Novec, Merck, Milan, Italy) was used for gas–liquid displacement measurements for pore size and overall porosity estimation. Ultra-pure water (filtered by a USF ELGA plant) was used to investigate membrane anti-wetting properties. NaCl with a degree of purity of 100% was purchased from VWR Chemicals. Humic acid (HA, Sigma Aldrich, Milan, Italy) was mixed with NaCl for hypersaline solutions. All materials were used as received.

2.2. Membrane Preparation

PVDF powder (12 wt.%) was dissolved and stirred for 24 at 40 °C in NMP. The homogeneous solution was cast on a glass support through the use of a micrometric film applicator (Elcometer) with a gap size of 250 μm. The casting solution was then coagulated in 2-propanol and washed with deionized water. After drying at room temperature, the membrane was further treated at 40 °C for 1 h and then sprayed with a solution of N-ethyl-o/p-toluenesulphonamide dissolved in ethanol at 1.5 w%. PNIPAM-NH$_2$ (N) and PNIPAM-COOH (A) were dissolved in water (10^{-2} M based on the repeat unit molecular weight) at 16 °C while MIL-140 (M) and graphene (G) were dispersed in PNIPAM-NH$_2$ under ultrasound treatment for 24 h at a concentration of 1.0 mgmL^{-1}. The pH value of each solution was adjusted to 5.7, which is typical of solutions containing humic acid.

A PVDF membrane was used as the substrate for discrete and random deposition of ionic PNIPAM/nanofillers complexes. The solutions were alternatively sprayed five times starting from the amine-terminated hydrogel (N) using a commercial glass spray bottle held 50 cm from the sublayer; a waiting period of 10 min was allowed between each layer to facilitate adsorption. For simplicity, these hybrid complexes were named as LAN in absence of 2D materials, LANG and LANM with graphene and Zr-MOF, respectively, and LANMG with graphene alternated with Zr-MOF (Scheme 1). All samples were dried in the air before testing.

2.3. Methods

Membrane morphology and topography features were examined using Scanning Electronic Microscopy (SEM, Zeiss EVO MA10, Oberkochen, Germany) and Atomic Force Microscopy (AFM, Nanoscope III Digital Instruments, VEECO Metrology Group, Santa Barbara, CA, USA). The latter was operated in tapping mode at a rate of 1 Hz across sample surfaces for 512 points. The pore size of the membranes was measured by gas–liquid displacement using a porosimeter (Capillary Flow Porometer-CFP 1500 AXEL, Porous Materials Inc., Ithaca, NY, USA). Overall porosity was measured by filling the membranes with FC-40. Membrane weight was estimated before and after filling and porosity was expressed in a percentage as the ratio between the volume occupied by the fluorine liquid and the volume of the membrane. The amount of materials deposited was estimated after drying through the use of a balance with five digits. Deposition, stability, and the thermo-responsive behavior of the ionic PNIPAM hydrogels were investigated by infrared spectroscopy in ATR mode (Spectrum One System, Perkin Elmer, Milan, Italy). MOF adsorption was also inspected by EDX (Zeiss EVO MA10, Oberkochen, Germany). Waterproofness was estimated by measuring the contact angle values throughout the functionalized surfaces (Cam200 KSV instruments, LTD, Helsinki, Finland).

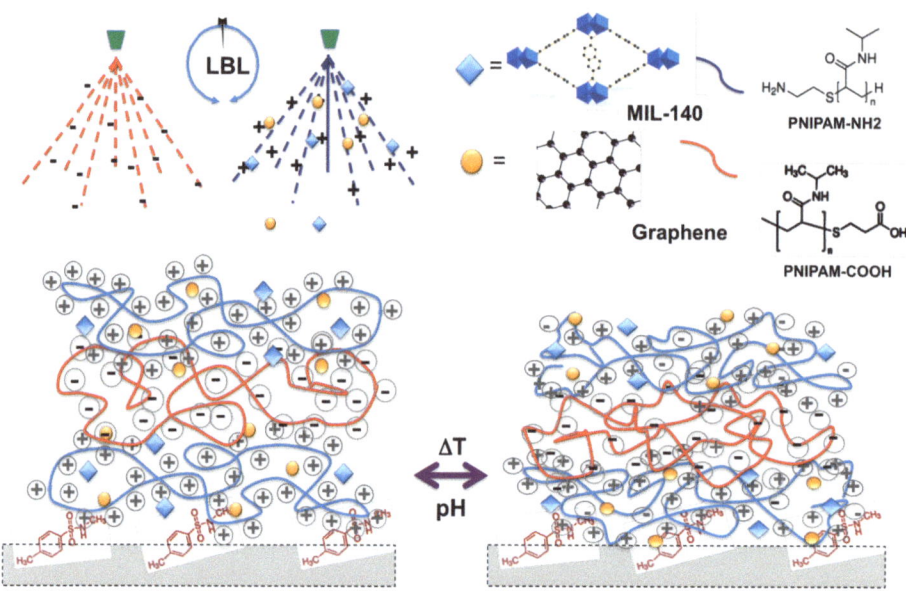

Scheme 1. Representative sequential deposition of ionic PNIPAM hydrogels and Zr-MOF and graphene on a PVDF substrate via an LBL spray technique.

2.4. Membrane Distillation Tests

Thermally driven DCMD experiments were executed using NaCl (35 gL^{-1}) and mixtures of NaCl/HA (35 gL^{-1}/1.0 mgmL^{-1}), flow rates of 6 Lh^{-1} (feed side) and 4.98 Lh^{-1} (permeate side), and temperatures of T_{feed} = 45 °C and T_{perm} = 16 °C. Retentate and distillate streams were converged in a counter-current way toward the membrane module containing the membrane, where the liquid water was evaporated. On the retentate side, a pump was taking and sending the heated feed to the membrane module. Additionally, on the distillate side, a second pump ensured the counter-current recycling of the cold stream in order to remove the vapor diffusing through the pores of the membrane from the solution. The trans-membrane fluxes were calculated by weighing the variations in the distillate tanks. The experiments were run from 6 to 18 h continuously. The salt conductivity of the feed and permeate streams was measured at the end of every single experiment through the use of a conductive meter (HI 2300 bench meter supplied by Hanna Instruments, Woonsocket, RI, USA).

3. Results and Discussion

3.1. Membrane Fabrication

On the assumption that each membrane is crucial for molecular separation, new concept membranes are here proposed to make water purification more eco-friendly when a DCMD operation is performed. Engineered membrane surfaces have been designed through combined phase inversion, WS, and LbL spray procedures. The intent is to provide a greener route for the fabrication of new functional membranes designed for more sustainable management of water.

Figure 1 displays the surface and a cross section of PVDF membranes before and after surface engineering. It is pertinent to consider the spherulitic-like structure of the PVDF membrane induced by the exchange of solvents. This kind of morphology is due to crystallization events, which take place during phase inversion. Because PVDF is a semi-crystalline polymer, well-sized spherulites interlinked by polymeric filaments are formed during the delayed solid–liquid demixing of the polymer solution, as displayed in the image included in Figure 1a'. Free gaps are generated between the polymeric particles through the overall symmetrically structured film and work as effective pores of the membranes (Table 1). This singular topography is also somewhat attractive for its high resistance to wetting (θ = 130 ± 2°), which can be regarded as an effect of the high irregularity of the surface (Rq = 112 nm). Highly interconnected open paths together with high waterproofness make this kind of PVDF membrane a suitable candidate for MD applications.

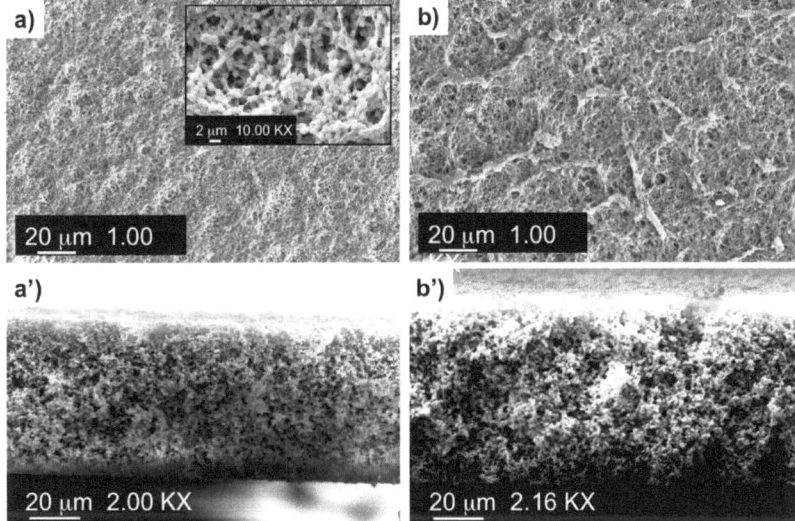

Figure 1. SEM images collected from the surface and along cross sections of pristine PVDF (**a,a'**) and engineered PVDF-LANM (**b,b'**) membranes.

To make this PVDF membrane a well-suited sublayer for PNIPAM hydrogels, the film was subjected to WS treatment through the use of a solution of N-ethyl-o/p-toluenesulphonamide in ethanol. This organic compound has typical amphiphilic properties and very low water solubility, with subsequent good affinity towards materials with different chemistry and resistance to aqueous media. Separately, colloidal dispersions of MIL-140 and graphene were prepared in the hydrogel-NH_2. The pH value of all solutions was adjusted to 5.7 so that the density of charge generated through the hydrogel segment chains allowed ion pairs to be formed by electrostatic attraction. It is relevant to observe how no continuous nano-films, but rather discrete aggregations of the complexes, were randomly deposited through the surfaces without affecting the bulk of the membrane (Figure 1b,b'). SEM images reveal the formation of fibrotic cords without occlusion of surface pores or penetration inside the membrane structure. All porosity, pore size, and pore distribution values calculated for the engineered PVDF membranes almost overlap with those measured for pristine PVDF so as not to offer further resistance to mass transfer (Figure 2a,a'). Indeed, all membranes exhibit two clusters of pores with a mean pore size of around 0.40–0.47 µm and a smallest pore size of around 0.19 µm, while the bubble point is between 0.85 and 0.9 µm (Table 1).

Table 1. Structural parameters estimated for all pristine and engineered membranes.

Complex	Overall Porosity (%)	Largest Pore Size (μm)	Mean Pore Size (μm)	Smallest Pore Size (μm)
PVDF	68 ± 2	0.88 ± 0.02	0.47 ± 0.01	0.197 ± 0.002
PVDF-LAN	67 ± 3	0.85 ± 0.03	0.40 ± 0.02	0.193 ± 0.007
PVDF-LANG	62 ± 1	0.85 ± 0.04	0.42 ± 0.03	0.198 ± 0.003
PVDF-LANM	62 ± 1	0.90 ± 0.03	0.40 ± 0.01	0.195 ± 0.002
PVDF-LANMG	68 ± 1	0.86 ± 0.04	0.42 ± 0.05	0.197 ± 0.005

However, it is important to observe that the pores around 0.4 μm in size provide the major contribution to the flow measured through the membrane. The incremental filter flow (incr. %FF) reaches the maximum value at 0.4 μm (Figure 2b,b'), providing an indication of the predominance of this pore cluster.

A morphological comparative analysis between PVDF membranes before and after functionalization (Figure 2 shows a comparison between pristine PVDF and PVDF-LANM membranes) reveals that the intrinsic structural features of the membranes do not undergo substantial changes. All of this is convenient if we consider that high interfacial area, which is crucial to water vapor diffusion during MD operations, is preserved.

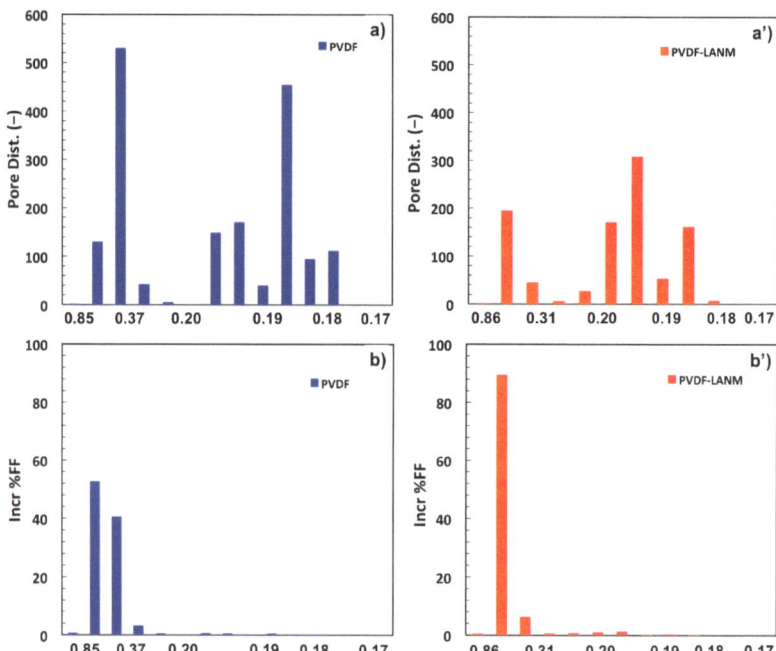

Figure 2. Pore size distribution and incremental filter flow (incr. %FF) estimated for pristine PVDF (a,a') and engineered PVDF-LANM (b,b') membranes.

3.2. Characterization of LAN Complexes

The formation, chemical stability, and thermal responsiveness of polycation/polyanion (LAN) complexes were investigated through infrared analysis (Figure 3). Firstly, ATR spectra collected onto neat hydrogels yielded a clear indication of the typical frequencies associated to the amide group, with band I located at 1639 cm^{-1} and band II positioned at 1538 cm^{-1}, while C-N stretching was detected at 1458 cm^{-1} (Figure 2a). For PNIPAM-COOH, an additional very weak absorption associated with carboxylic acid is detected at 1710 cm^{-1}, while a broad absorbance between 3600 and 3200 with a maximum intensity at 3291 cm^{-1} is ascribed to the overlapping of O-H and N-H stretching modes.

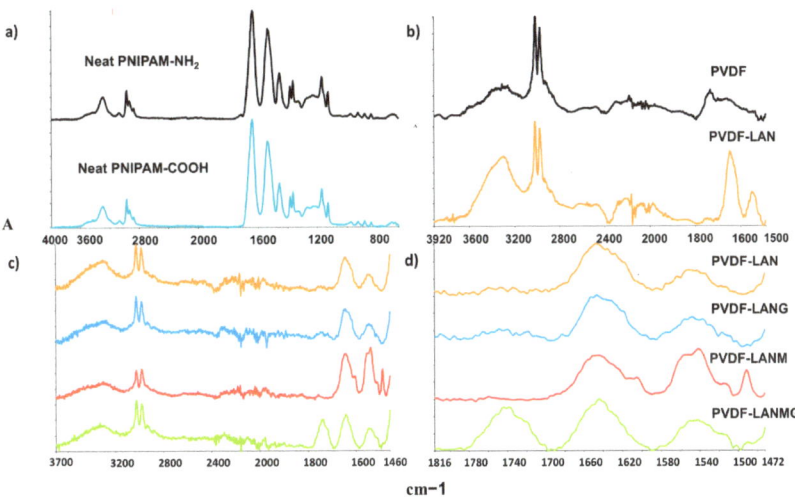

Figure 3. ATR spectra collected on (**a**) neat PNIPAM hydrogels; (**b**) PVDF functionalized with LAN; (**c**) PVDF functionalized with LANG, LAM, and LAMG complexes; (**d**) enlargement of the carbonyl regions related to LANG, LAM, and LAMG complexes.

Figure 3b shows how changes in the carbonyl amide vibrations—ν_s 1647 cm^{-1} and 1542 cm^{-1}—take place as the LAN complex is deposited onto the PVDF surface. Shifted carbonyl frequencies as well as broad flattening of O-H and N-H vibrations can be detected. The asset of the carbonyl bands is further modified when Zr-MOF is deposited on the surface of PVDF membranes through the use of ionic hydrogels (Figure 3c,d). In this case, an intensification of the carbonyl bands along with the appearance of various shoulders and a new peak around 1492 cm^{-1} can be appreciated. In this region of the spectrum, these overlapped and distinct new vibrations can be ascribed to the asymmetric and symmetric stretching of the carboxylate groups $\nu(COO^-)$, which are part of the $ZrO(O_2C-C_{10}H_6-CO_2)$ structure. Additionally, the intensity and broadness of the O-H stretching mode appears to be more intense around 3200 cm^{-1}. In the case of graphene, no substantial modifications are appreciated, while the spectrum collected on the membrane surface functionalized with the complex containing alternated aliquots of graphene and Zr-MOF (LANMG) shows three carbonyl stretching patterns, one of which is located at 1738 cm^{-1} (Figure 3c,d). This band is typical of the $\nu(C=O)$ mode of non-coordinated and free carboxylic acid groups. It should be noted that amine-terminated hydrogel bringing Zr-MOF and graphene was alternated with acid-terminated hydrogel. This obtains amine- and acid-terminated hydrogels with greater ability to neutralize the counter charges during deposition of the practically inert graphene, thus yielding more available COOH moieties of Zr-MOF in the successive step.

EDX confirms the significant presence of Zr in the chemical composition of the MIL-140 particles adhered to the PVDF membrane surface (Figure 4), while SEM and AFM micrographs distinguish the presence of the nanofillers on the membrane surfaces (Figures 4 and 5). In the case of AFM, topographical bright and edged regions are well distinct from the sublayer, also yielding an indication of the irregularity of the surface (Figure 5a). The random deposition of discrete objects produces an increase in the roughness factor, with a subsequent improvement in water repellence (Figure 5b). At equilibrium, membranes functionalized with complexes containing graphene (LANG) show a contact angle value of $136 \pm 4°$ against the $134 \pm 3°$ estimated for membranes containing Zr-MOF alone (LANM). For membranes with mixed Zr-MOF and graphene (LANMG), the contact angle value undergoes a negligible decrease ($133 \pm 3°$) due to the higher availability of free carboxyl groups, while a value of $129 \pm 4°$ was measured on membranes functionalized with ionic hydrogels without 2D materials (LAN) (Table 1). While the liquid water entry pressure (LEP$_w$) is less than 1 bar for all membrane samples, the singular topography of these hybrid membranes provides good resistance to liquid spreading due to the fact that the morphological component moves the surface properties towards higher irregularity and a subsequent reduced contact line between liquid and polymer (Figure 5c). In this case, 2D materials amplify the irregularity of the section profile, leading to a good fit between the root mean square deviation (R_q) and contact angle values (Table 2). Generally, increasing the value of surface roughness improved the anti-wetting behavior of engineered surfaces, thus leading to contact angle values of up to $137 \pm 4°$. Regarding this point, a short premise is required. MD requires operating hydrostatic pressure lower than that of pressure-driven processes such as reverse osmosis (RO). It works at pressures near that of atmospheric pressure since small differences in vapor partial pressure are enough to promote mass transfer. Thus, the high degree of waterproofness estimated for all proposed membranes is mostly expected to balance the low values of LEP, thereby resulting in a suitable resistance to liquid intrusion.

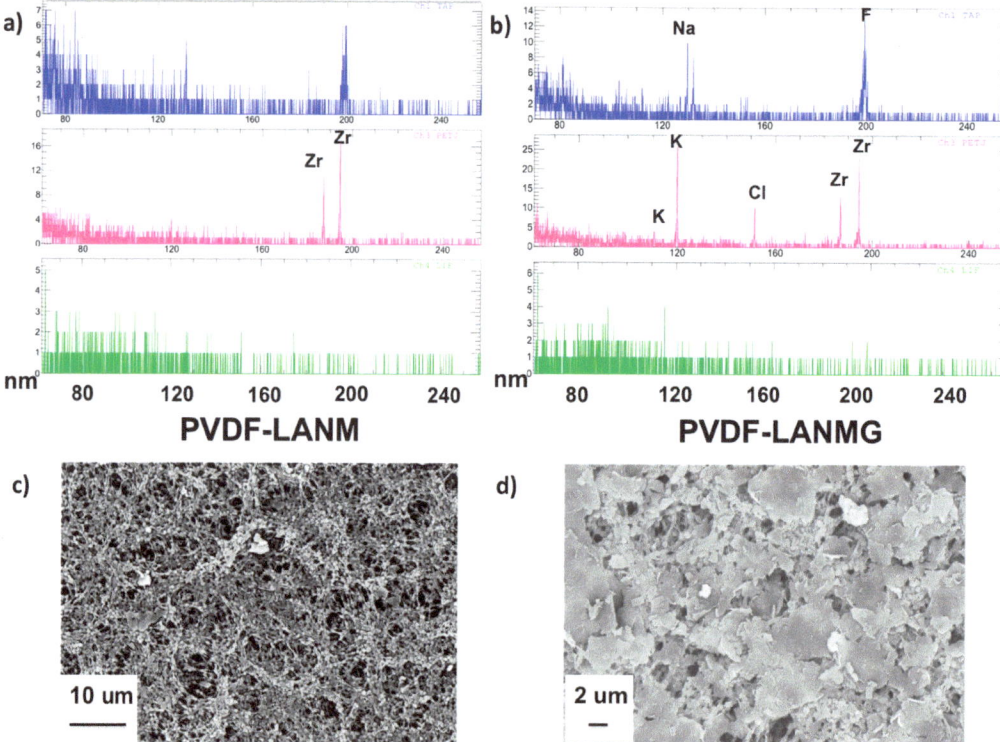

Figure 4. EDX spectra (**a,b**) and SEM images (**c,d**) collected on PNIPAM mixed hydrogels containing Zr-MOF (PVDFLANM) and Zr-MOF mixed with graphene (PVDF-LANMG).

Table 2. Contact angle values and surface roughness parameters related to the ion pairs adhered.

Complex	CA$_{t=0}$ (°)	CA$_{eq}$ (°)	R$_a$ (nm)	R$_q$ (nm)
PVDF-LAN	130 ± 4	129 ± 4	141 ± 66	184 ± 83
PVDF-LANMG	135 ± 3	133 ± 3	215 ± 72	262 ± 64
PVDF-LANM	136 ± 3	134 ± 3	223 ± 65	280 ± 77
PVDF-LANG	137 ± 4	136 ± 4	236 ± 59	299 ± 73

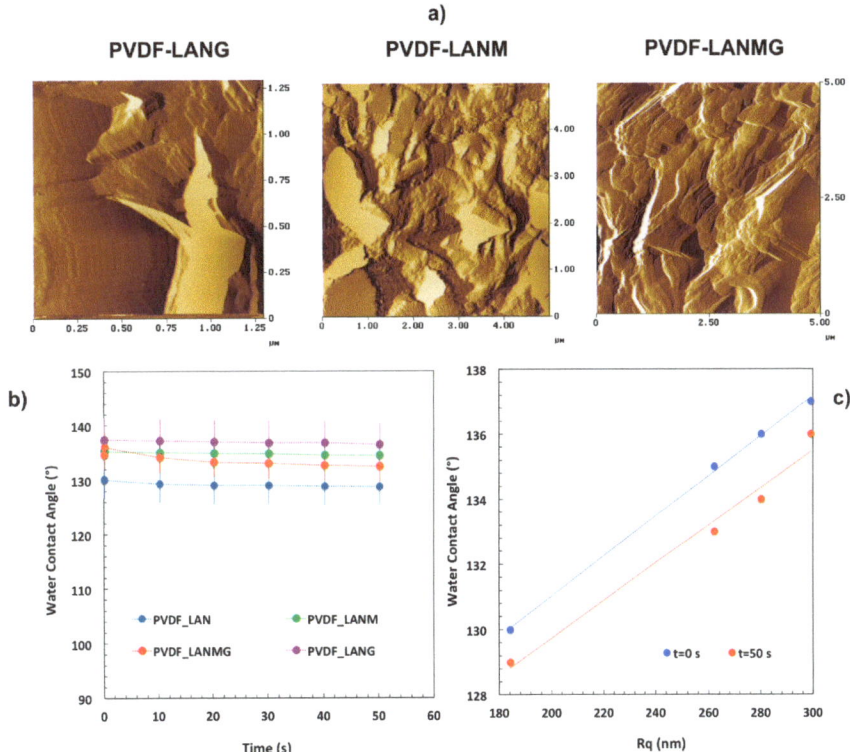

Figure 5. AFM topographies of the engineered membrane surfaces (**a**), contact angle values with time (**b**), and the relationships between water contact angles and surface roughness measured through all engineered membranes (**c**).

To evaluate the stability of the hierarchical materials in aqueous medium, membrane samples were submerged in water and stirred vigorously for 6 h at 40 °C. The films were then dried overnight and inspected again by infrared spectroscopy (Figure 6).

Comparison of the spectra collected from samples before and after treatment does not reveal substantial modifications to the intensity and vibration of the typical carbonyl bands, thereby suggesting the hydrophobic state of materials when operated at temperatures above 40 °C rather no leakage. It is known that PNIPAM hydrogels exhibit hydrophobic properties in aqueous solution at temperatures higher than the Low Critical Solution Temperature (LCST, around 32–33 °C) [82–84]. Above the LCST, a rigid water-soluble 'ice-like structure' is generated from the effect of intermolecular aggregations. Water molecules are released and hydrophobic groups are exposed to each other. In this state, the hydrogel exhibits typical hydrophobic behavior and insolubility in water. It is also known that functional side groups in the polymer, as well as changes in pH, and salts dissolved in water, may affect the LCST in the order of 0.1 to 5 °C [85–87]. As an example, values of 33.1 to 37.7 °C have been detected at pH 5 for some synthesized poly(NIPAM-co-AAD) copolymers and acid-terminated hydrogels. Our experiments give a clear indication about the stability of the hydrogels complexes in aqueous solutions at 40 °C and pH 5.7, denoting affinity towards wet PVDF sublayers rather than water.

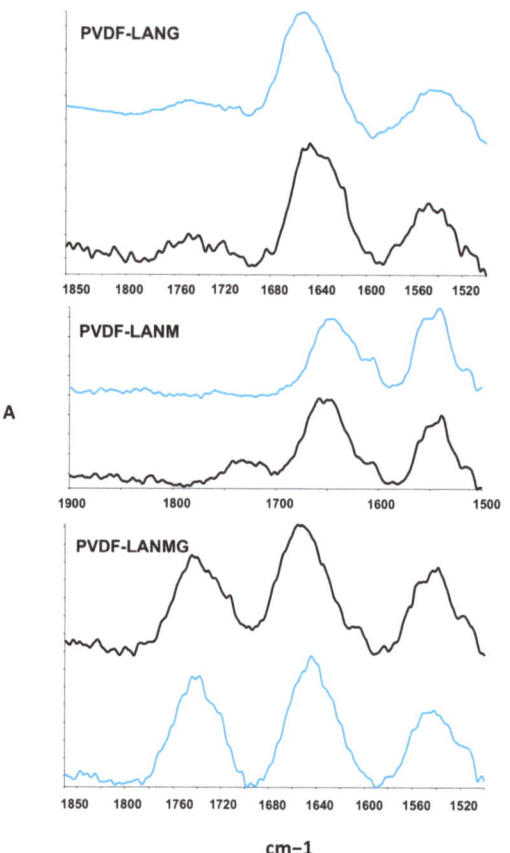

Figure 6. ATR spectra collected on the engineered PVDF membranes before (blue line) and after (black line) treatment at 40 °C under magnetic stirring.

3.3. Membrane Distillation Testing

Based on the previous results, MD experiments were carried out, with the composite–hybrid membranes exposed to saline streams at 45 °C. Under this temperature, the hydrogels were in a hydrophobic state and, hence, insoluble in aqueous media. Further, the pH value of the saline streams was around 5.7 and comparable to that of the LBL deposition. Thus, no changes in charge density are expected and the stability of the supramolecular complexes is preserved during MD. Experiments with a solution of NaCl at a concentration comparable to that of seawater (35 gL^{-1}) were conducted for 18 running hours.

Figure 7a displays the trend of the average flux measured through all membranes with time. As compared to the pristine PVDF, enhanced flux is observed through all functional membranes. A gain up to 17 and 19% is estimated for membranes adhered with LANG and LANM, respectively. The increase is around 10% for membranes functionalized with the LANGM complex. This is not surprising if we consider that MOFs are rich in carbonyl sites wherein water molecules can be temporarily adsorbed and transported through additional ordered nanoporous pathways. Similarly, defective graphene has been envisaged to have a great ability to assist water diffusion [26,35]. As expected, a slight decrease in flux is detected after 18 running hours, even if mass transfer through LANM continues to be the most consistent. All engineered membranes exhibit a selectivity of 99.99% against the 99.95% estimated for the pristine PVDF after the first six hours. At the end of the test, values higher than 99.9% can be estimated for all graphene-engineered membranes, while pristine PVDF and PVDF-LANM exhibit a rejection value of 99.8% (Figure 7b).

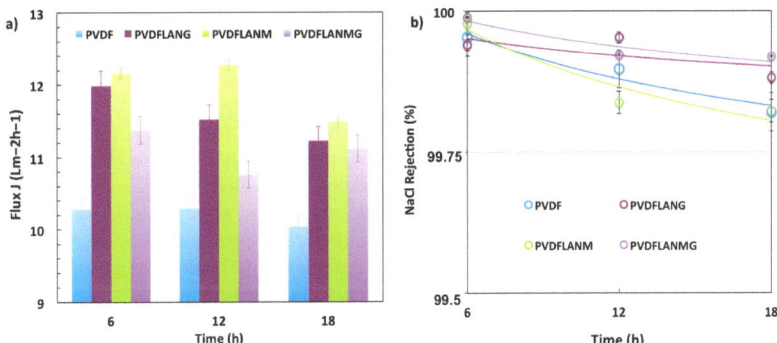

Figure 7. Averaged flux (**a**) and NaCl rejection (**b**) measured through pristine and engineered PVDF membranes during 18 continuous running hours with NaCl solution (35 gL^{-1}).

Mixtures such as NaCl/HA (35 gL^{-1}/1 mgmL^{-1}) were also treated with the engineered PVDF membranes (Figure 8a). After the first two hours of operation, fluxes of up to 11.7 Lm^{-2}h^{-1} were measured, with an increase of up to 58% with respect to the pristine PVDF. An increase of 30% continued to persist after six hours of operation. Figure 8b shows how the flux ratio is four to five times greater than that of the pristine PVDF during the first hour of the process. Successively, membranes with Zr-MOF show a flux ratio 1.5–1.2 times greater, indicating better performance (Figure 8b). It is relevant to observe that humic acid is strongly negatively charged at a pH greater than 4.7 [87]. Because the MD operation was performed under a pH of 5.7, membranes functionalized with aliquots of LAN and Zr-MOF exhibit a further negative charge due to the deprotonation of -COOH groups of the acid-terminated hydrogel and Zr-MOF structures. The same negative charge is consequently expected to produce higher repulsive electrostatic forces, thus leading to lower adhesion of the acid to the membrane surface. Initially, for the functional membranes, the removal rate of acid from the surface is higher than that on the neutral pristine membrane. As time progresses, the capability to contrast the acid adhesion on the surface continues to be more evident for membranes with a larger density of carboxylate groups (LANM, LANMG).

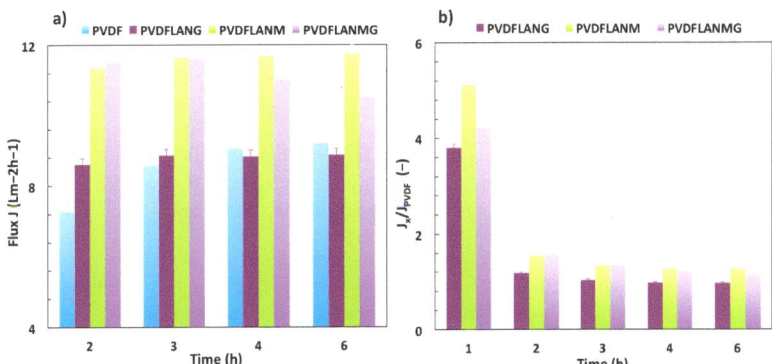

Figure 8. Averaged flux (**a**) through pristine and engineered membranes coming in contact with mixtures of NaCl/HA (35 gL^{-1}/1 mgmL^{-1}) and incremental ratio (**b**) with respect to the pristine PVDF.

Electrostatic repulsion has been also demonstrated to be crucial for cleaning. Reversible switching of the ionic charge throughout the surface has been proven to be enough to remove foulants from the functionalized surface quickly. To preserve the hydrophobic state of the supramolecular complexes, the membranes were washed with water at 46 °C, while the pH value was switched to a pH of 6.5 for 30 min. After a successive washing with pure water for a further 30 min, the pH was adjusted to 5.7 again and washing was continued for a further 30 min. To evaluate the recovery capability of the membranes, MD tests with pure water were further carried out. As displayed in Figure 9a, factors of recovery of 98–99% were obtained with LANM and LANMG, while a percentage of 95% was assessed for LANG against an estimated 92% for pristine PVDF. This means that strongly negatively charged membrane surfaces exhibit a better ability to remove adhered HA as an effect of repulsive forces, whereas the neutral membranes lose this capacity in the absence of additional chemical reagents. This result has to be regarded as another wide-ranging aspect of sustainability for MD processes because no additional harsh chemical agents are required to restore the quality of the engineered membranes after cleaning. Additionally, the stability and integrity of the supramolecular structures was assessed after cleaning and water permeation. Figure 9b shows no changes in intensity and carbonyl frequencies, with a more pronounced contribution of the band associated with protonated carboxylic groups observed instead due to the slight effect of the pH value. In this way, the stability and chemical steadiness of the complexes deposited on PVDF surfaces are fully satisfactory.

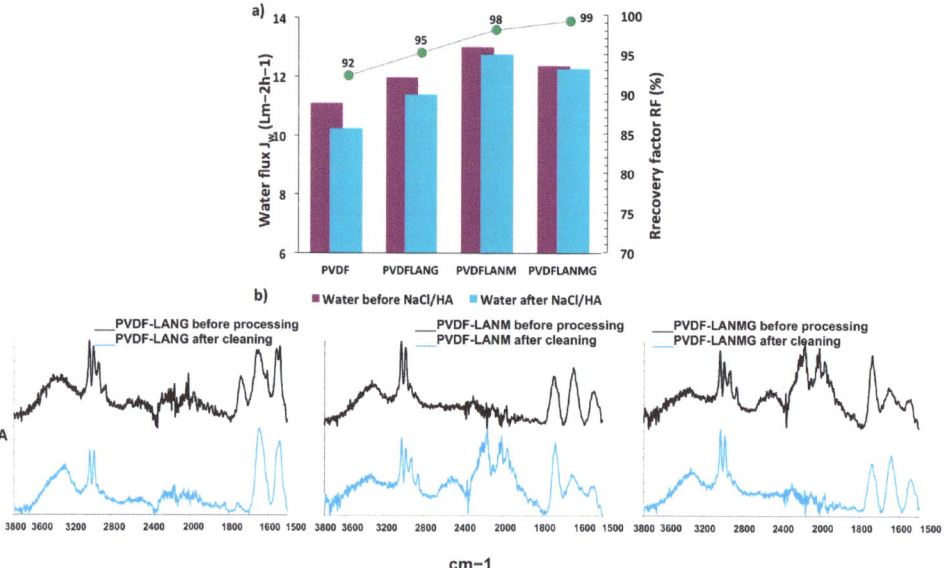

Figure 9. Water flux and recovery factor (**a**) estimated for all pristine and functional membranes before and after purification processes; ATR spectra (**b**) collected onto membranes before processing (blue line) and after the processing and cleaning procedure (black line).

In summary, we demonstrated how an ionic hydrogel-based nanoenvironment (0.5 mg/cm^2), where aliquots of 2D materials are confined, yields proficient cooperative mechanisms at the membrane surface–solution interface. Interfacial forces control water diffusion through the membranes, as comparative analyses confirm (Table 3). A practice currently in wide use is the filling of electrospun PVDF membranes with MOFs at concentrations of 0.1 to 5% in order to increase permeability under hard fluid dynamics conditions [72–74,88–91]. Herein, aliquots of hybrid responsive MOF supramolecular structures are demonstrated to already produce positive effects on flux under much softer operating conditions. Concerning graphene, many studies have also focused on the incorporation of few-layer, nanoplatelet, nanosheet, and quantum dot graphene in membranes [26,28,61,92,93]; however, only a few attempts have been made to adhere this material on the membrane surface for DCMD applications [35,92]. Herein, we demonstrate how the deposition of random and discrete aliquots of graphene on PVDF membrane surfaces is already beneficial to consistent productivity under mild conditions without the necessity of forming extensive coatings through more expensive procedures.

Table 3. Comparative data estimated for membranes functionalized with MOFs and graphene and operated in DCMD mode under different working conditions.

Membrane	Nanofiller	Amount [%]/Mode	ΔT [°C]	Feed/Permeate Flow Rate [mLmin^{-1}]	Prist./Func. Membr. Flux [Lm^{-2}h^{-1}]	Reference
PVDF nanofibrous membranes	Iron 1,3,5-benzenetricarboxylate MOF	5 (bulk)	32	1500/1500	2.1/3.3	[72]
Triple-layer PVDF/PAN/PVDF ENMs	Hydrophobic SiO$_2$/MOF/hydrophilic SiO$_2$	5/1.5/1 (bulk)	30	1500/1500	2.68/4.4	[73]
PVDF hollow fiber membrane	AlFu MOF	1 (bulk)	30	450/450	6.1/8.0	[74]
PVDF-HFP ENMs *	AlFu MOF	0.1 (bulk)	40	500/500	10.6/17.0	[91]
PVDF-f-G	Graphene	CVD (surface)	50	1000/1000	~0 **/3	[92]
PVDF/G0.005	Graphene coating	WF *** (surface)	24	100/80	5.4/7.5	[61]
PVDF-LANM	[ZrO(O$_2$C-C$_{10}$H$_6$-CO$_2$)]	LBL (surface)	30	100/80	10.2/12.3	In this work
PVDF-LANG	Graphene	LBL (surface)	30	100/80	10.2/11.5	In this work
PVDF-LANMG	[ZrO(O$_2$C-C$_{10}$H$_6$-CO$_2$)] Graphene	LBL (surface)	30	100/80	10.2/11.0	In this work

* ENMs: electrospun nanofiber membranes; ** PVDF: membrane with a densified skin layer; *** WF: Wet Filtration.

4. Conclusions

An eco-friendly strategy was proposed to realize new functional responsive interfaces that make membrane distillation processes much more sustainable. Zr-MOF and graphene were entangled with double responsive PNIPAM hydrogels, with discrete complexes randomly deposited on a PVDF sublayer via an LBL spray procedure. The stability and reliability of the adhered hybrid–composite structures were tuned, while the thermal and pH responsiveness of the PNIPAM was exploited to provide a hydrophobic state and a suitable repulsive environment during MD operations. The hydrophobicity of the surfaces was preserved when working at temperatures higher than the LCST traditionally observed for PNIPAM, while negative charge distribution through the membrane surface limited fouling due to HA. The hydrogels' pH responsiveness further addressed cleaning procedures in a safe and eco-friendly way, leading to a full recovery (RF = 98–99%) of initial water permeation properties for membranes mainly containing Zr-MOF complexes. Nanoscale confinement of 2D materials was hence demonstrated to be proficient and suitable for controlling interfacial events that result in beneficial changes in the overall productivity and eco-friendliness of MD processes. The effectiveness of membranes was proven through MD testing under somewhat soft conditions. An increase of up to 17–19% was measured through membranes functionalized with aliquots of PNIPAMs/graphene and PNIPAMs/MOFs coming in contact with synthetic seawater (NaCl 35 gL^{-1}). Flux through all engineered membranes was four to five times higher than in pristine PVDF membranes during the first hour of treatment with a mixture of sodium chloride (35 gL^{-1}) and humic acid (1 mgmL^{-1}), while better antifouling action was detected with increased time. This work demonstrates how the functionalization of PVDF membrane surfaces with aliquots of functional materials can be a more sustainable and fruitful route than incorporation of larger amounts of nanofiller into the bulk.

These novel responsive membranes can be regarded as the interplay of cooperative functions, which may well address the rationalization of new properties for functional PVDF membranes. This approach is aimed at paving the way for the realization of new families of active PVDF interfaces for more efficient, reliable, and sustainable recovery of reusable water by DCMD operations.

Author Contributions: Conceptualization, A.G.; Investigation, G.D.L. and G.C.; Data Analysis, A.G.; Writing—Draft Preparation, A.G.; Writing—Review and Editing, A.G. and W.J.; Funding Acquisition, A.G. All authors have read and agreed to the published version of the manuscript.

Funding: This research was funded by the Italian Ministry of Foreign Affairs and International Cooperation grant number MAE00694722021-05-20—Great Relevance International Project Italy (MAECI)–China (NSFC) 2018–2020; New Materials, with particular reference to Two-Dimensional Systems and Graphene (2DMEMPUR).

Institutional Review Board Statement: Not applicable.

Data Availability Statement: Not applicable.

Acknowledgments: We acknowledge a financial grant from the Italian Ministry of Foreign Affairs and International Cooperation within the framework of the Great Relevance International Project Italy (MAECI)–China (NSFC) 2018–2020—New Materials, with particular reference to Two-Dimensional Systems and Graphene (2DMEMPUR), MAE00694722021-05-20.

Conflicts of Interest: The authors declare that they have no known competing financial interests or personal relationships that could have appeared to influence the work reported in this paper.

References

1. Muluneh, M.G. Impact of climate change on biodiversity and food security: A global perspective. *Agric. Food Secur.* **2021**, *10*, 36. [CrossRef]
2. Adejumoke, A.I.; Babatunde, O.A.; Abimbola, P.O.; Tabitha, A.A.-A.; Adewumi, O.D.; Toyin, A.O. Water Pollution: Effects, Prevention, and Climatic Impact. In *Water Challenges of an Urbanizing World*; Glavan, M., Ed.; Intechopen: London, UK, 2019.
3. Qu, X.; Shi, L.; Qu, X.; Qiu, M.; Gao, W.; Wang, J. Evaluation of Groundwater Resources and Exploitation Potential: A Case from Weifang City of Shandong Province in China. *ACS Omega* **2021**, *6*, 10592–10606. [CrossRef]
4. Macknick, J.; Newmark, R.; Heath, G.; Hallett, K.C. *A Review of Operational Water Consumption and Withdrawal Factors for Electricity Generating Technologies*; Technical Report NREL/TP-6A20-50900; IOPscience: Bristol, UK, 2011.
5. Boretti, A.; Rosa, L. Reassessing the projections of the World Water Development Report. *Npj Clean Water* **2019**, *2*, 15. [CrossRef]
6. Fu, G.; Jin, Y.; Sun, S.; Yuan, Z.; Butler, D. The role of deep learning in urban water management: A critical review. *Water Res.* **2022**, *223*, 118973. [CrossRef] [PubMed]
7. Gugliuzza, A.; Basile, A. *Membranes for Clean and Renewable Power Applications*; Gugliuzza, A., Basile, A., Eds.; Woodhead Publishing: Cambridge, UK, 2014; pp. 1–410.
8. Cosgrove, W.J.; Loucks, D.P. Water management: Current and future challenges and research directions. *Water Resour. Res.* **2015**, *51*, 4823–4839. [CrossRef]
9. González, D.; Amigo, J.; Suáre, F. Membrane distillation: Perspectives for sustainable and improved desalination. *Renew. Sustain. Energy Rev.* **2017**, *80*, 238–259. [CrossRef]
10. Khayet, M. Solar desalination by membrane distillation: Dispersion in energy consumption analysis and water production costs (a review). *Desalination* **2013**, *308*, 89–101. [CrossRef]
11. Ashoor, B.B.; Mansour, S.; Giwa, A.; Dufour, V.; Hasan, S.W. Principles and applications of direct contact membrane distillation (DCMD): A comprehensive review. *Desalination* **2016**, *398*, 222–246. [CrossRef]
12. Speranza, V.; Trotta, F.; Drioli, E.; Gugliuzza, A. High-definition polymeric membranes: Construction of 3d lithographed channel arrays through control of natural building blocks dynamics. *ACS Appl. Mater. Interf.* **2010**, *2*, 459–466. [CrossRef]
13. Perrotta, M.L.; Saielli, G.; Casella, G.; Macedonio, F.; Giorno, L.; Drioli, E.; Gugliuzza, A. An ultrathin suspended hydrophobic porous membrane for high-efficiency water desalination. *Appl. Mater. Today* **2017**, *9*, 1–9. [CrossRef]
14. Rasool, M.; Banat, Q.F. Desalination by solar powered membrane distillation systems. *Desalination* **2013**, *308*, 186–197.
15. Usman, H.S.; Touati, K.; Rahaman, M.S. An economic evaluation of renewable energy-powered membrane distillation for desalination of brackish water. *Renew. Energy* **2021**, *169*, 1294–1304. [CrossRef]
16. Olatunji, S.O.; Camacho, L.M. Heat and Mass Transport in Modeling Membrane Distillation Configurations: A Review. *Front. Energy Res.* **2018**, *6*, 130. [CrossRef]
17. Abdallah, S.B.; Frikha, N.; Gabsi, S. Design of an autonomous solar desalination plant using vacuum membrane distillation, the MEDINA project. *Chem. Eng. Res. Des.* **2013**, *91*, 2782–2788. [CrossRef]

18. Khayet, M.; Matsuura, T. Preparation and Characterization of Polyvinylidene Fluoride Membranes for Membrane Distillation. *Ind. Eng. Chem. Res.* **2001**, *40*, 5710–5718. [CrossRef]
19. Francis, L.; Ejaz, F.; Nidal Hilal, A. Electrospun membranes for membrane distillation: The state of play and recent advances. *Desalination* **2022**, *526*, 115511. [CrossRef]
20. Yeszhanov, A.B.; Korolkov, I.V.; Dosmagambetova, S.S.; Zdorovets, M.V.; Güven, O. Recent Progress in the Membrane Distillation and Impact of Track-Etched Membranes. *Polymers* **2021**, *13*, 2520. [CrossRef]
21. Gugliuzza, A.; Basile, A. Membrane contactors: Fundamentals, membrane materials and key operations. In *Handbook of Membrane Reactors*; Basile, A., Ed.; Elsevier Ltd.: Amsterdam, The Netherlands; Woodhead Publishing Limited: Cambridge, UK, 2013; Volume 2, p. 54.
22. García-Fernández, L.; Khayet, M.; García-Payo, M.C. Membranes used in membrane distillation: Preparation and characterization. In *Pervaporation, Vapour Permeation and Membrane Distillation Principles and Applications*; Basile, A., Figoli, A., Khayet, M., Eds.; Woodhead Publishing Series in Energy; Woodhead Publishing Limited: Cambridge, UK, 2015; p. 317.
23. Frappa, M.; Castillo del Rio, A.E.; Macedonio, F.; Di Luca, G.; Drioli, E.; Gugliuzza, A. Exfoliated Bi_2Te_3-enabled membranes for new concept water desalination: Freshwater production meets new routes. *Water Res.* **2021**, *203*, 117503. [CrossRef]
24. Liu, L.; Xiao, Z.; Liu, Y.; Li, X.; Yin, H.; Volkov, A.; He, T. Understanding the fouling/scaling resistance of superhydrophobic/omniphobic membranes in membrane distillation. *Desalination* **2021**, *499*, 114864. [CrossRef]
25. Bilad, M.R.; Al Marzooqi, F.A.; Arafat, H.A. New Concept for Dual-Layer Hydrophilic/Hydrophobic Composite Membrane for Membrane Distillation. *J. Membr. Sepr. Technol.* **2015**, *4*, 122–133.
26. Frappa, M.; Del Rio Castillo, A.E.; Macedonio, F.; Politano, A.; Drioli, E.; Bonaccorso, F.; Pellegrini, V.; Gugliuzza, A. Few-layer graphene for advanced composite PVDF membranes dedicated to water distillation: A comparative study. *Nanoscale Adv.* **2020**, *2*, 4728–4739. [CrossRef] [PubMed]
27. Gong, B.; Yang, H.; Wu, S.; Xiong, G.; Yan, J.; Cen, K.; Bo, Z.; Ostrikov, K. Graphene Array-Based Anti-fouling Solar Vapour Gap Membrane Distillation with High Energy Efficiency. *Nano-Micro Lett.* **2019**, *11*, 51. [CrossRef] [PubMed]
28. Mao, Y.; Huang, Q.; Meng, B.; Zhou, K.; Liu, G.; Gugliuzza, A.; Drioli, E.; Jin, W. Roughness-enhanced hydrophobic graphene oxide membrane for water desalination via membrane distillation. *J. Membr. Sci.* **2020**, *611*, 118364. [CrossRef]
29. Liao, X.; Goh, K.; Liao, Y.; Wang, R.; Razaqpur, A.G. Bio-inspired super liquid-repellent membranes for membrane distillation: Mechanisms, fabrications and applications. *Adv. Coll. Interf. Sci.* **2021**, *297*, 102547. [CrossRef]
30. Gugliuzza, A.; Drioli, E. Role of additives in the water vapor transport through block co-poly(amide/ether) membranes: Effects on surface and bulk polymer properties. *Eur. Polym. J.* **2004**, *40*, 2381–2389. [CrossRef]
31. Wang, W.; Du, X.; Vahabi, H.; Zhao, S.; Yin, Y.; Kota, A.K.; Tong, T. Trade-off in membrane distillation with monolithic omniphobic membranes. *Nat. Comm.* **2019**, *10*, 3220. [CrossRef]
32. Zhang, Y.; Shena, F.; Cao, W.; Wan, Y. YHydrophilic/hydrophobic Janus membranes with a dual-function surface coating for rapid and robust membrane distillation desalination. *Desalination* **2020**, *491*, 114561. [CrossRef]
33. Öjemyr, L.N.; Lee, H.J.; Gennis, R.B.; Brzezinski, P. Functional interactions between membrane-bound transporters and membranes. *Proc. Natl. Acad. Sci. USA* **2010**, *107*, 15763. [CrossRef]
34. Georgouvelas, D.; Abdelhamida, H.N.; Li, J.; Edlund, U.; Mathew, A.P. All-cellulose functional membranes for water treatment: Adsorption of metal ions and catalytic decolorization of dyes. *Carbohydr. Polym.* **2021**, *264*, 118044. [CrossRef]
35. Gontarek-Castro, E.; Di Luca, G.; Lieder, M.; Gugliuzza, A. Graphene-Coated PVDF Membranes: Effects of Multi-Scale Rough Structure on Membrane Distillation Performance. *Membranes* **2022**, *12*, 511. [CrossRef]
36. De Luca, G.; Gugliuzza, A.; Drioli, E. Competitive hydrogen-bonding interactions in modified polymer membranes: A density functional theory investigation. *J. Phys. Chem.* **2009**, *113*, 5473–5477. [CrossRef]
37. Teng, J.; Shen, L.; He, Y.; Liao, B.-Q.; Wu, G.; Lin, H. Novel insights into membrane fouling in a membrane bioreactor: Elucidating interfacial interactions with real membrane surface. *Chemosphere* **2018**, *210*, 769–778. [CrossRef]
38. Pingitore, V.; Miriello, D.; Drioli, E.; Gugliuzza, A. Integrated carboxylic carbon nanotubes pathway with membranes for voltage-activated humidity detection and microclimate regulation. *Soft Matter* **2015**, *11*, 4461–4468. [CrossRef]
39. Zielinska, D.; Radecka, H.; Radecki, J. Contribution of membrane surface charge in the interaction of lead and tin derivatives with model lipid membrane. *Chemosphere* **2000**, *40*, 327–330. [CrossRef]
40. Perrotta, M.L.; Macedonio, F.; Tocci, E.; Giorno, L.; Drioli, E.; Gugliuzza, A. Graphene stimulates the nucleation and growth rate of NaCl crystals from hypersaline solution via membrane crystallization. *Environ. Sci. Water Res. Technol.* **2020**, *6*, 1723–1736. [CrossRef]
41. Hoek, E.M.V.; Bhattacharjee, S.; Elimelech, M. Effect of Membrane Surface Roughness on Colloid-Membrane DLVO Interactions. *Langmuir* **2003**, *19*, 4836–4847. [CrossRef]
42. Curcio, S.; Petrosino, F.; Morrone, M.; De Luca, G. Interactions between Proteins and the Membrane Surface in Multiscale Modeling of Organic Fouling. *J. Chem. Inf. Model.* **2018**, *58*, 1815–1827. [CrossRef]
43. Gugliuzza, A.; Fabiano, R.; Garavaglia, M.G.; Spisso, A.; Drioli, E. Study of the surface character as responsible for controlling interfacial forces at membrane-feed interface. *J. Coll. Interf. Sci.* **2006**, *303*, 388–403. [CrossRef]
44. Tanaka, M. Interplays of Interfacial Forces Modulate Structure and Function of Soft and Biological Matters in Aquatic Environments. *Front. Chem.* **2020**, *8*, 165. [CrossRef]

45. Titus, A.R.; Ferreira, L.A.; Belgovskiy, A.I.; Kooijman, E.E.; Mann, E.K.; Mann Jr, J.A.; Meyer, W.V.; Smart, A.E.V.; Uverskygh, N.; Zaslavsky, B.Y. Interfacial tension and mechanism of liquid–liquid phase separation in aqueous media. *Phys. Chem. Chem. Phys.* **2020**, *22*, 4574–4580. [CrossRef]
46. Fengler, C.; Arens, L.; Horn, H.; Wilhelm, M. Desalination of Seawater Using Cationic Poly(acrylamide) Hydrogels and Mechanical Forces for Separation. *Macromol. Mater. Eng.* **2020**, *305*, 2000383. [CrossRef]
47. Saravia, F.; Zwiener, C.; Frimmel, F.H. Interactions between membrane surface, dissolved organic substances and ions in submerged membrane filtration. *Desalination* **2006**, *192*, 280–287. [CrossRef]
48. Rana, D.; Matsuura, T. Surface Modifications for Antifouling Membranes. *Chem. Rev.* **2010**, *110*, 2448–2471. [CrossRef] [PubMed]
49. Dakheel, A.; Darwish, N.A.; Hilal, N. A Review of Colloidal Interactions in Membrane Separation. *Iran. J. Energy Environ.* **2010**, *1*, 144–159.
50. Decher, G.; Eckle, M.; Schmitt, J.; Struth, B. Layer-by-layer assembly multicomposite films. *Curr. Opin. Coll. Interf. Sci.* **1998**, *3*, 32–39. [CrossRef]
51. Pingitore, V.; Gugliuzza, A. Fabrication of porous semiconductor interfaces by pH-driven assembly of carbon nanotubes on honeycomb structured membranes. *J. Phys. Chem.* **2013**, *117*, 26562–26572. [CrossRef]
52. Yilong, H.; Giorno, L.; Gugliuzza, A. Photoactive gel for assisted cleaning during olive mill wastewater membrane microfiltration. *Membranes* **2017**, *7*, 66.
53. Cho, K.L.; Lomas, H.A.J.; Hill, F.; Caruso, S.E. Kentish, Spray Assembled, Cross-Linked Polyelectrolyte Multilayer Membranes for Salt Removal. *Langmuir* **2014**, *29*, 8784–8790. [CrossRef]
54. Arslan, M.; Dönmez, G.; Ergün, A.; Okutan, M.; Arı, G.A.; Deligöz, H. Preparation, Characterization, and Separation Performances of Novel Surface Modified LbL Composite Membranes from Polyelectrolyte Blends and MWCNT. *Polym. Eng. Sci.* **2020**, *60*, 346–351. [CrossRef]
55. Xu, G.R.; Wang, S.H.; Zhao, H.L.; Wu, S.B.; Xu, J.L.; Liu, X.Y. Layer-by-layer (LBL) assembly technology as promising strategy for tailoring pressure-driven desalination membranes. *J. Membr. Sci.* **2015**, *493*, 428–443. [CrossRef]
56. Qi, S.; Tang, C.Y. Cross-Linked Layer-by-Layer Membranes. In *Encyclopedia of Membranes*; Drioli, E., Giorno, L., Eds.; Springer: Berlin/Heidelberg, Germany, 2016.
57. Krizak, D.; Abbaszadeh, M.; Kundu, S. Desalination membranes by deposition of polyamide on polyvinylidene fluoride supports using the automated layer-by-layer technique. *Sep. Sci. Technnol.* **2022**, *57*, 1119–1127. [CrossRef]
58. Starr, B.J.; Tarabara, V.V.; Herrera-Robledo, M.; Zhou, M.; Roualdes, S.A.; Ayral, A. Coating porous membranes with a photocatalyst: Comparison of LbL self-assembly and plasma-enhanced CVD techniques. *J. Membr. Sci.* **2016**, *514*, 340–349. [CrossRef]
59. Gu, L.; Xie, M.-Y.; Jin, Y.; He, M.; Xing, X.-Y.; Yu, Y.; Wu, Q.-Y. Construction of Antifouling Membrane Surfaces through Layer-by-Layer Self-Assembly of Lignosulfonate and Polyethyleneimine. *Polymers* **2019**, *11*, 1782. [CrossRef]
60. Wågberg, L.; Erlandsson, J. The Use of Layer-by-Layer Self-Assembly and Nanocellulose to Prepare Advanced Functional Materials. *Adv. Mater.* **2021**, *33*, 2001474. [CrossRef]
61. Gontarek, E.; Macedonio, F.; Militano, F.; Giorno, L.; Lieder, M.; Politano, A.; Drioli, E.; Gugliuzza, A. Adsorption-assisted transport of water vapour in super-hydrophobic membranes filled with multilayer graphene platelets. *Nanoscale* **2019**, *11*, 11521–11529. [CrossRef]
62. Safaei, J.; Xiong, P.; Wang, G. Progress and prospects of two-dimensional materials for membrane-based water desalination. *Mater. Today Adv.* **2020**, *8*, 10010. [CrossRef]
63. Seo, D.H.; Pineda, S.; Woo, Y.C.; Xie, M.; Murdock, A.T.; Ang, E.Y.M.; Jiao, Y.; Park, M.J.; Lim, S.I.; Lawn, M.; et al. Anti-fouling graphene-based membranes for effective water desalination. *Nat. Comm.* **2018**, *9*, 683. [CrossRef]
64. Yang, J.; Li, Z.; Wang, Z.; Yuan, S.; Li, Y.; Zhao, W.; Zhan, X. 2D Material Based Thin-Film Nanocomposite Membranes for Water Treatment. *Adv. Mater. Technol.* **2021**, *6*, 2000862. [CrossRef]
65. Fatima, J.; Noor Shah, A.; Bilal Tahir, M.; Mehmood, T.; Ali Shah, A.; Tanveer, M.; Nazir, R.; Jan, B.L.; Alansi, S. Tunable 2D Nanomaterials; Their Key Roles and Mechanisms in Water Purification and Monitoring. *Front. Environ. Sci.* **2022**, *10*, 766743. [CrossRef]
66. Aqra, M.W.; Ramanathan, A.A. Graphene and related 2D materials for desalination: A review of recent patents. *Jordan J. Phys.* **2020**, *13*, 233–242. [CrossRef]
67. Cao, Z.; Liu, V.; Farimani, A.B. Water Desalination with Two-Dimensional Metal–Organic Framework Membranes. *Nano Lett.* **2019**, *19*, 8638–8643. [CrossRef] [PubMed]
68. Kurniawan, A.; Ismadji, S.; Soetaredjo, F.E.; Santoso, S.P.; Yuliansa, M.; Anggorowati, A.A. Metal-organic framework-based processes for water desalination: Current development and future prospects. In *Aquananotechnology Applications of Nanomaterials for Water Purification*; Abd-Elsalam, K.A., Zahid, M., Eds.; Micro Nano Technol Series; Elsevier: Amsterdam, The Netherlands, 2021; p. 491.
69. Wan, L.; Zhou, C.; Xu, K.; Feng, B.; Huang, A. Synthesis of highly stable UiO-66-NH$_2$ membranes with high ions rejection for seawater desalination. *Microporous Mesoporous Mater.* **2017**, *252*, 207–213. [CrossRef]
70. Zhu, J.; Qin, L.; Uliana, A.; Hou, J.; Wang, J.; Zhang, Y.; Li, X.; Yuan, S.; Li, J.; Tian, M.; et al. Elevated performance of thin film nanocomposite membranes enabled by modified hydrophilic MOFs for nanofiltration. *ACS Appl. Mater. Interf.* **2017**, *9*, 1975–1986. [CrossRef] [PubMed]

71. Werber, J.R.; Osuji, C.O.; Elimelech, M. Materials for next-generation desalination and water purification membranes. *Nat. Rev. Mater.* **2016**, *1*, 16018. [CrossRef]
72. Yang, F.; Efome, J.E.; Rana, D.; Matsuura, T.; Lan, C. Metal−Organic Frameworks Supported on Nanofiber for Desalination by Direct Contact Membrane Distillation. *ACS Appl. Mater. Interf.* **2018**, *10*, 11251–11260. [CrossRef]
73. Efome, J.E.; Rana, D.; Matsuura, T.; Yang, F.; Cong, Y.; Lan, C.Q. Triple-Layered Nanofibrous Metal−Organic Framework-Based Membranes for Desalination by Direct Contact Membrane Distillation. *ACS Sustain. Chem. Eng.* **2020**, *8*, 6601–6610. [CrossRef]
74. Cheng, D.; Zhao, L.; Li, N.; Smith, S.J.D.D.; Wu, J.; Zhang, J.; Ng, D.; Wu, C.; Martinez, M.R.; Batten, M.P.; et al. Aluminum fumarate MOF/PVDF hollow fiber membrane for enhancement of water flux and thermal efficiency in direct contact membrane distillation. *J. Membr. Sci.* **2019**, *588*, 117204. [CrossRef]
75. Liu, X.; Zhou, Y.; Zhang, J.; Tang, L.; Luo, L.; Zeng, G. Iron containing metal-organic frameworks: Structure, synthesis, and applications in environmental remediation. *ACS Appl. Mater. Interf.* **2017**, *9*, 20255–20275. [CrossRef]
76. Duan, J.; Pan, Y.; Liu, G.; Jin, W. Metal-organic framework adsorbents and membranes for separation applications. *Curr. Opin. Chem. Eng.* **2018**, *20*, 122–131. [CrossRef]
77. Yang, D.; Gates, B.C. Catalysis by metal organic frameworks: Perspective and suggestions for future research. *ACS Catal.* **2019**, *9*, 1779–1798. [CrossRef]
78. Xia, W.; Mahmood, A.; Zou, R.; Xu, Q. Metal organic frameworks and their derived nanostructures for electrochemical energy storage and conversion. *Energy Environ. Sci.* **2015**, *8*, 1837–1866. [CrossRef]
79. Chen, X.; Tong, R.; Shi, Z.; Yang, B.; Liu, H.; Ding, S.; Wang, X.; Lei, Q.; Wu, J.; Fang, W. MOF nanoparticles with encapsulated autophagy inhibitor in controlled drug delivery system for antitumor. *ACS Appl. Mater. Inter.* **2018**, *10*, 2328–2337. [CrossRef]
80. Zuo, J.; Chung, T.-S. Metal−Organic Framework-Functionalized Alumina Membranes for Vacuum Membrane Distillation. *Water* **2016**, *8*, 586. [CrossRef]
81. Dedecker, K.; Pillai, R.S.; Nouar, F.; Pires, J.; Steunou, N.; Dumas, E.; Maurin, G.; Serre, C.; Pinto, M.L. Metal-organic frameworks for cultural heritage preservation: The case of acetic acid removal. *J. ACS Appl. Mater. Interf.* **2018**, *10*, 13886–13894. [CrossRef]
82. Nakayama, M.; Okano, T.; Winnik, F.M. Poly(N-isopropylacrylamide)-based smart surfaces for cell sheet tissue engineering. *Mater. Matters* **2010**, *5*, 56.
83. Jain, K.; Vedarajan, R.; Watanabe, M.; Ishikiriyama, M.; Matsumi, N. Tunable LCST behavior of poly(N-isopropylacrylamide/ionic liquid) copolymers. *Polym. Chem.* **2015**, *6*, 6819–6825. [CrossRef]
84. Sarkar, N. Kinetics of thermal gelation of methylcellulose and hydroxypropylmethylcellulose in aqueous solutions. *Carbohydr. Polym.* **1995**, *26*, 195–203. [CrossRef]
85. Costa, M.C.M.; Silva, S.M.C.; Antunes, F.E. Adjusting the low critical solution temperature of poly(N-isopropylacrylamide) solutions by salts, ionic surfactants and solvents: A rheological study. *J. Mol. Liq.* **2015**, *210*, 113–118. [CrossRef]
86. Tauer, K.; Gau, D.; Schulze, S.; Völkel, A.; Dimova, R. Thermal property changes of poly(N-isopropylacrylamide) microgel particles and block copolymers. *Colloid Polym. Sci.* **2009**, *287*, 299–312. [CrossRef]
87. Gao, X.; Cao, Y.; Song, X.; Zhang, Z.; Xiao, C.; He, C.; Chena, X. pH- and thermo-responsive poly(N-isopropylacrylamideco-acrylic acid derivative) copolymers and hydrogels with LCST dependent on pH and alkyl side groups. *J. Mater. Chem. B* **2013**, *1*, 5578–5587. [CrossRef]
88. Stevenson, F.J. *Humus Chemistry*; John Wiley & Sons: Hoboken, NY, USA, 1982; pp. 1–443.
89. Hou, J.; Wang, H.; Zhang, H. Zirconium Metal–Organic Framework Materials for Efficient Ion Adsorption and Sieving. *Ind. Eng. Chem. Res.* **2020**, *59*, 12907–12923. [CrossRef]
90. Furukawa, H.; Gándara, F.; Zhang, Y.B.; Jiang, J.; Queen, W.L. Water Adsorption in Porous Metal−Organic Frameworks and Related Materials. *J. Am. Chem. Soc.* **2014**, *136*, 4369–4381. [CrossRef] [PubMed]
91. Wu, X.Q.; Mirza, N.R.; Huang, Z.; Zhang, J.; Zheng, Y.M.; Xiang, J.; Xi, Z. Enhanced desalination performance of aluminium fumarate MOF-incorporated electrospun nanofiber membrane with bead-on-string structure for membrane distillation. *Desalination* **2021**, *520*, 115338. [CrossRef]
92. Grasso, G.; Galiano, F.; Yoo, M.J.; Mancuso, R.; Park, H.B.; Gabriele, B.; Figoli, A.; Drioli, E. Development of graphene-PVDF composite membranes for membrane distillation. *J. Membr. Sci.* **2020**, *604*, 118017. [CrossRef]
93. Jafari, A.; Reza, M.; Kebria, S.; Rahimpour, A.; Bakeri, G. Graphene quantum dots modified polyvinylidenefluride (PVDF) nanofibrous membranes with enhanced performance for air Gap membrane distillation. *Chem. Eng. Process.-Process Intensif.* **2018**, *126*, 222. [CrossRef]

Disclaimer/Publisher's Note: The statements, opinions and data contained in all publications are solely those of the individual author(s) and contributor(s) and not of MDPI and/or the editor(s). MDPI and/or the editor(s) disclaim responsibility for any injury to people or property resulting from any ideas, methods, instructions or products referred to in the content.

Article

Application of Hybrid Electrobaromembrane Process for Selective Recovery of Lithium from Cobalt- and Nickel-Containing Leaching Solutions

Dmitrii Butylskii [1,*], Vasiliy Troitskiy [1], Daria Chuprynina [2], Lasâad Dammak [3], Christian Larchet [3] and Victor Nikonenko [1]

1 Membrane Institute, Kuban State University, 149 Stavropolskaya St., 350040 Krasnodar, Russia
2 Department of Analytical Chemistry, Kuban State University, 149 Stavropolskaya St., 350040 Krasnodar, Russia
3 CNRS, ICMPE, UMR 7182, Université Paris-Est Créteil, 2 Rue Henri Dunant, 94320 Thiais, France
* Correspondence: d_butylskii@bk.ru

Abstract: New processes for recycling valuable materials from used lithium-ion batteries (LIBs) need to be developed. This is critical to both meeting growing global demand and mitigating the electronic waste crisis. In contrast to the use of reagent-based processes, this work shows the results of testing a hybrid electrobaromembrane (EBM) method for the selective separation of Li^+ and Co^{2+} ions. Separation is carried out using a track-etched membrane with a pore diameter of 35 nm, which can create conditions for separation if an electric field and an oppositely directed pressure field are applied simultaneously. It is shown that the efficiency of ion separation for a lithium/cobalt pair can be very high due to the possibility of directing the fluxes of separated ions to opposite sides. The flux of lithium through the membrane is about 0.3 mol/($m^2 \times h$). The presence of coexisting nickel ions in the feed solution does not affect the flux of lithium. It is shown that the EBM separation conditions can be chosen so that only lithium is extracted from the feed solution, while cobalt and nickel remain in it.

Keywords: lithium extraction; spent lithium-ion battery; ion separation; electrobaromembrane separation; countercurrent electromigration

Citation: Butylskii, D.; Troitskiy, V.; Chuprynina, D.; Dammak, L.; Larchet, C.; Nikonenko, V. Application of Hybrid Electrobaromembrane Process for Selective Recovery of Lithium from Cobalt- and Nickel-Containing Leaching Solutions. *Membranes* **2023**, *13*, 509. https://doi.org/10.3390/membranes13050509

Academic Editors: Annarosa Gugliuzza and Cristiana Boi

Received: 12 April 2023
Revised: 28 April 2023
Accepted: 10 May 2023
Published: 11 May 2023

Copyright: © 2023 by the authors. Licensee MDPI, Basel, Switzerland. This article is an open access article distributed under the terms and conditions of the Creative Commons Attribution (CC BY) license (https://creativecommons.org/licenses/by/4.0/).

1. Introduction

The development of new approaches for the extraction of valuable components from aqueous solutions is an important task. For the extraction of alkali, alkaline-earth and transition metal compounds on an industrial scale, pyrometallurgy, reagent-based methods of hydrometallurgy, sorption, electrochemical reduction, etc., having high efficiency are often used. Multistage ion separation processes are used in industry for the production of lithium, cobalt, nickel, zinc, titanium, etc., with high purity [1–3]. All these processes are well developed when treating natural sources. However, the valuable metals listed above can also be obtained from secondary sources (e-waste, sludge, ash, tailing, etc.) [4–6]. The problem is acute when using, for example, lithium-ion batteries (LIBs). Tests of lithium materials newly obtained from spent LIBs have shown that they are not inferior in performance to materials made from lithium obtained from primary sources (natural brines and minerals) [7–9], but all of them cannot be called environmentally friendly.

The authors of review papers [1–3] note that the extraction of lithium and no-less-valuable cobalt from leachates of spent LIBs has clear advantages over their extraction from natural sources. First, the concentration of lithium and cobalt in leachates is usually quite high. Second, the composition of leachates is more predictable and less varied. Third, leachates do not contain multiply charged ions such as Ca^{2+} and Mg^{2+}, which have low value and which make it difficult to extract lithium from natural solutions. In addition, a significant advantage is the reduction of the environmental burden due to the neutralization of spent LIBs as hazardous waste [10].

However, it is difficult to organize a LIB processing plant. This is mainly due to the need to sort different types of batteries and the complexity of their disassembly, as well as the high reactivity of materials in spent LIBs, which can ignite and explode on contact with air [11]. For these reasons, the reuse of spent LIBs is insignificant. For example, in Australia (world leader in lithium mining) in 2017–2018, only 6% of LIBs that were out of service during this period were recycled [12]. The profit from the recycling of different types of spent LIBs also differs. Bhandari et al. [13] note that the processing of NMC batteries (the cathode is obtained from a mixture of lithium, nickel, manganese and cobalt) is more economically attractive than the recycling of cheaper LFP batteries (lithium iron phosphate cathode). This is due to the ability to extract valuable nickel and cobalt from leachates of NMC batteries in addition to lithium, as in the case of LFP.

In industry, the processing of spent LIBs is carried out using three traditional methods: pyrometallurgy, hydrometallurgy, or a combination of these. Pyrometallurgical processing does not allow the extraction of lithium from spent LIBs; it goes to the slag [10]. The target component is valuable Co and other metals [14]. Lithium can be recovered from spent LIBs using hydrometallurgical processes. According to the latest estimates [13], one ton of spent NMC811-type LIBs can bring a profit from the sale of materials of at least USD 6500 (in 2021). More than 50% of the profit will come from recovered lithium.

Membrane methods for separating the components of spent LIBs are still under development [1]. Selective electrodialysis (SED) is most commonly used to extract Li^+ ions from LIB leachates [15–18]. The SED technology differs from conventional electrodialysis in the use of multiple units of at least one special-grade ion-exchange membrane (cell pair). These membranes are commercially available [1–3]. They pass singly charged ions well and reject multiply charged ones. This makes it possible to efficiently separate Li^+ ions from Co^{2+}, Ni^{2+} and Mn^{2+} [18].

Another attractive membrane technology is the hybrid electrobaromembrane (EBM) method [1]. Unlike electrodialysis, EBM separation uses nonselective porous membranes. Separated ions of the same charge sign move in an electric field through the pores of this membrane to the corresponding electrode, while a commensurate counter convective flow is created in the pores. The selectivity of separation is achieved due to the difference in the mobility of the competing ions [19–22].

In recent studies on EBM devices, impressive results have been achieved in the separation of Li^+/K^+ ions when using feed solutions imitating natural waters [23–25]. It was shown that the ion separation coefficient for the Li^+/K^+ ions can be as high as 59 [23,26] or even 150 [24,25]. For the Li^+/Na^+ pair, the selective permeability coefficient is somewhat lower, reaching 30 [24]. However, the EBM method and porous membranes have not yet been tested in the recovery of lithium from secondary sources represented by leachates or liquors of spent LIBs.

In this regard, the purpose of this study is to expand the scope of the EBM method, as well as to study the possibility of using nanoporous membranes that do not have selectivity for a certain type of ions but ensure their selective separation. The paper presents the results of testing a hybrid EBM method for separating Li^+ and Co^{2+} ions, as well as Li^+, Co^{2+} and Ni^{2+} ions contained in leachates of spent LIBs. The efficiency of the EBM method is analyzed, and the parameters of ion separation are compared with other membrane methods.

2. Materials and Methods

Two types of feed solutions were used. A mixture of lithium and cobalt sulfates was used to determine the optimal separation parameters, and then nickel sulfate was added to this mixture to test the possibility of separating lithium from a more complex mixture. The main characteristics of the feed solution components that affect the efficiency of EBM separation are presented in Table 1.

Table 1. Some characteristics of ions (at 25 °C) in the feed solution.

Ion	Symbol	Diffusion Coefficient, 10^{-9} m^2/s [27]	Stokes Radius, Å [28]
Lithium	Li$^+$	1.04	2.38
Cobalt	Co^{2+}	0.73	3.35
Nickel	Ni^{2+}	0.66	2.92
Sulfate	SO$_4^{2-}$	1.06	2.30

A mixture of 0.05 M Li$_2$SO$_4$ and 0.05 M CoSO$_4$ was used as the first type of feed solution (pH = 4.2–4.4) and a mixture of 0.05 M Li$_2$SO$_4$, 0.025 M CoSO$_4$ and 0.025 M NiSO$_4$ was used as the second type of feed solution (pH = 5.2). The concentrations of the components were within wide limits, which are typical for solutions of spent LIB leachates.

In this work, a track-etched membrane (designated as TEM #811) was used as a nanoporous membrane. It was produced from a polyethylene terephthalate (PET) film at the Joint Institute for Nuclear Research (Dubna, Russia). The properties of the TEM are described in Table 2.

Table 2. Characteristics of the TEM#811 track-etched membrane.

Parameter	Value
Thickness	10 μm
Density (dry) *	1.10 ± 0.05 g/cm^3
Pore density **	5.0×10^9 pores/cm^2
Pore diameter	35 ± 3.0 nm **
	28 ± 2.0 nm ***
Surface porosity	$5.3\% \pm 1.0\%$
Water uptake	5%
Hydraulic permeability	0.10 ± 0.02 cm^3/(cm$^2 \times$ min $\times$ bar)
Functional groups	hydroxyl and carboxyl groups [29]
Exchange capacity *	0.064 ± 0.003 mmol/g$_{wet}$
Electrical conductivity (0.1 M NaCl) *	0.81 mS/cm
Integral diffusion permeability coefficient (0.1 M NaCl) *	2.8×10^{-7} cm^2/s

* The results are presented in Ref. [30] for sample #811. ** Estimated by scanning electron microscopy (SEM); *** estimated by hydraulic permeability.

On the left- and right-hand sides the TEM is surrounded by auxiliary anion-exchange (AEM) MA-41 heterogeneous membranes (JCC Shchekinoazot, Pervomayskiy, Russia) to form flow chambers. Solutions of the same composition and volume (0.15 L) were pumped through the left-hand (I) and right-hand (II) chambers, separated by a porous membrane, at the same flow rate (5.4 L/h) (Figure 1). A 0.1 M Na$_2$SO$_4$ solution was pumped through the electrode chambers (4 L). Separation experiments at given parameters were repeated at least four times. The duration of each experiment was 8 h; this amount of time was needed to measure the fluxes of competing ions at a given electric current and a pressure drop in a steady state of the system. A convective flow directed from chamber II to chamber I was created opposite to the electromigration of the competitive cations. This was achieved by increasing the pressure of the solution in the circuit passing through chamber II using an automatic nitrogen dosing system.

Samples of the solutions from chambers I and II were taken at the beginning and the end of the separation process to determine the concentration of Li$^+$-ions using a Dionex ICS-3000 ion-chromatograph with a conductometric detector (Dionex, Sunnyvale, CA, USA). The concentration of cobalt and nickel (if any) was determined using direct spectrophotometric analysis with a UV-1800 TM ECOVIEW (Shanghai Mapada Instruments Co., Shanghai, China) instrument.

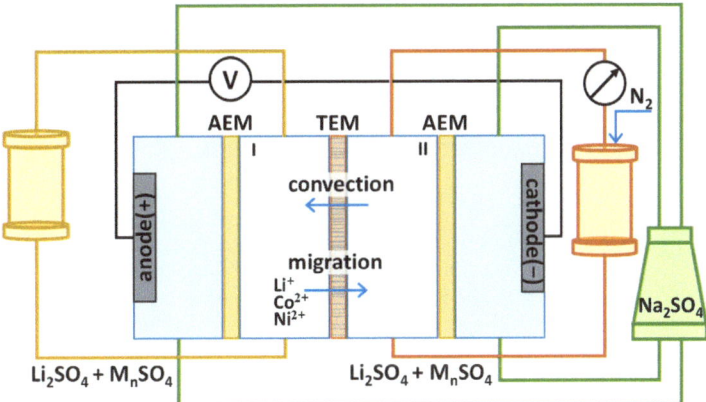

Figure 1. Schematic diagram of the setup for studying the parameters of the selective separation of Li^+/Co^{2+} and $Li^+/Co^{2+}/Ni^{2+}$ cations using the hybrid electrobaromembrane method.

3. Results

Let us consider the mechanism of ion separation by the hybrid electrobaromembrane method in more detail (Figure 2). When only an external electric field is applied, the separated Li^+ and Co^{2+} ions migrate through the pores of the TEM to the negative cathode and their velocities are proportional to their mobility u_k (or diffusion coefficient D_k, multiplied by the charge numbers z_k): $v_k^{migr} = u_k E = (D_k z_k F/RT)E$, where E is the electric field strength. Under the action of one driving force (electric field), Li^+ and Co^{2+} ions are freely transferred through the wide pores of the membrane.

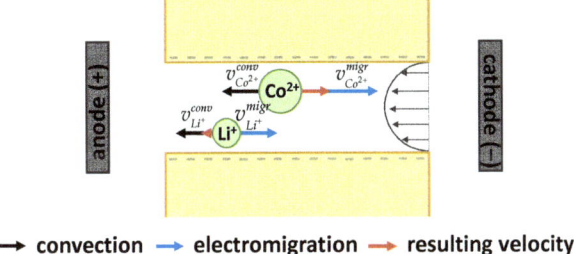

Figure 2. Scheme of ion velocities in the pore of a track-etched membrane. The velocity of electromigration (blue arrows) is proportional to the ion mobility and electric field; the velocity of convective transfer (black arrows) depends only on the pressure drop and is the same for both ions. The resulting velocities (red arrows) can be directed to different sides.

When a pressure field is applied to the system along with the electric field, a convective flow is created in the pores of the TEM. The convective flow is opposite to the electromigration flow of separated ions. The velocity of convection is the same for both separated ions. For efficient separation, it is necessary to choose the ratio of the rates of convection and electromigration so that the resulting rate of the least mobile ion (here, these are Li^+ ions) tends to zero. It is also possible to choose conditions such that the ions to be separated move in different directions, as shown in Figure 2.

3.1. Theory of EBM Separation

Separation of ions by the EBM method occurs under the action of the two external forces mentioned above. Additionally, a concentration difference may appear when the solutions are different on both sides of the membrane. Therefore, when calculating the fluxes of competing counterions ($k = 1, 2$), it is necessary to take into account electromigration, convection and diffusion contributions [23]:

$$j_1 = j_1^{migr} + j_1^{dif} + j_1^{conv} = \frac{i\tilde{t}_1}{z_1 F} + c_1 v^{conv} \gamma \tag{1}$$

$$j_2 = j_2^{migr} + j_2^{dif} + j_2^{conv} = \frac{i\tilde{t}_2}{z_2 F} + c_2 v^{conv} \gamma \tag{2}$$

where i is the current density and j_k, $\tilde{t}_k$, c_k and z_k are the flux density, effective transport number, concentration and charge number of cation k; v^{conv} is the convective velocity; and γ is the surface porosity (the fraction of the membrane surface occupied by the pore openings).

The contribution of diffusion is taken into account implicitly through the value of the effective transport number. The $\tilde{t}_k$ value characterizes the fraction of electric charge carried by ion k under the action of electric current and diffusion (if any). It is easy to see that in Equations (1) and (2), only the $\tilde{t}_k$ and v^{conv} values are unknown. However, assuming that the convective velocity depends only on the pressure drop and is independent of the given current, this can be estimated by the Hagen-Poiseuille equation:

$$v^{conv} = \frac{1}{32} \frac{\Delta p d^2}{\eta L} \tag{3}$$

where Δp is the pressure difference between chamber II and chamber I, d and L are the diameter and length of a pore, and η is the liquid viscosity.

Now only $\tilde{t}_k$ is unknown, and can be determined by the fit between the experimental and calculation results. At the same time, taking into account a relatively large pore diameter, $\tilde{t}_k$ cannot be much greater than the (electromigration) transport numbers in a free solution t_k:

$$\tilde{t}_k \approx t_k = \frac{z_k^2 D_k c_k}{\sum_{j=1,2,3} z_j^2 D_j c_j} \tag{4}$$

Due to the different mobility of competing ions in an electric field, it is possible to choose values of the set current and pressure drop such that the flux of one of them through the track-etched membrane tends to zero [1,23,24]. Let us suppose that $j_2 = 0$; Equation (2) can be used to express the value of the convective velocity:

$$v^{conv} = \frac{i\tilde{t}_2}{z_2 c_2 F \gamma} \tag{5}$$

Substituting Equation (3) in Equation (1), taking into account that the convective flow is opposed to the electromigration flow, the following expression can be obtained:

$$j_1 = \frac{i}{F} \left(\frac{\tilde{t}_1}{z_1} - \frac{\tilde{t}_2}{z_2} \frac{c_1}{c_2} \right) \tag{6}$$

If Equation (4) is taken into account, expression (6) can be rewritten in the following form:

$$j_1 \approx \frac{i\tilde{t}_1}{z_1 F} \left(1 - \frac{z_2 D_2}{z_1 D_1} \right) \tag{7}$$

Conducting a brief analysis of Equation (7), it should be noted that the flux density of the ions, which has the highest mobility in an electric field, increases as the $z_2 D_2 / z_1 D_1$ ratio decreases. If $z_2 D_2$ is close to $z_1 D_1$, the competing ions cannot be separated.

3.2. Separation of Li^+/Co^{2+}-Ions

Lithium and cobalt have different electromigration velocities v_k^{migr}, since this velocity is proportional to the electrical mobility of the ions. Cobalt ions are more mobile in an electric field than lithium ions. The z_2D_2/z_1D_1 ratio for the Li^+/Co^{2+} pair is about 0.71. This means that, from a theoretical point of view, the separation of these ions by the EBM method is possible. However, it is known that cobalt sulfate exists in neutral and acid aqueous solutions in two forms: Co^{2+} and $CoSO_4$ (Figure 3). According to the Medusa/Hydra software, at the pH value of the feed solution (4.2–4.4), the fraction of Co^{2+} ions is 0.6. For calculations, the values of the equilibrium constants from the Hydra built-in database were used [31].

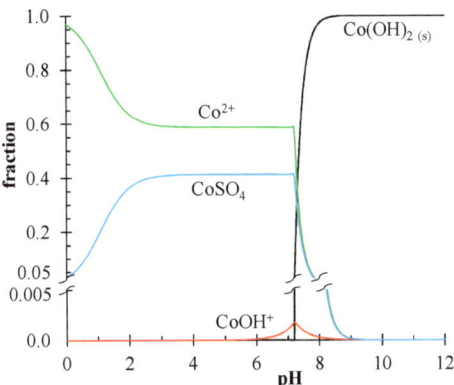

Figure 3. Equilibrium diagram of cobalt compounds in an aqueous solution obtained using the Medusa/Hydra software package [31].

Taking into account the proportion of Co^{2+} ions in the solutions pumped through chamber I and chamber II, the transport numbers of lithium and cobalt ions in free solution in Equation (4) are as follows: $t_{Li^+} = 0.17$ and $t_{Co^{2+}} = 0.14$, while $t_{SO_4^{2-}} = 0.69$. Due to the relatively large pore size ($d_{av} = 35 \pm 3$ nm), the transport numbers $\tilde{t}_k$ of separated ions in a nanoporous membrane should not differ significantly from their transport numbers in solution. This means that the flux of cobalt ions through the membrane determined by electromigration should be lower than the lithium flux, despite the cobalt ions having higher mobility in an electric field.

To determine the experimental flux of cation k under separation ($k = Li^+$ or Co^{2+}), it is necessary to measure the quasi-stationary rate of change of the cation concentration dc_k/dt in chamber I or chamber II:

$$j_k = \frac{V}{s}\frac{dc_k}{dt} \qquad (8)$$

where V is the volume of solution in chamber I or chamber II, s is the membrane surface area, and t is the duration of an experiment.

The efficiency of ion separation is characterized by the ion separation coefficient $S_{Li^+/Co^{2+}}$ (also called the permselectivity coefficient between two counterions) [32,33]:

$$S_{Li^+/Co^{2+}} = \frac{j_{Li^+}/j_{Co^{2+}}}{c^0_{Li^+}/c^0_{Co^{2+}}} = \frac{\Delta c_{Li^+}/\Delta c_{Co^{2+}}}{c^0_{Li^+}/c^0_{Co^{2+}}} \qquad (9)$$

where j_{Li^+} and $j_{Co^{2+}}$ are the flux densities of Li^+ and Co^{2+} through the membrane, respectively; $c^0_{Li^+}$ and $c^0_{Co^{2+}}$ are the concentrations of these ions in chamber I or chamber II (in our case, the solutions in these chambers are the same); and Δc_{Li^+} and $\Delta c_{Co^{2+}}$ are the changes in concentrations of these ions in chamber I or chamber II. Note that when calculating the value of $S_{Li^+/Co^{2+}}$, the total concentration of cobalt was used, and this can exist in two forms: as a doubly charged cation and in the composition of cobalt sulfate. The method of cobalt concentration measurement (see experimental section) also determines its total concentration. A similar approach has been previously used in other studies [34,35].

Figure 4 shows the results of the separation of Li$^+$ and Co^{2+} ions using the EBM method, as well as the results of the calculation using Equations (1)–(3). The experiments were carried out at the constant pressure drop of 0.3 bar between chambers II and I, since this value provided optimal separation without significant overflow of solution from the chamber under pressure II to the adjacent chamber I.

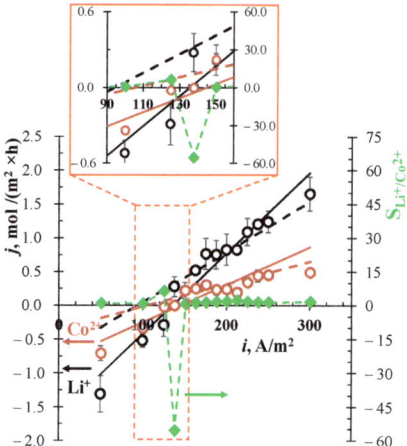

Figure 4. Flux densities of Li$^+$ (black circles) and Co^{2+} ions (dark red circles) through the TEM #811 membrane, and the ion separation coefficient $S_{Li^+/Co^{2+}}$ (diamonds with dashed green line) vs. the current density for the EBM system at a constant pressure drop of 0.3 bar. Experimental data are shown by markers; lines are calculated using Equations (1)–(3), taking into account d = 32 nm, $\tilde{t}_{Li^+}$ = 0.32 and $\tilde{t}_{Co^{2+}}$ = 0.15 (black and dark red solid lines) and d = 26 nm, with $\tilde{t}_{Li^+}$ = 0.20 and $\tilde{t}_{Co^{2+}}$ = 0.09 (black and dark red dashed lines) as fitting parameters.

When a low current value (50 A/m^2) is set in the system, the experimental values of the fluxes of both separated ions are negative at Δp = 0.3 bar. This means that the process is controlled by the convection. Since both separated ions experience the same effect of convection, one should not expect high separation selectivity with these parameters. When current density is in the range 125–137.5 A/m^2, the flux of cobalt ions through the membrane tends to zero. In this case, the flux of lithium ions is significant, and is determined either by convection (at 125 A/m^2) or migration (at 137.5 A/m^2). This leads to an increase in selectivity, with a value ranging from 8 up to −55. It is important to note that at 125 A/m^2 the fluxes of lithium and cobalt are negative and codirectional. At 137.5 A/m^2, the lithium flux becomes positive, while the cobalt flux remains negative. Taking into account the fact that only lithium ions can leave the feed solution, i.e., the outgoing fluxes of the competing ions are zero, the separation coefficient should formally be set to infinity. However, since the feed solution and receiving solution in experiments were identical, $S_{Li^+/Co^{2+}}$ cannot be evaluated as infinity. Hereafter, when the fluxes of lithium and competing ions have opposite signs, "not available for calculation" will be written and abbreviated as "n/a".

This feature makes it possible to effectively separate Li$^+$ and Co^{2+} ions despite the obtained value of the ion separation coefficient. With a further increase in current at Δp = 0.3 bar, the ion separation coefficient does not differ much from 1.

The values of the transport numbers $\tilde{t}_k$ of competing cations in a nanoporous membrane were determined by the best fit between the experiment and the theory. When

d = 32 nm, the fitted values were 0.32 and 0.15 for lithium and cobalt, respectively. The difference from their transport numbers in solution (t_{Li^+} = 0.17 and $t_{Co^{2+}}$ = 0.14) is probably due to the error in determining the average pore diameter (35 ± 3 nm according to SEM results). It was recently found that a loose cation-conductive intermediate gel layer between the pore solution and the nonconductive membrane bulk material can form in the pores of TEMs during track etching [36,37]. Some of the cations driven by the electric field pass through this loose layer, which leads to an increase in their transport numbers.

On the other hand, the formation of a loose layer can lead to narrowing in the middle part of the pores. Indeed, if the hydraulic permeability of TEM #811 is taken into account, then using the Hagen-Poiseuille Equation (3), an average pore size of 28 ± 2 nm can be obtained [23,36]. Figure 4 shows good agreement between the theoretical dependence (dashed lines) and positive values of the fluxes corresponding to the dominant migration. Here, the average diameter calculated from the hydraulic permeability (26 nm) and the values of the fitted transport numbers becomes closer to those in free solution ($\tilde{t}_{Li^+}$ = 0.20 and $\tilde{t}_{Co^{2+}}$ = 0.09 were used).

Since at the beginning of the experiment, the solutions on both sides of the TEM were the same, and the composition of these solutions changed only slightly (by less than 20%) during the experiment, the contribution of diffusion to the ion transport can be ignored. Therefore, the effective transport numbers in Equations (1) and (2) should be close to the electromigration transport numbers.

In addition, the presented theoretical analysis does not take into account the interaction of the separated cations with the hydroxyl and carboxyl groups of the polyethylene terephthalate TEM [38]. The negative charge of the pore walls will attract the lithium and cobalt ions inside the pore (mainly in the electrical double layer on the pore walls). As a result, the cation transport numbers in the pore should be slightly greater than in the free solution. Moreover, a fixed charge will affect doubly charged cobalt ions to a greater extent than lithium ions [39].

3.3. Separation of $Li^+/Co^{2+}/Ni^{2+}$-Ions

In addition to lithium and cobalt, the leachates of spent LIBs also contain nickel and manganese ions, which are less valuable [1–3,18]. The literature analysis allows us to conclude that the presence of coexisting ions in the feed solution significantly affects the efficiency of separation by electrodialysis [15,18,40]. For example, Ji et al. [40] showed that the presence of Na^+, K^+ and Ca^{2+} ions in natural water (used as a feed solution) negatively affects the efficiency of separation of Mg^{2+} and Li^+ by selective electrodialysis. With an increase in the c_{Na^+}/c_{Li^+} ratio in the feed solution from 1 to 20, S_{Mg^{2+}/Li^+} decreases from 8.7 to 1.8. The negative effect is associated with a decrease in the flux of Li^+ ions through the membrane in the presence of an excess of Na^+ ions. The presence of Ni^{2+} and Mn^{2+} ions in leachates as a feed solution similarly affects the flux of Co^{2+} ions through the membrane [15,18].

To evaluate the effect of coexisting Ni^{2+} ions on the separation efficiency of Li^+ and Co^{2+} ions through the TEM #811 membrane, optimal parameters selected from the dependence in Figure 4 (Δp = 0.3 bar; i = 137.5 A/m^2) were used. The concentration of cobalt ions was reduced by half (usually the ratio of Co^{2+} and Ni^{2+} ions in leachates is approximately 1:1 [16,18]), due to which the ionic strength of the solution did not change. The composition of the feed solution was as follows: 0.05 M Li_2SO_4, 0.025 M $CoSO_4$ and 0.025 M $NiSO_4$ (pH = 5.2). Table 3 shows the results of the separation of Li^+/Co^{2+}-ions by the EBM method in the presence of Ni^{2+} ions.

Under selected conditions, in the presence of Ni^{2+} ions, the selectivity of Li^+/Co^{2+}-ion separation is significantly reduced ($S_{Li^+/Co^{2+}}$ = 4). The flux of lithium ions through the membrane remains the same, within experimental error, and the flux of cobalt ions increases (0.02 mol/(m^2 × h)).

Table 3. Comparison of separation efficiency of Li^+/Co^{2+}-ions and $Li^+/Co^{2+}/Ni^{2+}$-ions by EBM method at $\Delta p = 0.3$ bar.

Current Density, i, A/m²	Ions in the Feed Solution	$c^0_{M^{n+}}$, g/L	$j_{M^{n+}}$, mol/(m² × h)	$S_{Li^+/M^{n+}}$
Li^+/Co^{2+}-containing feed solution				
137.5	Li^+	0.69	0.28	–
	Co^{2+}	2.95	−0.0025	n/a
125	Li^+	0.69	−0.29	–
	Co^{2+}	2.95	−0.022	8
$Li^+/Co^{2+}/Ni^{2+}$-containing feed solution				
137.5	Li^+	0.69	0.33	–
	Co^{2+}	1.47	0.02	4
	Ni^{2+}	1.47	0.04	2
125	Li^+	0.69	0.30	–
	Co^{2+}	1.47	−0.02	n/a
	Ni^{2+}	1.47	−0.005	n/a

The flux of Ni^{2+} ions through the membrane is 0.04 mol/(m² × h). The difference between the fluxes of Co^{2+} and Ni^{2+} is explained by the fact that $NiSO_4$ dissociates in aqueous solutions better than $CoSO_4$. The fraction of Ni^{2+} ions in the feed solution is approximately 0.61 (0.39 in the form of its sulfate), while that of Co^{2+} ions is 0.58 (pH = 5.2). The transport number (and hence the flux) of Ni^{2+} ions in pore solution is higher than the transport number of Co^{2+} ions ($t_{Ni^{2+}} = 0.07$ and $t_{Co^{2+}} = 0.06$). The $z_2 D_2 / z_1 D_1$ ratio for the Li^+/Ni^{2+} pair is higher than for the Li^+/Co^{2+} pair (0.76 and 0.71, respectively). This means that it is more difficult to separate Li^+ and Ni^{2+} ions than Li^+ and Co^{2+}. Along with the measurement error, this probably explains the high flux of these ions through the membrane compared to the experiment with the same parameters without the addition of Ni^{2+} ions. To estimate competing ion fluxes, the change in concentration over time in the chambers of the EBM device is determined against the background of a high concentration of the analyte in the feed solution.

Due to the fact that the fluxes of both Co^{2+} ions and Ni^{2+} ions were positive at 137.5 A/m² (controlled by migration), in order to increase the efficiency of lithium extraction, the current density was reduced to 125 A/m² at the same pressure value ($\Delta p = 0.3$ bar) (Table 3). This caused the fluxes of cobalt and nickel to become negative, which means that the dominant transport mechanism is convection. However, the lithium flux changed insignificantly, from 0.33 to 0.30 mol/(m² × h). This allowed lithium to be fractionated from the mixed solution. Taking into account that the fluxes of cobalt and nickel ions are negative, the separation coefficient is not available for calculation. The energy consumption ranged from 0.55 to 0.92 kWh/mol Li^+ for the entire cell (depending on $\tilde{t}_{Li^+}$), which is comparable with the energy consumption of conventional electrodialysis. The methodology for calculating energy consumption is presented in Ref. [23].

3.4. Analysis of Obtained Results

Let us make a brief analysis of the obtained separation characteristics and compare them with similar characteristics found by different authors using other membrane methods. In Table 4, the results from some recent papers on the selective recovery of lithium from leachates of spent LIBs using membrane technologies were compiled. The fluxes of Li^+ ions through the membrane, as well as the fluxes of competing cations, were calculated using the published data presented in the relevant articles [15–18,41,42]. Calculations were made using Equations (8) and (9).

Table 4. Comparison of recovery/rejection of used lithium using membrane methods.

Method	Membrane	Feed Solution	Experiment Details	j_{Li^+}, mol/(m² × h)	Competing Cation, C^+	$j_{M^{n+}}$, mol/(m² × h)	$S_{Li^+/M^{n+}}$
Selective electrodialysis Ref. [18]	Cell with monovalent selective Selemion CSO (Asahi Glass, Tokyo, Japan) or Neosepta CIMS membranes, as well as Neosepta AMX (Astom, Shunan, Japan)	Leach solution of NMC111 cathodic materials: 2.60 g/L Li⁺, 7.88 g/L Co²⁺, 8.01 g/L Ni²⁺, 4.40 g/L Mn²⁺, 51.45 g/L SO₄²⁻ (pH = 2.8)	125 A/m²	1.92 over the Selemion CSO; 3.05 over the Neosepta CIMS	Co^{2+} Ni^{2+} Mn^{2+} Co^{2+} Ni^{2+} Mn^{2+}	0.56 0.54 0.29 0.20 0.18 0.12	1.25 1.3 1.4 5.6 6.1 5.4
Selective electrodialysis Ref. [17]	5 cell pair with monovalent selective PC-MVK & PC-MVA membranes (PCA GmbH, Heusweiler, Germany)	0.1 g/L Li⁺ & 0.3 g/L Co²⁺	5 V (1 V/cell) ~15 A/m²	0.1	Co^{2+}	8.8×10^{-3}	4
Selective electrodialysis Ref. [15]	Cell with laboratory-made selective PAN-5C8Q membrane as well as Neosepta ASE (Astom, Shunan, Japan)	0.027 g/L Li⁺, 0.108 g/L Co²⁺, 0.049 g/L Ni²⁺	5 V	0.047	Co^{2+} Ni^{2+}	0.044 1.5×10^{-3}	0.5 7
Conventional electrodialysis + complexation & selective electrodialysis Ref. [16]	Cell with monovalent selective Neosepta CMS membrane, as well as Neosepta AMX, Neosepta CMX (Astom, Japan) and PCA PC 400D (PCA GmbH, Heusweiler, Germany)	Leach solution of NMC111 cathodic materials: 0.07 g/L Li⁺, 0.2 g/L Co²⁺, 0.2 g/L Ni²⁺, 0.18 g/L Mn²⁺ & SO₄²⁻ (pH ~ 1.5)	18 V (Stage 1); 18 V (Stage 2); 3 V (Stage 3)	0.165 over the Neosepta CMS (Stage 3)	Ni-EDTA⁻ over the PCA PC 400D (Stage 1); Co-EDTA⁻ over the PCA PC 400D (Stage 2); Mn²⁺ over the Neosepta CMS (Stage 3)	0.057 0.042 5.7×10^{-4}	n/a n/a 92
Bipolar membrane electrodialysis + complexation Ref. [41]	Cell with Neosepta BP-1E bipolar membrane (Astom, Japan), as well as Selemion CMV and Selemion AMV (Asahi Glass, Tokyo, Japan)	0.14 g/L Li⁺, 1.18 g/L Co²⁺, 3.72 g/L NO₃⁻, 1.84 g/L Na⁺ (pH = 7.0)	20 V	0.77 over the Selemion CMV	Co-EDTA⁻ over the Selemion AMV	0.33	n/a
Nanofiltration Ref. [42]	Ccell with VNF2 nanofiltration membrane (Vontron Membrane Technology Ltd., Beijing, China)	Leach liquor of lithium-iron-phosphate spent LiBs: 23.9 g/L Li⁺, 0.78 g/L Ni²⁺, 0.58 g/L Co²⁺, 0.67 g/L Mn²⁺, 27.3 g/L Fe³⁺, 0.18 g/L Al³⁺, 0.28 g/L Cr³⁺, 0.059 g/L Cu²⁺, 11.0 g/L PO₄³⁻ (pH = 2.2)	10 bar	0.67	Co^{2+} Ni^{2+} Mn^{2+}	3.0×10^{-4} 5.7×10^{-4} 3.0×10^{-4}	6.2 4.7 8.2

Table 4. Cont.

Method	Membrane	Feed Solution	Experiment Details	j_{Li^+}, mol/(m² × h)	Competing Cation, C⁺	$j_{M^{n+}}$, mol/(m² × h)	$S_{Li^+/M^{n+}}$
Hybrid electro-baromembrane (EBM) method [this study]	Cell with TEM #811 track-etched membrane, as well as two MA-41 (JCC Shchekinoazot, Pervomayskiy, Russia)	0.69 g/L Li⁺, 2.95 g/L Co²⁺, 9.6 g/L SO₄²⁻ (pH = 4.2–4.4) or 0.69 g/L Li⁺, 1.47 g/L Co²⁺, 1.47 g/L Ni²⁺, 9.6 g/L SO₄²⁻ (pH = 5.2)	137.5 A/m² 0.3 bar	0.36 * 0.33	Co²⁺ Co²⁺ Ni²⁺	0.01 * 0.02 0.04	18 * 4 2

* Results of one of the experiments with codirectional fluxes of separated ions at given parameters, in contrast to the results discussed above (Figure 4, Table 3), where the average values for four series of experiments are presented.

It is known that selective electrodialysis (SED) allows separate monovalent and multivalent ions of the same charge sign using special-grade monovalent-ion-selective ion-exchange membranes. In the literature, the examples of the successful application of SED can be found to separate lithium from doubly charged ions of cobalt, nickel, and manganese [15,17,18]. Competing ion fluxes through ion-exchange membranes are determined by the set current and the concentration of ions in the feed solution. The ion separation coefficient $S_{Li^+/M^{n+}}$ can reach 5–7 [15,17,18].

Another approach is to use conventional and special-grade ion-exchange membranes together, as well as complexation with EDTA [16,41]. This makes it possible to transfer multiply charged cations, in the form of anionic complexes, through conventional ion-exchange membranes. Lithium does not form a complex and is transported over special-grade membranes. Separation efficiency increases significantly, but the ion separation coefficient cannot always be calculated (is not available), since lithium and doubly charged coexisting ions are transported through different membranes. Only a small portion of doubly charged ions are transported together with singly charged ions, and the ion separation coefficient reaches very high values [16].

In the pressure-driven-membrane method, nanofiltration can be effectively used to separate singly and multiply charged ions. As in the case of electrodialysis, the fluxes of the ions to be separated depend primarily on the magnitude of the driving force (excess pressure) and the concentration of the ions to be separated in the feed solution. The parameters of the selective extraction of lithium from spent LIBs using nanofiltration [42] are close to the parameters obtained using electrodialysis methods (Table 4).

The EBM method can compete with known membrane methods. The flux of lithium through the membrane is comparable to the fluxes of this ion obtained in nanofiltration and electrodialysis. However, the method makes it possible to choose the parameters in such a way that the fluxes of the separated ions are directed in opposite directions (Figure 4). It is possible to choose such conditions when Li^+ is extracted from chamber I (a positive flux) while the direction of fluxes of competing cations is negative: they cannot pass from chamber I (the feed solution) to chamber II (the receiving solution). Therefore, the selectivity coefficient is theoretically equal to infinity. This opens possibilities for fractionation of the components.

Although the ion-separation coefficient with the EBM method can be much higher than with other membrane methods [1], like other membrane methods, this method is not without drawbacks. The main problem lies in the accuracy of selecting the separation parameters in order to bring the flux of one of the competing ions to zero as accurately as possible or direct it in the opposite direction.

4. Conclusions

In this work, the possibility of using the hybrid electrobaromembrane (EBM) method for the separation of Li^+ and Co^{2+} ions, as well as Li^+, Co^{2+} and Ni^{2+} contained in leachates of spent LIBs, was studied. For the separation, a track-etched membrane with a pore diameter d_{av} of 35 ± 3 nm was used. At this value of d_{av}, there are no steric hindrances for the ion transfer through the pores. However, the relatively wide pores of track-etched membranes are an important condition for creating oppositely directed convective transport and electromigration. This organization of these two fluxes allows very effective separation of ions with the same charge sign. Based on both the calculations and the results of the EBM separation experiment, the flux of lithium ions through the membrane can be expected to be about 0.3 mol/(m² × h). Under conditions where the feeding and receiving solutions were identical, the flux of cobalt ions was directed from the receiving to the feeding solution and was close to zero (−0.0025 mol/(m² × h)). Formally, through the calculation of the flux ratio, the ion separation coefficient of Li^+ and Co^{2+} ions was −55. But taking into account that only lithium ions can leave the feed solution, the separation coefficient should be equal to infinity. The presence of coexisting nickel ions in the feed solution leads to a decrease in the separation efficiency of lithium and cobalt if the fluxes of these cations are codirected.

However, the separation conditions can be chosen so that only lithium is extracted from the initial solution, while the fluxes of other ions are oppositely directed.

Author Contributions: Conceptualization, D.B., L.D. and V.N.; methodology, D.B. and V.N.; validation, D.B., L.D., C.L. and V.N.; formal analysis, D.B. and V.N.; investigation, D.B., V.T. and D.C.; resources, D.B., C.L. and V.N.; data curation, D.B.; writing—original draft preparation, D.B., L.D., C.L. and V.N.; writing—review and editing, D.B. and V.N.; visualization, D.B. and V.T.; supervision, D.B.; project administration, D.B.; funding acquisition, D.B. All authors have read and agreed to the published version of the manuscript.

Funding: This study was funded by the Russian Science Foundation, project № 22-79-00178, https://rscf.ru/en/project/22-79-00178/ (accessed on 11 April 2023).

Institutional Review Board Statement: Not applicable.

Data Availability Statement: Not applicable.

Acknowledgments: We thank P. Yu. Apel from the Joint Institute for Nuclear Research (Dubna, Russia) for providing us with the track-etched TEM #811 membrane.

Conflicts of Interest: The authors declare no conflict of interest.

References

1. Butylskii, D.Y.; Dammak, L.; Larchet, C.; Pismenskaya, N.D.; Nikonenko, V.V. Selective recovery and re-utilization of lithium: Prospects for the use of membrane methods. *Russ. Chem. Rev.* **2023**, *92*, RCR5074. [CrossRef]
2. Tabelin, C.B.; Dallas, J.; Casanova, S.; Pelech, T.; Bournival, G.; Saydam, S.; Canbulat, I. Towards a low-carbon society: A review of lithium resource availability, challenges and innovations in mining, extraction and recycling, and future perspectives. *Miner. Eng.* **2021**, *163*, 106743. [CrossRef]
3. Swain, B. Recovery and recycling of lithium: A review. *Sep. Purif. Technol.* **2017**, *172*, 388–403. [CrossRef]
4. Tuncuk, A.; Stazi, V.; Akcil, A.; Yazici, E.Y.; Deveci, H. Aqueous metal recovery techniques from e-scrap: Hydrometallurgy in recycling. *Miner. Eng.* **2012**, *25*, 28–37. [CrossRef]
5. Chen, Y.; Qiao, Q.; Cao, J.; Li, H.; Bian, Z. Precious metal recovery. *Joule* **2021**, *5*, 3097–3115. [CrossRef]
6. Coman, V.; Robotin, B.; Ilea, P. Nickel recovery/removal from industrial wastes: A review. *Resour. Conserv. Recycl.* **2013**, *73*, 229–238. [CrossRef]
7. Liu, Y.; Liu, M. Reproduction of Li battery LiNixMnyCo$_{1-x-y}$O$_2$ positive electrode material from the recycling of waste battery. *Int. J. Hydrogen Energy* **2017**, *42*, 18189–18195. [CrossRef]
8. Chan, K.H.; Anawati, J.; Malik, M.; Azimi, G. Closed-Loop Recycling of Lithium, Cobalt, Nickel, and Manganese from Waste Lithium-Ion Batteries of Electric Vehicles. *ACS Sustain. Chem. Eng.* **2021**, *9*, 4398–4410. [CrossRef]
9. Xu, P.; Dai, Q.; Gao, H.; Liu, H.; Zhang, M.; Li, M.; Chen, Y.; An, K.; Meng, Y.S.; Liu, P.; et al. Efficient Direct Recycling of Lithium-Ion Battery Cathodes by Targeted Healing. *Joule* **2020**, *4*, 2609–2626. [CrossRef]
10. Yu, M.; Bai, B.; Xiong, S.; Liao, X. Evaluating environmental impacts and economic performance of remanufacturing electric vehicle lithium-ion batteries. *J. Clean. Prod.* **2021**, *321*, 128935. [CrossRef]
11. Chen, Y.; Kang, Y.; Zhao, Y.; Wang, L.; Liu, J.; Li, Y.; Liang, Z.; He, X.; Li, X.; Tavajohi, N.; et al. A review of lithium-ion battery safety concerns: The issues, strategies, and testing standards. *J. Energy Chem.* **2021**, *59*, 83–99. [CrossRef]
12. Zhao, Y.; Rüther, T.; Staines, J. *Australian Landscape for Lithium-Ion Battery Recycling and Reuse in 2020*; Future Battery Industries CRC: Bentley, WA, Australia, 2021.
13. Bhandari, N.; Cai, A.; Yuzawa, K.; Zhang, J.; Joshi, V.; Fang, F.; Lee, G.; Harada, R.; Shin, S. *Global Batteries the Greenflation Challenge*; Goldman Sachs &, Co.: New York, NY, USA, 2022.
14. Velázquez-Martínez, O.; Valio, J.; Santasalo-Aarnio, A.; Reuter, M.; Serna-Guerrero, R. A Critical Review of Lithium-Ion Battery Recycling Processes from a Circular Economy Perspective. *Batteries* **2019**, *5*, 68. [CrossRef]
15. Siekierka, A.; Yalcinkaya, F. Selective cobalt-exchange membranes for electrodialysis dedicated for cobalt recovery from lithium, cobalt and nickel solutions. *Sep. Purif. Technol.* **2022**, *299*, 121695. [CrossRef]
16. Chan, K.H.; Malik, M.; Azimi, G. Separation of lithium, nickel, manganese, and cobalt from waste lithium-ion batteries using electrodialysis. *Resour. Conserv. Recycl.* **2022**, *178*, 106076. [CrossRef]
17. Afifah, D.N.; Ariyanto, T.; Suprapto, S.; Prasetyo, I. Separation of Lithium Ion from Lithium-Cobalt Mixture using Electrodialysis Monovalent Membrane. *Eng. J.* **2018**, *22*, 165–179. [CrossRef]
18. Gmar, S.; Chagnes, A.; Lutin, F.; Muhr, L. Application of Electrodialysis for the Selective Lithium Extraction Towards Cobalt, Nickel and Manganese from Leach Solutions Containing High Divalent Cations/Li Ratio. *Recycling* **2022**, *7*, 14. [CrossRef]
19. Brewer, A.K.; Madorsky, S.L.; Westhaver, J.W. The Concentration of 39K and 41K by Balanced Ion Migration in a Counterflowing Electrolyte. *Science* **1946**, *104*, 156–157. [CrossRef]

20. Forssell, P.; Kontturi, K. Experimental Verification of Separation of Ions Using Countercurrent Electrolysis in a Thin, Porous Membrane. *Sep. Sci. Technol.* **1983**, *18*, 205–214. [CrossRef]
21. Kontturi, K.; Forssell, P.; Ekman, A. Separation of Ions Using Countercurrent Electrolysis in a Thin, Porous Membrane. *Sep. Sci. Technol.* **1982**, *17*, 1195–1204. [CrossRef]
22. Butylskii, D.; Troitskiy, V.; Chuprynina, D.; Kharchenko, I.; Ryzhkov, I.; Apel, P.; Pismenskaya, N.; Nikonenko, V. Selective Separation of Singly Charged Chloride and Dihydrogen Phosphate Anions by Electrobaromembrane Method with Nanoporous Membranes. *Membranes* **2023**, *13*, 455. [CrossRef]
23. Butylskii, D.Y.; Pismenskaya, N.D.; Apel, P.Y.; Sabbatovskiy, K.G.; Nikonenko, V.V. Highly selective separation of singly charged cations by countercurrent electromigration with a track-etched membrane. *J. Memb. Sci.* **2021**, *635*, 119449. [CrossRef]
24. Tang, C.; Bondarenko, M.P.; Yaroshchuk, A.; Bruening, M.L. Highly selective ion separations based on counter-flow electromigration in nanoporous membranes. *J. Memb. Sci.* **2021**, *638*, 119684. [CrossRef]
25. Tang, C.; Yaroshchuk, A.; Bruening, M.L. Ion Separations Based on Spontaneously Arising Streaming Potentials in Rotating Isoporous Membranes. *Membranes* **2022**, *12*, 631. [CrossRef] [PubMed]
26. Kislyi, A.G.; Butylskii, D.Y.; Mareev, S.A.; Nikonenko, V. V Model of Competitive Ion Transfer in an Electro-Baromembrane System with Track-Etched Membrane. *Membr. Membr. Technol.* **2021**, *3*, 131–137. [CrossRef]
27. Lide, D.R. *CRC Handbook of Chemistry and Physics*; CRC Press: Boca Raton, FL, USA, 2005; ISBN 0-8493-0485-7.
28. Nightingale, E.R. Phenomenological Theory of Ion Solvation. Effective Radii of Hydrated Ions. *J. Phys. Chem.* **1959**, *63*, 1381–1387. [CrossRef]
29. Apel, P.Y. Track-Etching. In *Encyclopedia of Membrane Science and Technology*; Hoek, E.M.V., Tarabara, V.V., Eds.; John Wiley & Sons, Inc.: Hoboken, NJ, USA, 2013; pp. 332–355.
30. Sarapulova, V.V.; Pasechnaya, E.L.; Titorova, V.D.; Pismenskaya, N.D.; Apel, P.Y.; Nikonenko, V.V. Electrochemical Properties of Ultrafiltration and Nanofiltration Membranes in Solutions of Sodium and Calcium Chloride. *Membr. Membr. Technol.* **2020**, *2*, 332–350. [CrossRef]
31. Puigdomenech, I. HYDRA (Hydrochemical Equilibrium-Constant Database) and Medusa (Make Equilibrium Diagrams Using Sophisticated Algorithms). Available online: https://www.kth.se/che/medusa/ (accessed on 11 April 2023).
32. Sata, T.; Sata, T.; Yang, W. Studies on cation-exchange membranes having permselectivity between cations in electrodialysis. *J. Memb. Sci.* **2002**, *206*, 31–60. [CrossRef]
33. Wang, W.; Liu, R.; Tan, M.; Sun, H.; Niu, Q.J.; Xu, T.; Nikonenko, V.; Zhang, Y. Evaluation of the ideal selectivity and the performance of selectrodialysis by using TFC ion exchange membranes. *J. Memb. Sci.* **2019**, *582*, 236–245. [CrossRef]
34. Karpenko, T.V.; Kovalev, N.V.; Kirillova, K.R.; Achoh, A.R.; Melnikov, S.S.; Sheldeshov, N.V.; Zabolotsky, V.I. Competing Transport of Malonic and Acetic acids across Commercial and Modified RALEX AMH Anion-Exchange Membranes. *Membr. Membr. Technol.* **2022**, *4*, 118–126. [CrossRef]
35. Kontturi, K.; Ojala, T.; Forssell, P. Transport of acetate and chloroacetate weak electrolytes through a thin porous membrane in counter-current electrolysis. *J. Chem. Soc. Faraday Trans. 1 Phys. Chem. Condens. Phases* **1984**, *80*, 3379. [CrossRef]
36. Nichka, V.S.; Mareev, S.A.; Apel, P.Y.; Sabbatovskiy, K.G.; Sobolev, V.D.; Nikonenko, V.V. Modeling the Conductivity and Diffusion Permeability of a Track-Etched Membrane Taking into Account a Loose Layer. *Membranes* **2022**, *12*, 1283. [CrossRef]
37. Apel, P.; Koter, S.; Yaroshchuk, A. A Time-resolved pressure-induced electric potential in nanoporous membranes: Measurement and mechanistic interpretation. *J. Membr. Sci.* **2022**, *653*, 120556. [CrossRef]
38. Apel, P. Track etching technique in membrane technology. *Radiat. Meas.* **2001**, *34*, 559–566. [CrossRef]
39. Tang, C.; Yaroshchuk, A.; Bruening, M.L. Flow through negatively charged, nanoporous membranes separates Li^+ and K^+ due to induced electromigration. *Chem. Commun.* **2020**, *56*, 10954–10957. [CrossRef]
40. Ji, P.-Y.; Ji, Z.-Y.; Chen, Q.-B.; Liu, J.; Zhao, Y.-Y.; Wang, S.-Z.; Li, F.; Yuan, J.-S. Effect of coexisting ions on recovering lithium from high Mg^{2+}/Li^+ ratio brines by selective-electrodialysis. *Sep. Purif. Technol.* **2018**, *207*, 1–11. [CrossRef]
41. Iizuka, A.; Yamashita, Y.; Nagasawa, H.; Yamasaki, A.; Yanagisawa, Y. Separation of lithium and cobalt from waste lithium-ion batteries via bipolar membrane electrodialysis coupled with chelation. *Sep. Purif. Technol.* **2013**, *113*, 33–41. [CrossRef]
42. Kumar, R.; Liu, C.; Ha, G.-S.; Park, Y.-K.; Ali Khan, M.; Jang, M.; Kim, S.-H.; Amin, M.A.; Gacem, A.; Jeon, B.-H. Downstream recovery of Li and value-added metals (Ni, Co, and Mn) from leach liquor of spent lithium-ion batteries using a membrane-integrated hybrid system. *Chem. Eng. J.* **2022**, *447*, 137507. [CrossRef]

Disclaimer/Publisher's Note: The statements, opinions and data contained in all publications are solely those of the individual author(s) and contributor(s) and not of MDPI and/or the editor(s). MDPI and/or the editor(s) disclaim responsibility for any injury to people or property resulting from any ideas, methods, instructions or products referred to in the content.

Article

On the Performance of a Ready-to-Use Electrospun Sulfonated Poly(Ether Ether Ketone) Membrane Adsorber

Niki Joosten [1,2,3], Weronika Wyrębak [1], Albert Schenning [2], Kitty Nijmeijer [1] and Zandrie Borneman [1,*]

[1] Membrane Materials and Processes, Department of Chemical Engineering and Chemistry, Eindhoven University of Technology, 5600 MB Eindhoven, The Netherlands; d.c.nijmeijer@tue.nl (K.N.)
[2] Stimuli-responsive Functional Materials and Devices, Department of Chemical Engineering and Chemistry, Eindhoven University of Technology, 5600 MB Eindhoven, The Netherlands
[3] Wetsus, European Centre of Excellence for Sustainable Water Technology, Oostergoweg 9, 8911 MA Leeuwarden, The Netherlands
* Correspondence: z.borneman@tue.nl

Abstract: Motivated by the need for efficient purification methods for the recovery of valuable resources, we developed a wire-electrospun membrane adsorber without the need for post-modification. The relationship between the fiber structure, functional-group density, and performance of electrospun sulfonated poly(ether ether ketone) (sPEEK) membrane adsorbers was explored. The sulfonate groups enable selective binding of lysozyme at neutral pH through electrostatic interactions. Our results show a dynamic lysozyme adsorption capacity of 59.3 mg/g at 10% breakthrough, which is independent of the flow velocity confirming dominant convective mass transport. Membrane adsorbers with three different fiber diameters (measured by SEM) were fabricated by altering the concentration of the polymer solution. The specific surface area as measured with BET and the dynamic adsorption capacity were minimally affected by variations in fiber diameter, offering membrane adsorbers with consistent performance. To study the effect of functional-group density, membrane adsorbers from sPEEK with different sulfonation degrees (52%, 62%, and 72%) were fabricated. Despite the increased functional-group density, the dynamic adsorption capacity did not increase accordingly. However, in all presented cases, at least a monolayer coverage was obtained, demonstrating ample functional groups available within the area occupied by a lysozyme molecule. Our study showcases a ready-to-use membrane adsorber for the recovery of positively charged molecules, using lysozyme as a model protein, with potential applications in removing heavy metals, dyes, and pharmaceutical components from process streams. Furthermore, this study highlights factors, such as fiber diameter and functional-group density, for optimizing the membrane adsorber's performance.

Keywords: membrane adsorber; sulfonated poly(ether ether ketone) (sPEEK); electrospinning; lysozyme; dynamic adsorption capacity; fiber diameter; functional-group density; sulfonation degree; electrostatic interactions; specific surface area

Citation: Joosten, N.; Wyrębak, W.; Schenning, A.; Nijmeijer, K.; Borneman, Z. On the Performance of a Ready-to-Use Electrospun Sulfonated Poly(Ether Ether Ketone) Membrane Adsorber. *Membranes* **2023**, *13*, 543. https://doi.org/10.3390/membranes13060543

Academic Editors: Annarosa Gugliuzza and Cristiana Boi

Received: 26 April 2023
Revised: 17 May 2023
Accepted: 18 May 2023
Published: 23 May 2023

Copyright: © 2023 by the authors. Licensee MDPI, Basel, Switzerland. This article is an open access article distributed under the terms and conditions of the Creative Commons Attribution (CC BY) license (https://creativecommons.org/licenses/by/4.0/).

1. Introduction

Our resource consumption surpasses the earth's replenishment rate. To prevent depletion and ensure long-term sustainability, a transition from a linear to a circular economy is necessary [1]. Looking especially at the key element of water, a circular economy requires the purification of water to safeguard clean drinking water and the recovery of the valuable resources it contains [2]. Currently, many valuable resources are lost in discarded rest streams due to a lack of cost-effective recovery technologies [3]. In many cases, the concentration of the valuable resource is too low, resulting in high energy costs, or the rest streams are contaminated, increasing the purification process costs [3]. Additionally, every industry has unique resources in its rest stream, such as proteins in the dairy industry, ionic species (e.g., nitrogen, phosphorous, and potassium) in the agricultural industry, and dyes in the textile industry [4,5]. These factors make it challenging to find a single, efficient technology to recover and valorize these resources [6,7]. Therefore, an efficient, and especially versatile, purification technology is needed that can be tailored to the needs of various industries.

Traditionally, adsorption processes using packed-bed column technology are frequently applied to remove target components or impurities from aqueous streams [8]. However, significant limitations involving high pressure drop and low throughput separation (the amount of material processed per unit time), seriously impede the development and scale-up of this technology [8,9]. In addition, coarse impurities can easily clog the column, making this technology unsuitable for streams with high-mass components [10]. To overcome these limitations, several developments in column technology have been made.

One such development is reducing the particle size in the packed bed, which increases the adsorption capacity. By reducing the size of the porous particle, the diffusion length is shortened, making the adsorption sites more accessible and increasing the throughput [11]. However, this also intensifies the limitations of clogging and high pressure drop, leading to column deformation and channeling. This results in an early breakthrough, which implies noncomplete utilization of the adsorption capacity [12].

Nonporous and/or core-shell rigid particles, on the other hand, offer the advantage of greater robustness and lower pressure drop [13]. Solute diffusion is no longer a limiting factor due to the absence of pores [8]. Unfortunately, these particles have a lower surface area, which results in a lower capacity [13].

Perfusive or super-porous particles have been developed to increase convective mass transport and reduce the diffusive dependency of the purification step to obtain higher throughput. These particles allow solute molecules to pass through faster and at lower pressures compared to packed beds and the capacity is higher compared to nonporous particles [8,14].

Expanded beds were developed to prevent clogging of the column [10]. An upward flow is applied to increase the space between the particles and allows coarser impurities to flow through. However, the size and density of the particles and the flow rate must be carefully balanced; if the particles are small, the flow must be limited to avoid overexpansion of the bed, while if the particles are large, the flow must be high enough to prevent sedimentation of the particles [15]. In both situations, the throughput of the bed is limited, either due to low flow rates or due to restricted diffusion reducing the adsorption capacity.

Membrane adsorbers have been developed to overcome these limitations. In these adsorbers, the adsorbent particles are fixed in a porous matrix or the matrix itself acts as the adsorbent [16]. The target substances are adsorbed on the adsorptive moieties in the membrane adsorber while the solvent with nonbinding and coarse impurities permeates through the pores [17]. Within this porous structure, convective mass transport takes place, which allows operation at higher flow rates compared to diffusion-controlled packed-bed chromatography [8,18]. This leads to a reduced pressure drop and facilitates the scale-up of the membrane adsorber technology [9]. However, most membrane adsorbers made by modifying micro/macroporous membranes have a low adsorption capacity due to a low surface area and a large pore size distribution [12]. A variance in porosity causes a preferential flow of the solute molecules through the larger pores resulting in an early breakthrough [8,12].

Electrospinning was introduced to enhance the surface area and versatility of membrane adsorbers, allowing the creation of nanofibrous porous mats with tailor-made functionalities to selectively recover valuable components [19]. This technique uses an electrostatic force to overcome the surface tension of a polymer solution, converting it into a fiber structure that is deposited on a collector paper forming a porous mat (Figure 1) [20,21]. These mats can be stacked with a random overlay orientation of the fibers to decrease the effective pore size distribution and achieve an even flow dispersion [22]. Electrospun membrane adsorbers offer a promising cost-effective platform technology for resource recovery. This is because electrospinning gives opportunities to (1) tailor the selectivity by functionalization of the electrospun membranes through polymer blending, functional particle embedding during electrospinning or chemical post-functionalization; (2) control permeability by adjusting the bed height, porosity, and fiber diameter such that nonbinding and coarser impurities easily elute through the bed while the desired components can bind to the (functionalized) electrospun fibers; and (3) facilitate easy production and linear scale-up [19,23–25]. Recent progress made in the use of electrospun nanofibers for membrane adsorbers is discussed in several review papers [9,19,23,24]. So far, the versatility of electrospun membrane adsorbers has been studied primarily in terms of design, fabrication, and type of functionalization [19]. However, many of these adsorbers require multistep synthesis for functionalization [26–32], which limits their entrance into industry and the market. To overcome this, alternative fabrication routes, such as the use of pre-functionalized polymers should be explored [33]. Systematic studies to tailor the performance by controlling electrospinning conditions are limited [19].

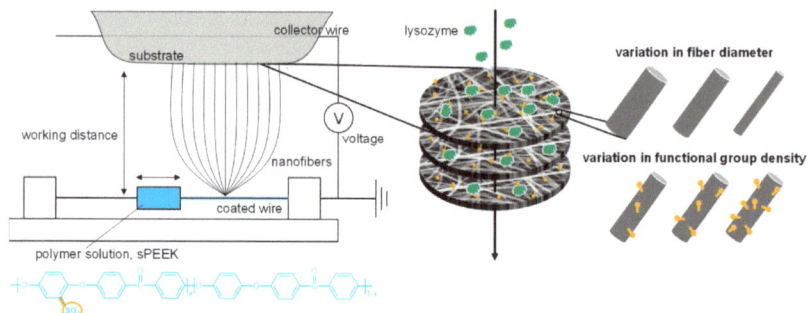

Figure 1. Schematic representation of the fabrication process for sPEEK membrane adsorbers using a wire-electrospinning device, along with a visualization of the factors studied for their impact on performance.

This study develops electrospun sulfonated poly(ether ether ketone) (sPEEK) membrane adsorbers that eliminate the need for any post-functionalization steps due to the inherent presence of the functional sulfonate groups in the polymer (Figure 1). The sulfonate groups allow selective binding of the model protein lysozyme under neutral pH through electrostatic interactions. The negatively charged strong acidic sulfonic acid groups and the positively charged lysozyme (with an isoelectric point of 11.35) are attracted to each other by Coulombic interactions [34,35]. Other interactions, such as hydrogen bonding, hydrophobic interactions, and van der Waals forces, also contribute to the binding affinity between the protein and the membrane adsorber [35–37]. The choice of lysozyme as the model protein in this study is due to its stability, antimicrobial properties, and is a natural preservative, which makes lysozyme an ideal test molecule and is therefore widely studied [38]. Most important, the versatility of electrospinning is explored with a focus on the effect of fiber thickness or functional-group density on the adsorber performance (Figure 1). The sPEEK fiber thickness was tailored by varying the concentration of the polymer solution and fine-tuning the process parameters of the electrospinner [20,39,40]. sPEEK membrane adsorbers with thicknesses of 90 ± 18 nm, 132 ± 27 nm, and 166 ± 18 nm were investigated. Additionally, the effect of the functional-group density on the binding capacity was studied by creating sPEEK-based membrane adsorbers from sPEEK with different sulfonation degrees (52%, 62%, and 72%).

2. Materials and Methods

2.1. Materials

Sulfonated poly(ether ether ketone) (sPEEK) was purchased from FumaTech-BWT GmbH, Bietigheim-Bissingen, Germany, Fumion® with sulfonation degrees of 52% (sPEEK-52, x = 0.52 in Figure 1), 62% (sPEEK-62), and 72% (sPEEK-72). Dimethylacetamide (DMAc) was supplied by Sigma–Aldrich, Darmstadt, Germany and N-methylpyrrolidone (NMP) by Biosolve B.V., Valkenswaard, The Netherlands. To dry the solvents molecular sieves were used (4 Å, Sigma–Aldrich, Darmstadt, Germany). Hydrochloric acid (HCl, Supelco® from Sigma–Aldrich, Darmstadt, Germany), sodium hydroxide (NaOH, VWR Chemicals, Boxmeer, The Netherlands), and sodium chloride (NaCl, Sanal® P, AkzoNobel, Deventer, The Netherlands) were used for membrane pretreatment and/or characterization. Demineralized water was obtained from an Elga water purification system from Veolia, Weert, The Netherlands. Lysozyme (LZ) from hen egg white (Mw ~ 14,600, Fluka analytical), phosphate buffers saline (PBS) tablets (pH 7.4, 0.01 M PBS, total ionic strength 0.15 M, Sigma–Aldrich, Darmstadt, Germany), and syringe filter holders (25 mm, Sartorius, Goettingen, Germany) were used to measure the membrane performance.

2.2. Preparation of Electrospun Membranes

For electrospinning, the polymer is dissolved in a solvent. During the spinning process, the polymer solidifies, thereby forming a fiber. Both the polymer, solution, and process conditions define, e.g., the dimensions of the fiber. The driving force in electrospinning is the electrical field built between the polymer supply and the collector. This electric charge causes instability in the polymer solution because of the induction of charges on the polymer and the charge builds up mainly at the surface of the liquid, destabilizing the meniscus of the droplet on the wire. When the electric charge overcomes the surface tension, a jet is formed.

Before solution preparation, the sPEEK polymers were dried in the vacuum oven at 80 °C for six hours. The dried sPEEK-52 was used to prepare a 22 weight-% (wt %) solution using NMP as a solvent, which had been dried using molecular sieves. The solution was placed on the roller bench for at least 24 h. Additionally, the polymer solution was placed in an ultrasonic bath at 25 °C for at least 4 h to break up any gel particles that may be present in the solution. Polymer solutions of 17–25 wt % sPEEK-62 and sPEEK-72, with solubility properties distinct from sPEEK-52, were prepared using dried DMAc as a solvent and placed on the roller bench for at least 15 h. While the polymer solution was still hazy, the solution was sonicated to obtain a homogeneous transparent solution. Electrospinning was performed using a wire-electrospinning device (Nanospider NS LAB, Elmarco, Liberec, Czech Republic). The relative humidity and temperature of the electrospinning chamber were controlled (desiccant dehumidifier system, ML270PLUS, Munters, Den Haag, The Netherlands).

The polymer solutions were electrospun from a carrier with an orifice of 0.8 mm moving along the working wire electrode at a speed of 150 mm/s. The applied voltage between the working and collecting electrode (working distance was set to 150 mm) was set at 80 kV. The substrate was not moving and its distance to the collecting electrode was set at 25 mm. Nanofibers were produced at 22 ± 0.5 °C under $25 \pm 1\%$ relative humidity, except for 23.4 wt % sPEEK-72, which was produced under $20 \pm 1\%$ relative humidity. The obtained spunbound membranes were placed for conditioning in 1 M HCl on a shaking plate for one hour to ensure that all sulfonate groups have an H^+ as their counterion. Then, the membranes were rinsed in demineralized water by refreshing the water multiple times until a neutral pH was obtained. Next, the membranes were dried in a vacuum oven at 80 °C for at least 6 h.

2.3. Membrane Characterization

2.3.1. Scanning Electron Microscopy (SEM)

The morphology of the fabricated membranes was evaluated using SEM (JEOL IT-100, Nieuw-Vennep, The Netherlands) with 10 kV accelerating voltage and probe current setting 32. All measured samples were platinum coated for 60 s at 40 mA using a sputtercoater (JFC-2300HR, JEOL, Nieuw-Vennep, The Netherlands). Fiber dimensions were measured on at least 100 spots at 10.000× magnification using ImageJ software.

2.3.2. BET Surface Area

The Brunauer–Emmett–Teller (BET) specific surface area of the electrospun membranes was determined by N_2 physisorption at liquid N_2 temperature ($-196\ °C$) with a Micromeretics TriStar II (Eindhoven, The Netherlands) using the Plus 3.03 software with optimized BET calculation. Prior to the measurement, samples of ~0.1 g were outgassed for 20 h at 80 °C under vacuum.

2.3.3. Capillary Liquid Porometry

The pore size distribution was studied by porometry (Porolux 500, Porometer, Nazareth, Belgium) on a sample with a diameter of 25 mm. Nitrogen gas was used as the pressurizing agent, wetting was done with Porofil® (15.9 dyn/cm, supplied by Porometer, Nazareth, Belgium). From each membrane, two samples were measured with the following settings: shape factor 0.715, pressure increasing slope 120 s/bar, final pressure 6 bar, number of measurements steps wet curve 50, and number of measurements steps dry curve 25.

2.3.4. Water Uptake and Swelling

The sPEEK water uptake and swelling (%) were measured on cast sPEEK films in duplicates. Hereto the SPEEK solutions that were prepared for the electrospinning were cast on a glass plate using a 500 μm casting knife for sPEEK-52 and sPEEK-62, and a 300 μm casting knife for sPEEK-72. Then the films were dried for two days in a nitrogen box and six days in a nitrogen oven at 120 °C. Then the films were immersed in water for three days to ensure a fully saturated water uptake. Subsequently, the films were carefully wiped with paper to remove excess solution and weighed. The films were put in a vacuum oven at 60 °C for 20 h. Once again, the films were weighed, and the water uptake was calculated using Equation (1) and the thickness using Equation (2).

$$\text{Water uptake} = \frac{m_{wet} - m_{dry}}{m_{dry}} \cdot 100\% \tag{1}$$

where m_{wet} is the weight of the wet membrane (g) and m_{dry} is the weight of the dried membrane (g).

$$\text{Swelling thickness} = \frac{t_{wet} - t_{dry}}{t_{dry}} \cdot 100\% \tag{2}$$

where t_{wet} is the thickness of the wet membrane (μm) and t_{dry} is the thickness of the dried membrane (μm).

2.3.5. Ion-Exchange Capacity

The ion-exchange capacity reflects the number of functional cationic groups that are available for ion exchange. The ion-exchange capacity of the electrospun membranes was determined through acid–base titration as reported by Park et al. [41]. First, the membranes were immersed overnight in 1 M HCl to convert them into the H^+ form. After, the membranes were thoroughly rinsed with demineralized water to remove the unbound H^+ ions from the spunbound membranes. Subsequently, the membranes were soaked three times for one hour in 15 mL 2 M NaCl to exchange Na^+ for H^+. For each membrane sample, the combined salt solutions were titrated with 0.01 M NaOH using a titrator from Metler Toledo with sensor DGi115-SC. The ion-exchange capacity (meq/g dry membrane) was calculated using the following Equation (3):

$$\text{Ion-exchange capacity} = \frac{M_{NaOH} \cdot V_{NaOH}}{m_{dry}} \tag{3}$$

where M_{NaOH} is the molar concentration of the sodium hydroxide solution (M), V_{NaOH} the volume of sodium hydroxide needed to titrate the acid (mL) and m_{dry} the dry mass of the membrane (g). All ion-exchange capacity measurements were executed in triplicate.

2.4. Membrane Performance

2.4.1. Static Lysozyme Adsorption

The maximum adsorption capacity of lysozyme on the membrane adsorbers was evaluated through static adsorption experiments. First, lysozyme solutions with concentrations ranging from 0–2.5 mg/mL in PBS buffer were made and measured by UV-vis spectroscopy at 280 nm (Shimadzu UV-1280, 's-Hertogenbosch, The Netherlands). The sPEEK membranes (~0.02 g) were immersed in 2 mL lysozyme solution with the predetermined concentrations (in duplicates). The adsorption experiments were carried out at room temperature for 20 h on a shaking plate to ensure equilibrium and the concentrations were measured again. The amount of adsorbed lysozyme (mg/g) was calculated using the following Equation (4):

$$\text{Adsorbed lysozyme} = \frac{(C_0 - C_t) \cdot V}{m} \tag{4}$$

where C_0 is the initial concentration and C_t the equilibrium concentration of lysozyme in solution (mg/mL), V is the volume of the solution (mL), and m the mass of the membrane (g). The adsorption isotherm follows the Langmuir isotherm, which is described by the following Equation (5):

$$q_e = \frac{Q_m \cdot K_d \cdot C_e}{(1 + K_d \cdot C_e)} \tag{5}$$

where q_e is the equilibrium adsorption capacity (mg/g), C_e is the equilibrium concentration (mg/mL), and K_d is the equilibrium constant (mL/mg). Q_m is the maximum adsorption capacity using a curve fitting (mg/g).

2.4.2. Dynamic Lysozyme Adsorption

For the determination of the dynamic adsorption capacity, 5–20 membrane discs with a diameter of 25 mm each were cut and stacked in a filter holder (total mass membrane 0.05–0.15 g). The membrane mass available for adsorption was determined as the mass enclosed within the o-ring. The filter holder was connected to a syringe filled with PBS solution to flush the system and eliminate any potential contaminants. The flow velocity was controlled with a syringe pump (Chemyx Inc. Fusion 200, Stafford, TX, USA) and set at 1.0 mL/min for the adsorption step. The syringe was filled with 0.5 mg/mL lysozyme in PBS solution and the permeate was collected in fractions of ~0.7 mL. The concentration of lysozyme was determined by UV-vis spectroscopy at 280 nm. When the concentration in the permeate exceeded 10% of the feed concentration (breakthrough point) the pump was stopped. The adsorption capacity was determined by interpolation of the adsorption curve at 10% breakthrough. The dynamic adsorption was executed for most samples in duplicates, and the adsorption capacity showed an error margin of ≤11%. For the washing step, the lysozyme in the syringe was replaced by PBS buffer solution to remove the unbound lysozyme. The flow velocity in this washing step was set at 0.5 mL/min for practical reasons. For desorption, the syringe was filled with 0.5 M NaCl in PBS solution and the flow velocity was set at 1 mL/min. The amount of desorbed lysozyme in the desorption buffer was determined by UV-vis spectroscopy at 280 nm and the recovery is calculated using the following Equation (6):

$$\text{Recovery} = \frac{P_D}{P_L - P_{AW}} \cdot 100\% \tag{6}$$

where P_D is the amount of protein removed from the membrane stack in the desorption step (mg), P_L is the amount of protein loaded on the membrane stack (mg), and P_{AW} is the amount of unbound protein eluted in the adsorption and washing step (mg).

3. Results and Discussion

3.1. Electrospun sPEEK Membrane Adsorbers

Membrane adsorbers with sulfonic acid functional groups were fabricated by wire-electrospinning a 19 wt % sPEEK-62 in DMAc solution. SEM images of the nanofibrous mats after the conditioning step show a uniform fiber morphology with an average fiber diameter of 132 ± 27 nm (Figure 2) with an associated BET surface area of 12.3 ± 2.1 m^2/g. The obtained fiber diameter is relatively small compared to the data provided in the review paper by Yang et al. (ranging from 150–15,000 nm) [9]. As a result, the surface area is relatively high compared to values reported in literature (4–7 m^2/g) [9].

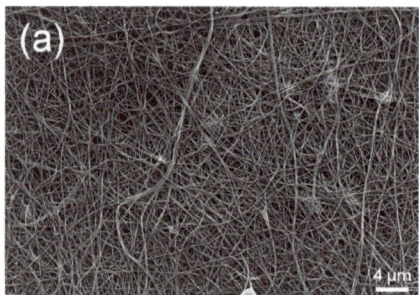

Figure 2. SEM images of sPEEK-62 membrane adsorbers in the H$^+$ form made by wire-electrospinning using a 19 wt % polymer solution; (**a**) 2500× magnification and (**b**) 10,000× magnification.

Achieving a uniform plug-flow velocity through the membrane adsorber and utilizing its complete adsorption capacity relies on a narrow pore size distribution. The pore size distribution of this sPEEK-62 membrane is shown in Figure 3. The pores of the membrane are almost 50 times larger than the size of a lysozyme molecule (4.5 × 3 × 3 nm), allowing convective transport of the lysozyme without clogging the pores [42].

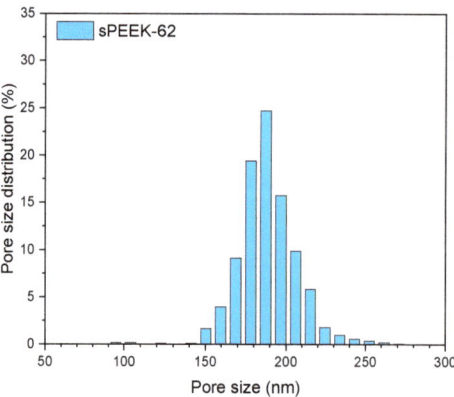

Figure 3. Pore size distribution measured with capillary liquid porometry of sPEEK-62 membrane adsorbers in the H$^+$ form made by wire-electrospinning using a 19 wt % polymer solution.

The membrane adsorber performance is studied by measuring the static adsorption capacity using lysozyme as a model protein (Equation (4), Figure 4). Lysozyme is positively charged at neutral pH enabling electrostatic binding with the negatively charged sulfonic acid groups of sPEEK [34]. The experimental results in Figure 4 have been fitted using the Langmuir adsorption isotherm expression from Equation (5), which gives a maximum equilibrium adsorption capacity Q_m of 72 mg/g.

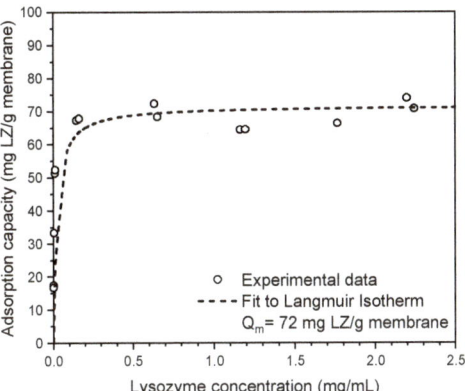

Figure 4. Static adsorption isotherm (at equilibrium concentration, C_e) of lysozyme (LZ) in PBS buffer (pH = 7.4) on an electrospun sPEEK-62 membrane adsorber with a maximum adsorption capacity (Q_m) of 72 mg LZ/g membrane.

The dynamic adsorption capacity of the membrane adsorber is studied by measuring the breakthrough curve, showing the lysozyme load (permeate volume in the adsorption step) versus the eluent concentration (Figure 5). Initially, no lysozyme is detected in the eluent, indicating full adsorption on the membrane adsorber. When the adsorber becomes saturated, with the majority of the adsorption sites occupied, the first lysozyme molecules start to break through, causing a lysozyme increase in the eluent. By convention, the dynamic adsorption capacity is determined at 10% breakthrough ($q_{10\%}$) to minimize product loss. Dynamic adsorption experiments of a stack of 20 membranes loaded with 0.5 mg/mL lysozyme at a flow of 0.1 mL/min show an adsorption capacity at 10% breakthrough of 59.3 mg/g (Figure 5a). This corresponds to a lysozyme adsorption capacity that is comparable to the static adsorption capacity at C_e of 0.05 mg/mL, being the concentration at 10% breakthrough. This means full utilization of the lysozyme adsorption capacity and indicates an almost ideal plug flow in the membrane stack. Furthermore, the dynamic adsorption capacity at varying flow velocity (from 0.1 mL/min to 1 mL/min) showed only small deviations (within the expected-error margin, as reported in the experimental section) and was independent of the set flow velocities, confirming that convective mass transport is dominant in the membrane adsorber stack (Figure 5a–c).

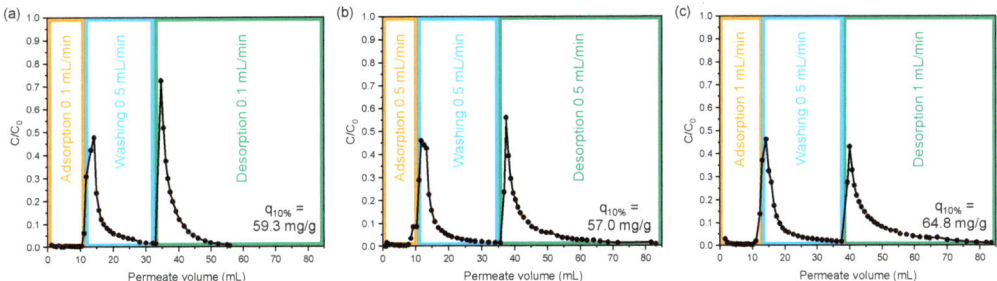

Figure 5. Dynamic lysozyme adsorption and desorption curves of electrospun sPEEK-62 membrane adsorbers with flow velocities of (**a**) 0.1 mL/min, (**b**) 0.5 mL/min, and (**c**) 1 mL/min (0.5 mg LZ/mL, pH 7.4); $q_{10\%}$ is the adsorption capacity at 10% breakthrough.

The peak in the desorption step shows the amount of lysozyme removed from the membrane adsorber using 0.5 M NaCl in PBS buffer, which is 31 ± 2% of the total adsorbed lysozyme, regardless of the flow velocity. This suggests that lysozyme adsorbs on the membrane by two modes of adsorption, namely reversible electrostatic interactions, and irreversible hydrophobic interactions. Adsorption with these two modes is supported by the study of Dismer et al. who showed that in the case of a resin with a hydrophobic backbone (polystyrene) functionalized with sulfonic acid groups (such as our sulfonated PEEK), lysozyme binding occurs through electrostatic and hydrophobic interactions [43,44]. Furthermore, computer simulations by Yu et al. indicate that when the lysozyme binds with its hydrophobic region with only four positive residues in its surroundings (in contrast to the other hydrophobic region surrounded by 13 positive residues), desorption is inhibited [45]. Once the hydrophobic binding sites are covered with lysozyme only the reversible electrostatic binding sites remain available, which leads to an overall decrease in lysozyme adsorption capacity in the second and subsequent cycles (Figure 6). The recovery, normalized with the recovery of the first cycle, shows a small increase in the second cycle. This could be attributed to a binding rearrangement from hydrophobic to electrostatic binding, likely due to the change in ionic strength between the desorption step of cycle 1 (0.65 M) and the adsorption step of cycle 2 (0.15 M). At low ionic strength, electrostatic interactions are dominant in protein adsorption, while at higher ionic strength, the charges are screened by the ions in solution, and hydrophobic interactions become dominant in protein adsorption [46,47]. From the third cycle onward, the adsorption capacity and recovery remained fairly constant, indicating that the hydrophobic binding sites are occupied and interactions in these cycles primarily occur at the reversible electrostatic binding sites, i.e., sulfonic acid groups. These findings demonstrate that sPEEK-62 membrane adsorbers are reusable and capable of operating at a constant efficiency after the third cycle.

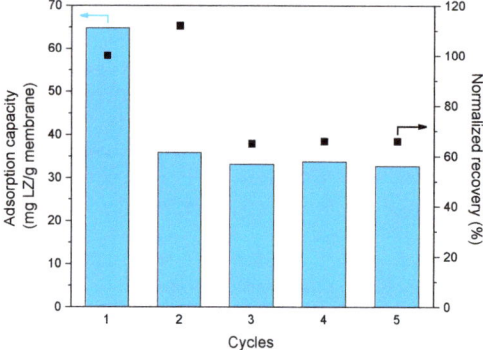

Figure 6. Dynamic lysozyme adsorption capacity and recovery of sPEEK-62 membrane adsorber for 5 cycles (flow velocity 1.0 mL/min, 0.5 mg LZ/mL, pH 7.4). The recovery is normalized with the recovery of the first cycle.

3.2. The Effect of Fiber Diameter

The versatility of electrospinning is explored with a focus on the relationship between fiber diameter and performance. The sPEEK-62 membrane adsorbers with different fiber diameters were fabricated by varying the concentration of the polymer solution in the wire-electrospinning process. The SEM images of the fabricated sPEEK membranes after proton exchange show a variation in fiber diameter from 90 ± 18 nm to 166 ± 18 nm with increasing polymer concentration (Figure 7). Based on our experience, this is the widest range of fiber diameters achievable through wire-electrospinning with this material because lower concentrations of polymer solution yield fibers with beads (as already observed in SEM with 17 wt % sPEEK-62), and higher concentrations are too viscous and prone to gelation, making them unsuitable for electrospinning [25,40]. Small optimizations in the electrospinning process parameters could be made to enlarge the variations in fiber diameter, although it is known that the polymer concentration has the largest effect on fiber diameter [39]. Despite the considerable variations in fiber diameter observed with

SEM, only minor variations in the surface area were measured with BET. Specifically, the measured surface areas were 11.2 ± 0.7 m^2/g, 12.3 ± 2.1 m^2/g, and 16.5 ± 1.3 m^2/g for the sPEEK-62 membrane adsorbers created with 17 wt %, 19 wt %, and 25 wt % polymer solution, respectively. Surprisingly, the membrane adsorber with the largest fiber diameter has the highest surface area, which could be due to the submicron-scale surface roughness of the fibers. This surface roughness is likely a result of buckling instability during the electrospinning process, where the skin layer formed on the polymer fibers collapses as the solvent evaporates, leading to a wrinkled surface. This phenomenon occurs more frequently with thicker fibers, which have a reduced surface-to-volume ratio and therefore longer drying times, increasing the likelihood of skin-layer formation [48–50]. Consequently, it is more probable that the thickest fiber has the greatest surface roughness, even though it cannot be observed with SEM. In contrast, the membrane adsorber with the smallest fiber diameter could have a reduced surface area due to the formation of beads.

Figure 7. SEM images of sPEEK-62 membrane adsorbers in the H$^+$ form made by wire-electrospinning using polymer concentrations of (**a**) 17 wt %, (**b**) 19 wt %, and (**c**) 25 wt %. Average fiber diameters are given in the bottom left corner of each image.

For a fair comparison of the dynamic adsorption capacity, it is important to take the pore size distribution into account as it influences the flow distribution through the membrane adsorbers. The porometry results show that the pore size distribution is in the same range for all three samples (Figure 8) and will probably be narrowed down when using a stack of membranes [26,28]. It is worth noting that larger fiber diameters correspond to larger pore sizes. This is because thicker fibers are created by pulling more material from the wire, which makes the space between the fibers wider resulting in larger pores [51–54].

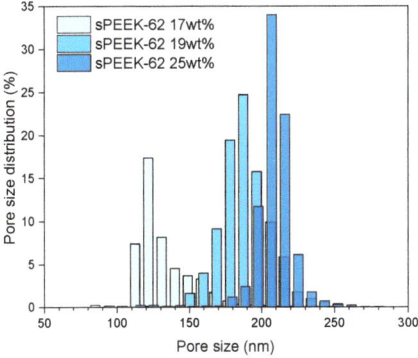

Figure 8. Pore size distribution of sPEEK-62 membrane adsorbers with different fiber diameters (made with polymer concentrations of 17 wt %, 19 wt %, and 25 wt %) as measured with porometry.

Dynamic adsorption measurements with lysozyme are performed to evaluate the performance of all three sPEEK-62 membrane adsorbers. The adsorption capacity results (in mg/g) show that there is no correlation between adsorption capacity and fiber diameter since the total surface area of the membrane did not change with a change in the fiber diameter (Table 1). When the adsorption capacity is normalized for the surface area, a slight decrease is observed for the sPEEK-62 membranes produced with higher polymer concentrations. As the material properties remain constant for these membranes and the average pore size is larger for those produced with higher polymer concentrations, this suggests that the amount of lysozyme adsorbed per surface area is slightly decreased due to the larger pores. This could be attributed to the longer time required for the lysozyme to reach the surface of the adsorber. Despite this, the recovery values for all three membrane adsorbers were similar, as the material properties, i.e., ratio electrostatic and hydrophobic interactions, are identical for all three sPEEK-62 membranes.

Table 1. Dynamic lysozyme adsorption capacity at 10% breakthrough and recovery of sPEEK-62 membrane adsorbers with different fiber diameters (made with different polymer concentrations of 17 wt %, 19 wt %, and 25 wt %).

	17 wt % sPEEK-62	19 wt % sPEEK-62	25 wt % sPEEK-62
BET surface area (m^2)	11.2 ± 0.7	12.3 ± 2.1	16.5 ± 1.3
Adsorption capacity (mg/g)	66.0 ± 6.4	60.1 ± 6.6	76.0 ± 3.0
Adsorption capacity (mg/m^2)	5.9 ± 0.7	4.9 ± 1.0	4.6 ± 0.4
Recovery (%)	37 ± 11%	35 ± 6%	36 ± 3%

3.3. The Effect of Sulfonation Degree

The versatility of sPEEK membrane adsorbers is explored by using the sPEEK of variable sulfonation degrees (52%, 62%, and 72%), as validated by H-NMR using the method of Zaidi et al. [55]. The sulfonation degree determines the number of functional groups present for adsorption. First, the properties of the different sPEEK polymers are studied by measuring the water uptake and swelling behavior of cast sPEEK films (Table 2). The water uptake and swelling thickness properties of sPEEK-52 and sPEEK-62 are rather similar, whereas sPEEK-72 shows a substantial increase in both the water uptake and the swelling thickness. This is supported by Zaidi et al. where they show that the water uptake increases linearly with a sulfonation degree up to 65% (with a reported water uptake of 33%), followed by a rapid increase above a sulfonation degree of 70% (with a water uptake of 47% at a sulfonation degree of 72%) [55]. The high density of SO$_3$H groups in the highly sulfonated sPEEK can form clusters that absorb more water, explaining the rapid increase in water uptake [55,56]. The high water uptake results in severe swelling and gelation affecting the dimensional stability; at higher sulfonation degree values (100%) the polymer becomes even water soluble. Therefore sPEEK-72 was selected as the upper limit in sulfonation degree.

Table 2. Water uptake and swelling thickness of cast sPEEK films and the ion-exchange capacity of electrospun sPEEK membranes.

	sPEEK-52	sPEEK-62	sPEEK-72
Water uptake (%)	34 ± 5%	39 ± 2%	51 ± 7%
Swelling thickness (%)	11 ± 1%	12 ± 9%	18 ± 8%
Ion-exchange capacity (meq/g)	1.30 ± 0.02	1.56 ± 0.05	1.89 ± 0.09

The experimentally determined ion-exchange capacity of the electrospun membranes is reported in Table 2. The quantity of accessible functional groups increases using sPEEK with a higher sulfonation degree. It should be noted that the ion-exchange capacity represents the overall number of functional cationic groups available for small ions that can penetrate inside the fiber. However, it is unlikely that lysozyme, with a dimension of approximately 4.5 × 3 × 3 nm [42], has access to the subsurface sulfonic acid groups, resulting in a lower quantity of functional groups available for lysozyme binding compared to the measured ion-exchange capacity.

The properties of the sPEEK polymer change with the sulfonation degree. In the electrospinning process, the charge mainly accumulates at the surface of the liquid, which destabilizes the meniscus of the droplet and changes the jet formation. For this reason, the electrospinning parameters had to be adjusted for every polymer solution (22 wt % sPEEK-52, 19 wt % sPEEK-62, and 23 wt % sPEEK-72), as described in the experimental section. Despite efforts to achieve uniform thickness in the sPEEK fibers of varying sulfonation degrees, variations in fiber diameter were observed in the SEM images (Figure 9). Additionally, the fusion of some fibers was observed in the sPEEK-72 membrane, which may be attributed to the high degree of swelling of this polymer. As a result, an increased average diameter is measured compared to the pristine membrane, which had an average diameter of 158 ± 35 nm (SEM image not shown).

Figure 9. SEM images of membrane adsorbers in the H^+ form made by wire-electrospinning using sPEEK with different sulfonation degrees of (**a**) 52%, (**b**) 62%, and (**c**) 72%. Average fiber diameters are given in the bottom left corner of each image.

The surface areas of the membranes, as measured with BET, were 14.9 ± 2.9 m^2/g, 12.3 ± 2.1 m^2/g, and 9.5 ± 2.4 m^2/g for sPEEK-52, sPEEK-62, and sPEEK-72, respectively. The BET measurements show a reduction in surface area for the sPEEK-72 membrane compared to the sPEEK-62 membrane, likely due to the fusion of fibers, as observed in SEM. Additionally, the sPEEK-52 membrane shows an even higher surface area than the sPEEK-62 membrane with the smallest fiber diameter (11.2 ± 0.7 m^2/g, 17 wt % sPEEK-62, discussed in the previous section), despite having a slightly higher fiber diameter. This can be attributed to a more homogenous fiber formation with sPEEK-52 resulting in a higher surface area.

Additionally, porometry showed that the sPEEK-52 membranes had a narrower pore size distribution, indicating more homogeneity of the fibrous structure (Figure 10). The average pore size of sPEEK-72 is slightly smaller than that of sPEEK-62, although sPEEK-72 has a higher fiber diameter. This reduction in pore size is likely due to the higher swelling of sPEEK-72.

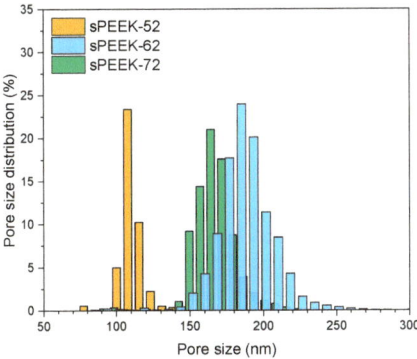

Figure 10. Pore size distribution of the sPEEK membrane adsorbers in the H^+ form with different sulfonation degrees (52%, 62%, and 72%) as measured with porometry.

The effect of sulfonation degree on the membrane performance is studied by measuring the dynamic lysozyme adsorption and desorption curves. The results show that the sulfonation degree, i.e., the number of functional groups per unit mass, did not directly influence the lysozyme adsorption capacity in mg/g, as sPEEK-52 with the lowest functional-group density exhibits the highest adsorption capacity (Table 3). However, when considering the adsorption capacity in mg/m^2, which normalizes for surface area, the adsorption capacity increased with increasing sulfonation degree.

Table 3. Dynamic lysozyme adsorption capacity at 10% breakthrough and recovery of sPEEK membrane adsorbers with different sulfonation degrees (52%, 62%, and 72%).

	sPEEK-52	sPEEK-62	sPEEK-72
BET surface area (m^2)	14.9 ± 2.9	12.3 ± 2.1	9.5 ± 2.4
Adsorption capacity (mg/g)	65.7 ± 5.4	60.1 ± 6.6	63.9 ± 3.4
Adsorption capacity (mg/m^2)	4.4 ± 0.9	4.9 ± 1.0	6.7 ± 1.7
Recovery (%)	35 ± 5%	35 ± 6%	43 ± 7%

The literature has shown that lysozyme with a net charge of +8 at neutral pH (17 positively charged residues and 9 negatively charged residues) binds with four positive key residues on the negatively charged surface of the membrane adsorber [44,46]. The orientation of lysozyme on the membrane surface depends on the ligand type, ionic strength, and surface interactions, and can be side-on (with its long axis parallel to the surface) or, more efficiently, end-on [44,46]. Depending on the orientation, the lysozyme monolayer surface coverage varies from 2–3 mg/m^2 [57]. The adsorption capacity in mg/m^2 shows that all our membranes are completely covered by lysozyme, without being limited in the number of functional groups. In fact, the coverage seems to exceed that of a monolayer. This is due to the discrepancy in measurement conditions. The BET is measured in the dry state, while the adsorption is measured in the wet state. The swelling of the fibers creates more spaces for lysozyme adsorption and increases the surface area beyond what is measured in the dry state using BET. This also explains the high adsorption capacity in mg/m^2 of the sPEEK-72 membrane adsorber, as severe swelling of this sample is observed increasing the surface area in the wet state. It is worth noting that the standard deviation for this sample is high, which may be attributed to the swelling resulting in fiber fusion which can vary between samples. Overall, the adsorption data, combined with the surface area, suggest that the swelling, which is linked to the sulfonation degree, is the main factor that increases the adsorption capacity.

However, this positive effect of swelling is not unlimited; with an increase in sulfonation degree, fiber swelling can increase the flow resistance. Pump stalling has been observed when the flow rates exceeded 1 mL/min, which limits the throughput and reduces the efficiency of the adsorption process. Moreover, excessive fiber swelling can create dead zones and promote channeling that induces a decrease in the dynamic adsorption capacity.

The lysozyme recovery is comparable for the sPEEK-52 and sPEEK-62 membranes, but it increases for the sPEEK-72 membrane. This can be attributed to the higher number of charged groups in the sPEEK-72 membrane, which also makes the surface more hydrophilic. As a result, the reversible electrostatic interactions are favored over the irreversible hydrophobic interactions, leading to easier desorption and a higher lysozyme recovery.

To put this work into perspective, the adsorption capacity of our membrane is similar to the adsorption capacity of Sartobind® S (52 mg/g), a commercial membrane adsorber with grafted polymers containing sulfonic acid groups [58]. However, in the literature, membrane adsorbers that surpass the adsorption capacity of our sPEEK are reported [59,60]. The drawback of these membrane adsorbers is that they require functionalization through multistep synthesis, whereas our sPEEK membranes can be directly used after electrospinning, making them easy to scale-up. Electrospinning offers the opportunity to tailor the fiber diameter; however, only small changes in surface area and, therefore, adsorption capacity are observed. To enhance the adsorption capacity, co-electrospinning of functionalized adsorptive/affinity particles can be explored, which offer additional selective binding sites [17,23,24,61]. Another approach is grafting polymer brushes, which enable multilayer stacking of proteins, leading to high effective surface areas [62–65]. Increasing the number of functional groups is also a strategy for enhancing adsorption capacity, as done in this study by increasing the number of functional groups. However, an increased degree of sulfonation did not result in an increased lysozyme adsorption capacity. This is because lysozyme covers many more charged groups than itself needs to bind on the membrane adsorber; therefore, in this specific case of lysozyme binding, there is no advantage in increasing the charge density of the membrane adsorber. Conversely, smaller target molecules that do not cover the excess charge and that do not experience steric hindrance can benefit from a higher sulfonation degree. From this point of view, testing sPEEK membrane adsorbers has provided valuable insights into the relationship between surface area, swelling, and adsorption capacity.

4. Conclusions

In this work, a wire-electrospun sPEEK membrane adsorber was developed without the need for any additional functionalization step. This sPEEK-based membrane adsorber has a dynamic lysozyme adsorption capacity (at 10% breakthrough) of 59.3 mg/g. This is comparable to the static adsorption capacity at the same concentration, demonstrating full utilization of the lysozyme adsorption capacity and indicating an almost ideal plug flow in the membrane stack. The dynamic adsorption capacity is independent of the flow rate (varying from 0.1 to 1 mL/min), indicating that convective mass transport is dominant, and adsorption is not limited by diffusion. Cycling the dynamic adsorption capacity experiment suggests that lysozyme binds by two modes of adsorption, namely reversible electrostatic interactions and irreversible hydrophobic interactions. Furthermore, this experiment demonstrates that the sPEEK membrane adsorbers are reusable and capable of operating at a constant efficiency after the third cycle. The versatility of electrospinning was explored by creating nanofibrous mats with variable diameters, ranging from 90 ± 18 to 166 ± 18 nm. The dynamic adsorption capacity and surface area in this range of fiber diameters were minimally affected by variations in fiber diameter. Therefore, electrospinning offers membrane adsorbers with consistent performance, even when the process yields variations in fiber diameter. Additionally, the effect of the functional-group density on the binding capacity was studied by creating membrane adsorbers from sPEEK with different sulfonation degrees (52%, 62%, and 72%). In all presented cases, at least a monolayer coverage was obtained. This suggests that there is already an abundant number of functional groups available within the area occupied by a lysozyme molecule, and, therefore, increasing the functional-group density does not enhance the adsorption capacity. In brief, our study has demonstrated the successful development of a membrane absorber that can be immediately used for binding positively charged molecules. However, the potential for performance tuning through adjustments to fiber diameter or functional-group density is limited.

Author Contributions: Conceptualization, N.J., A.S., K.N. and Z.B.; methodology, N.J. and W.W.; validation, N.J. and W.W.; formal analysis, N.J. and W.W.; investigation, N.J. and W.W.; resources, A.S. and K.N.; data curation, N.J. and W.W.; writing—original draft preparation, N.J.; writing—review and editing, N.J., A.S., K.N. and Z.B.; visualization, N.J.; supervision, A.S., K.N. and Z.B.; project administration, A.S. and K.N.; funding acquisition, A.S. and K.N. All authors have read and agreed to the published version of the manuscript.

Funding: This research received funding from the Netherlands Organisation for Scientific Research (NWO) in the framework of the "Sustainable water technology Partnership Programme NWO-Wetsus on the Impact of Water Research on Energy, Industry, Health, Economy and Environment 2016, project number ALWET.2016.001. This work was performed in the cooperation framework of Wetsus, European Centre of Excellence for Sustainable Water Technology. Wetsus is co-funded by the Dutch Ministry of Economic Affairs and Climate Policy, the European Union, the Province of Fryslân and the Northern Netherlands Provinces. The authors like to thank the participants of the research theme "Desalination & Concentrates" for their financial support.

Institutional Review Board Statement: Not applicable.

Data Availability Statement: Data are contained within the article.

Acknowledgments: The authors like to thank Renate Evers for her assistance and for making the SEM images.

Conflicts of Interest: The authors declare no conflict of interest. Additionally, the funders had no role in the design of the study; in the collection, analyses, or interpretation of data; in the writing of the manuscript.

References

1. Díaz, J.S.; Settele, E.S.; Brondízio, H.T.; Ngo, M.; Guèze, J.; Agard, A.; Arneth, P.; Balvanera, K.A.; Brauman, S.H.M.; Butchart, K.M.A.; et al. *IPBES, Summary for Policymakers of the Global Assessment Report on Biodiversity and Ecosystem Services of the Intergovernmental Science-Policy Platform on Biodiversity and Ecosystem Services*; IPBES: Bonn, Germany, 2019; ISBN 9783947851133.
2. Shannon, M.A.; Bohn, P.W.; Elimelech, M.; Georgiadis, J.G.; Mariñas, B.J.; Mayes, A.M. Science and Technology for Water Purification in the Coming Decades. *Nature* **2008**, *452*, 301–310. [CrossRef]
3. Kehrein, P.; Van Loosdrecht, M.; Osseweijer, P.; Garfí, M.; Dewulf, J.; Posada, J. A Critical Review of Resource Recovery from Municipal Wastewater Treatment Plants-Market Supply Potentials, Technologies and Bottlenecks. *Environ. Sci. Water Res. Technol.* **2020**, *6*, 877–910. [CrossRef]
4. Singh, R.P.; Singh, P. *Advances in Biological Treatment of Industrial Waste Water and Their Recylcing for a Sustainable Future*; Springer: Singapore, 2019; ISBN 9789811314674.
5. Rosemarin, A.; Macura, B.; Carolus, J.; Barquet, K.; Ek, F.; Järnberg, L.; Lorick, D.; Johannesdottir, S.; Pedersen, S.M.; Koskiaho, J.; et al. Circular Nutrient Solutions for Agriculture and Wastewater—A Review of Technologies and Practices. *Curr. Opin. Environ. Sustain.* **2020**, *45*, 78–91. [CrossRef]
6. Yadav, G.; Mishra, A.; Ghosh, P.; Sindhu, R.; Vinayak, V.; Pugazhendhi, A. Technical, Economic and Environmental Feasibility of Resource Recovery Technologies from Wastewater. *Sci. Total Environ.* **2021**, *796*, 149022. [CrossRef] [PubMed]
7. Udugama, I.A.; Petersen, L.A.H.; Falco, F.C.; Junicke, H.; Mitic, A.; Alsina, X.F.; Mansouri, S.S.; Gernaey, K.V. Resource Recovery from Waste Streams in a Water-Energy-Food Nexus Perspective: Toward More Sustainable Food Processing. *Food Bioprod. Process.* **2020**, *119*, 133–147. [CrossRef]
8. Avramescu, M.E.; Zandrie Borneman, M.W. Membrane Chromatography. In *Handbook of Membrane Separations*; Routledge Handbooks Online: London, UK, 2008; pp. 25–63.
9. Yang, X.; Merenda, A.; AL-Attabi, R.; Dumée, L.F.; Zhang, X.; Thang, S.H.; Pham, H.; Kong, L. Towards next Generation High Throughput Ion Exchange Membranes for Downstream Bioprocessing: A Review. *J. Memb. Sci.* **2022**, *647*, 120325. [CrossRef]
10. Mattiasson, B. *Expanded Bed Chromatography*; Kluwer Academic Publishers: Amsterdam, The Netherlands, 1999; Volume 8, pp. 1–271. [CrossRef]
11. McDonald, P.D.; Bidlingmeyer, B.A. Strategies for Successful Preparative Liquid Chromatography. *J. Chromatogr. Libr.* **1987**, *38*, 1–103. [CrossRef]
12. Ghosh, R. Protein Separation Using Membrane Chromatography: Opportunities and Challenges. *J. Chromatogr. A* **2002**, *952*, 13–27. [CrossRef]
13. Unger, K.K.; Lamotte, S.; Machtejevas, E. *Column Technology in Liquid Chromatography*; Elsevier: Amsterdam, The Netherlands, 2013; ISBN 9780124158078.
14. Liapis, A.I.; McCoy, M.A. Theory of Perfusion Chromatography. *J. Chromatogr. A* **1992**, *599*, 87–104. [CrossRef]
15. Mattiasson, B.; Nandakumar, M.P. Physicochemical Basis of Expanded-Bed Adsorption for Protein Purification. *Sep. Sci. Technol.* **2000**, *2*, 417–430. [CrossRef]
16. Orr, V.; Zhong, L.; Moo-Young, M.; Chou, C.P. Recent Advances in Bioprocessing Application of Membrane Chromatography. *Biotechnol. Adv.* **2013**, *31*, 450–465. [CrossRef] [PubMed]
17. Hao, S.; Jia, Z.; Wen, J.; Li, S.; Peng, W.; Huang, R.; Xu, X. Progress in Adsorptive Membranes for Separation—A Review. *Sep. Purif. Technol.* **2021**, *255*, 117772. [CrossRef]
18. Boi, C.; Malavasi, A.; Carbonell, R.G.; Gilleskie, G. A Direct Comparison between Membrane Adsorber and Packed Column Chromatography Performance. *J. Chromatogr. A* **2020**, *1612*, 460629. [CrossRef] [PubMed]
19. Fu, Q.; Duan, C.; Yan, Z.; Si, Y.; Liu, L.; Yu, J.; Ding, B. Electrospun Nanofibrous Composite Materials: A Versatile Platform for High Efficiency Protein Adsorption and Separation. *Compos. Commun.* **2018**, *8*, 92–100. [CrossRef]
20. Liao, Y.; Loh, C.H.; Tian, M.; Wang, R.; Fane, A.G. Progress in Electrospun Polymeric Nanofibrous Membranes for Water Treatment: Fabrication, Modification and Applications. *Prog. Polym. Sci.* **2018**, *77*, 69–94. [CrossRef]
21. Yin, J.Y.; Boaretti, C.; Lorenzetti, A.; Martucci, A.; Roso, M.; Modesti, M. Effects of Solvent and Electrospinning Parameters on the Morphology and Piezoelectric Properties of PVDF Nanofibrous Membrane. *Nanomaterials* **2022**, *12*, 962. [CrossRef]

22. Menkhaus, T.J.; Fong, H. *Electrospun Nanofibers for Protein Adsorption*; Elsevier: Amsterdam, The Netherlands, 2018; ISBN 9780323512701.
23. HMTShirazi, R.; Mohammadi, T.; Asadi, A.A.; Tofighy, M.A. Electrospun Nanofiber Affinity Membranes for Water Treatment Applications: A Review. *J. Water Process Eng.* **2022**, *47*, 102795. [CrossRef]
24. Nayl, A.A.; Abd-Elhamid, A.I.; Awwad, N.S.; Abdelgawad, M.A.; Wu, J.; Mo, X.; Gomha, S.M.; Aly, A.A.; Bräse, S. Review of the Recent Advances in Electrospun Nanofibers Applications in Water Purification. *Polymers* **2022**, *14*, 1594. [CrossRef]
25. Chen, H.; Huang, M.; Liu, Y.; Meng, L.; Ma, M. Functionalized Electrospun Nanofiber Membranes for Water Treatment: A Review. *Sci. Total Environ.* **2020**, *739*, 139944. [CrossRef] [PubMed]
26. Liu, B.; Wei, C.; Ng, I.; Loke, P.; Lin, K.; Chang, Y. Effective Purification of Lysozyme from Chicken Egg White by Tris(Hydroxymethyl Aminomethane Affinity Nanofiber Membrane. *Food Chem.* **2020**, *327*, 127038. [CrossRef]
27. Thi, D.; Huong, M.; Liu, B.; Siong, W.; Loke, P. Highly Efficient Dye Removal and Lysozyme Purification Using Strong and Weak Cation-Exchange Nanofiber Membranes. *Int. J. Biol. Macromol.* **2020**, *165*, 1410–1421. [CrossRef]
28. Zhang, L.; Menkhaus, T.J.; Fong, H. Fabrication and Bioseparation Studies of Adsorptive Membranes/Felts Made from Electrospun Cellulose Acetate Nanofibers. *J. Memb. Sci.* **2008**, *319*, 176–184. [CrossRef]
29. Ye, J.; Wang, X.; Chu, J.; Yao, D.; Zhang, Y.; Meng, J. Electrospun Poly(Styrene-Co-Maleic Anhydride) Nanofibrous Membrane: A Versatile Platform for Mixed Mode Membrane Adsorbers. *Appl. Surf. Sci.* **2019**, *484*, 62–71. [CrossRef]
30. Ma, Z.; Lan, Z.; Matsuura, T.; Ramakrishna, S. Electrospun Polyethersulfone Affinity Membrane: Membrane Preparation and Performance Evaluation. *J. Chromatogr. B Anal. Technol. Biomed. Life Sci.* **2009**, *877*, 3686–3694. [CrossRef] [PubMed]
31. Schneiderman, S.; Zhang, L.; Fong, H.; Menkhaus, T.J. Surface-Functionalized Electrospun Carbon Nanofiber Mats as an Innovative Type of Protein Adsorption/Purification Medium with High Capacity and High Throughput. *J. Chromatogr. A* **2011**, *1218*, 8989–8995. [CrossRef] [PubMed]
32. Ng, I.S.; Song, C.P.; Ooi, C.W.; Tey, B.T.; Lee, Y.H.; Chang, Y.K. Purification of Lysozyme from Chicken Egg White Using Nanofiber Membrane Immobilized with Reactive Orange 4 Dye. *Int. J. Biol. Macromol.* **2019**, *134*, 458–468. [CrossRef]
33. Yang, X.; Hsia, T.; Merenda, A.; AL-Attabi, R.; Dumee, L.F.; Thang, S.H.; Kong, L. Constructing Novel Nanofibrous Polyacrylonitrile (PAN)-Based Anion Exchange Membrane Adsorber for Protein Separation. *Sep. Purif. Technol.* **2022**, *285*, 120364. [CrossRef]
34. Wetter, L.R.; Deutsch, H.F. Immunological Studies on Egg White Proteins. *J. Biol. Chem.* **1951**, *192*, 237–242. [CrossRef]
35. Guélat, B.; Ströhlein, G.; Lattuada, M.; Morbidelli, M. Electrostatic Model for Protein Adsorption in Ion-Exchange Chromatography and Application to Monoclonal Antibodies, Lysozyme and Chymotrypsinogen A. *J. Chromatogr. A* **2010**, *1217*, 5610–5621. [CrossRef]
36. Abu-Thabit, N.Y. Near-Infrared PH Sensor Based on a SPEEK–Polyaniline Polyelectrolyte Complex Membrane. *Proceedings* **2019**, *3*, 11. [CrossRef]
37. Abu-Thabit, N.Y.; Ali, S.A.; Zaidi, S.M.J.; Mezghani, K. Novel Sulfonated Poly(Ether Ether Ketone)/Phosphonated Polysulfone Polymer Blends for Proton Conducting Membranes. *J. Mater. Res.* **2012**, *27*, 1958–1968. [CrossRef]
38. Nawaz, N.; Wen, S.; Wang, F.; Nawaz, S.; Raza, J.; Iftikhar, M.; Usman, M. Lysozyme and Its Application as Antibacterial Agent in Food Industry. *Molecules* **2022**, *27*, 6305. [CrossRef] [PubMed]
39. Zhu, G.; Zhao, L.Y.; Zhu, L.T.; Deng, X.Y.; Chen, W.L. Effect of Experimental Parameters on Nanofiber Diameter from Electrospinning with Wire Electrodes. *IOP Conf. Ser. Mater. Sci. Eng.* **2017**, *230*, 012043. [CrossRef]
40. Boaretti, C.; Roso, M.; Lorenzetti, A.; Modesti, M. Synthesis and Process Optimization of Electrospun PEEK-Sulfonated Nanofibers by Response Surface Methodology. *Materials* **2015**, *8*, 4096–4117. [CrossRef] [PubMed]
41. Park, J.W.; Wycisk, R.; Pintauro, P.N. Nafion/PVDF Nanofiber Composite Membranes for Regenerative Hydrogen/Bromine Fuel Cells. *J. Memb. Sci.* **2015**, *490*, 103–112. [CrossRef]
42. Blake, C.; Koenig, D.; Mair, G.; North, A.; Phillips, D.; Sarma, V. The Three-Dimensional Structure of Hen Eggwhite Lysozyme. *Nature* **1965**, *206*, 757–761. [CrossRef]
43. Dismer, F.; Petzold, M.; Hubbuch, J. Effects of Ionic Strength and Mobile Phase PH on the Binding Orientation of Lysozyme on Different Ion-Exchange Adsorbents. *J. Chromatogr. A* **2008**, *1194*, 11–21. [CrossRef]
44. Dismer, F.; Hubbuch, J. A Novel Approach to Characterize the Binding Orientation of Lysozyme on Ion-Exchange Resins. *J. Chromatogr. A* **2007**, *1149*, 312–320. [CrossRef]
45. Yu, G.; Liu, J.; Zhou, J. Mesoscopic Coarse-Grained Simulations of Hydrophobic Charge Induction Chromotography (HCIC) for Protein Purification. *AIChE J.* **2015**, *61*, 2035–2047. [CrossRef]
46. Yu, G.; Liu, J.; Zhou, J. Mesoscopic Coarse-Grained Simulations of Lysozyme Adsorption. *J. Phys. Chem. B* **2014**, *118*, 4451–4460. [CrossRef]
47. Xie, Y.; Zhou, J.; Jiang, S. Parallel Tempering Monte Carlo Simulations of Lysozyme Orientation on Charged Surfaces. *J. Chem. Phys.* **2010**, *132*, 065101. [CrossRef] [PubMed]
48. Pai, C.; Boyce, M.C.; Rutledge, G.C. Morphology of Porous and Wrinkled Fibers of Polystyrene Electrospun from Dimethylformamide. *Macromolecules* **2009**, *42*, 2102–2114. [CrossRef]
49. Wang, L.; Pai, C.L.; Boyce, M.C.; Rutledge, G.C. Wrinkled Surface Topographies of Electrospun Polymer Fibers. *Appl. Phys. Lett.* **2009**, *94*, 6–9. [CrossRef]

50. Tan, Y.; Hu, B.; Song, J.; Chu, Z.; Wu, W. *Bioinspired Multiscale Wrinkling Patterns on Curved Substrates: An Overview*; Springer: Singapore, 2020; Volume 12, ISBN 0123456789.
51. Bagherzadeh, R.; Najar, S.S.; Latifi, M.; Tehran, M.A.; Kong, L. A Theoretical Analysis and Prediction of Pore Size and Pore Size Distribution in Electrospun Multilayer Nanofibrous Materials. *J. Biomed. Mater. Res.-Part A* **2013**, *101*, 2107–2117. [CrossRef] [PubMed]
52. Krifa, M.; Yuan, W. Morphology and Pore Size Distribution of Electrospun and Centrifugal Forcespun Nylon 6 Nanofiber Membranes. *Text. Res. J.* **2016**, *86*, 1294–1306. [CrossRef]
53. Li, D.; Frey, M.W.; Joo, Y.L. Characterization of Nanofibrous Membranes with Capillary Flow Porometry. *J. Membr. Sci.* **2006**, *286*, 104–114. [CrossRef]
54. Eichhorn, S.J.; Sampson, W.W. Statistical Geometry of Pores and Statistics of Porous Nanofibrous Assemblies. *J. R. Soc. Interface* **2005**, *2*, 309–318. [CrossRef]
55. Zaidi, S.M.J.; Mikhailenko, S.D.; Robertson, G.P.; Guiver, M.D.; Kaliaguine, S. Proton Conducting Composite Membranes from Polyether Ether Ketone and Heteropolyacids for Fuel Cell Applications. *J. Memb. Sci.* **2000**, *173*, 17–34. [CrossRef]
56. Roelofs, K.S. Sulfonated Poly (Ether Ether Ketone) Based Membranes for Direct Ethanol Fuel Cells. Ph.D. Thesis, University of Stuttgart, Stuttgart, Germany, 2010; pp. 1–199.
57. Xu, K.; Ouberai, M.M.; Welland, M.E. A Comprehensive Study of Lysozyme Adsorption Using Dual Polarization Interferometry and Quartz Crystal Microbalance with Dissipation. *Biomaterials* **2013**, *34*, 1461–1470. [CrossRef]
58. Chiu, H.T.; Lin, J.M.; Cheng, T.H.; Chou, S.Y.; Huang, C.C. Direct Purification of Lysozyme from Chicken Egg White Using Weak Acidic Polyacrylonitrile Nanofiber-Based Membranes. *J. Appl. Polym. Sci.* **2012**, *125*, E616–E621. [CrossRef]
59. Rajesh, S.; Schneiderman, S.; Crandall, C.; Fong, H.; Menkhaus, T.J. Synthesis of Cellulose-Graft-Polypropionic Acid Nanofiber Cation-Exchange Membrane Adsorbers for High-Efficiency Separations. *ACS Appl. Mater. Interfaces* **2017**, *9*, 41055–41065. [CrossRef] [PubMed]
60. Menkhaus, T.J.; Varadaraju, H.; Zhang, L.; Schneiderman, S.; Bjustrom, S.; Liu, L.; Fong, H. Electrospun Nanofiber Membranes Surface Functionalized with 3-Dimensional Nanolayers as an Innovative Adsorption Medium with Ultra-High Capacity and Throughput. *Chem. Commun.* **2010**, *46*, 3720–3722. [CrossRef] [PubMed]
61. Sanaeepur, H.; Ebadi Amooghin, A.; Shirazi, M.M.A.; Pishnamazi, M.; Shirazian, S. Water Desalination and Ion Removal Using Mixed Matrix Electrospun Nanofibrous Membranes: A Critical Review. *Desalination* **2022**, *521*, 115350. [CrossRef]
62. Keating, J.J.; Imbrogno, J.; Belfort, G. Polymer Brushes for Membrane Separations: A Review. *ACS Appl. Mater. Interfaces* **2016**, *8*, 28383–28399. [CrossRef]
63. Dai, J.; Bao, Z.; Sun, L.; Hong, S.U.; Baker, G.L.; Bruening, M.L. High-Capacity Binding of Proteins by Poly (Acrylic Acid) Brushes and Their Derivatives. *Langmuir* **2006**, *22*, 4274–4281. [CrossRef] [PubMed]
64. Jain, P.; Sun, L.; Dai, J.; Baker, G.L.; Bruening, M.L. High-Capacity Purification of His-Tagged Proteins by Affinity Membranes Containing Functionalized Polymer Brushes. *Biomacromolecules* **2007**, *8*, 3102–3107. [CrossRef]
65. Janzen, R.; Unger, K.K.; Müller, W.; Hearn, M.T.W. Adsorption of Proteins on Porous and Non-Porous Poly(Ethyleneimine) and Tentacle-Type Anion Exchangersa. *J. Chromatogr. A* **1990**, *522*, 77–93. [CrossRef]

Disclaimer/Publisher's Note: The statements, opinions and data contained in all publications are solely those of the individual author(s) and contributor(s) and not of MDPI and/or the editor(s). MDPI and/or the editor(s) disclaim responsibility for any injury to people or property resulting from any ideas, methods, instructions or products referred to in the content.

Article

Bactericide Activity of Cellulose Acetate/Silver Nanoparticles Asymmetric Membranes: Surfaces and Porous Structures Role

Ana Sofia Figueiredo [1,2,3,*], Ana Maria Ferraria [4,5,6,*], Ana Maria Botelho do Rego [4,5,6], Silvia Monteiro [7], Ricardo Santos [7], Miguel Minhalma [1,2,3], María Guadalupe Sánchez-Loredo [8], Rosa Lina Tovar-Tovar [8] and Maria Norberta de Pinho [1,2,6]

1. CeFEMA-Center of Physics and Engineering of Advanced Materials, Instituto Superior Técnico, Universidade de Lisboa, 1049-001 Lisbon, Portugal
2. LaPMET-Associate Laboratory of Physics for Materials and Emergent Technologies, Instituto Superior Técnico, Universidade de Lisboa, 1049-001 Lisbon, Portugal
3. Instituto Superior de Engenharia de Lisboa, Instituto Politécnico de Lisboa, 1959-007 Lisbon, Portugal
4. BSIRG-iBB-Institute for Bioengineering and Biosciences, Universidade de Lisboa, 1049-001 Lisbon, Portugal
5. Associate Laboratory i4HB—Institute for Health and Bioeconomy at Instituto Superior Técnico, Universidade de Lisboa, 1049-001 Lisbon, Portugal
6. Chemical Engineering Department, Instituto Superior Técnico, Universidade de Lisboa, 1049-001 Lisbon, Portugal
7. Laboratório de Análises, Instituto Superior Técnico, Universidade de Lisboa, 1049-001 Lisbon, Portugal
8. Instituto de Metalurgia, Facultad de Ingeniería, Universidad Autónoma de San Luis Potosí, San Luis Potosí 78210, Mexico
* Correspondence: ana.figueiredo@isel.pt (A.S.F.); ana.ferraria@tecnico.ulisboa.pt (A.M.F.); Tel.: +351-218-317-000 (A.S.F.); +351-218-419-258 (A.M.F.)

Citation: Figueiredo, A.S.; Ferraria, A.M.; Botelho do Rego, A.M.; Monteiro, S.; Santos, R.; Minhalma, M.; Sánchez-Loredo, M.G.; Tovar-Tovar, R.L.; de Pinho, M.N. Bactericide Activity of Cellulose Acetate/Silver Nanoparticles Asymmetric Membranes: Surfaces and Porous Structures Role. *Membranes* **2023**, *13*, 4. https://doi.org/10.3390/membranes13010004

Academic Editor: Annarosa Gugliuzza

Received: 13 October 2022
Revised: 14 December 2022
Accepted: 15 December 2022
Published: 21 December 2022

Copyright: © 2022 by the authors. Licensee MDPI, Basel, Switzerland. This article is an open access article distributed under the terms and conditions of the Creative Commons Attribution (CC BY) license (https://creativecommons.org/licenses/by/4.0/).

Abstract: The antibacterial properties of cellulose acetate/silver nanoparticles (AgNP) ultrafiltration membranes were correlated with their integral asymmetric porous structures, emphasizing the distinct features of each side of the membranes, that is, the active and porous layers surfaces. Composite membranes were prepared from casting solutions incorporating polyvinylpyrrolidone-covered AgNP using the phase inversion technique. The variation of the ratio acetone/formamide and the AgNP content resulted in a wide range of asymmetric porous structures with different hydraulic permeabilities. Comprehensive studies assessing the antibacterial activity against *Escherichia coli* (cell death and growth inhibition of bacteria in water) were performed on both membrane surfaces and in *E. coli* suspensions. The results were correlated with the surface chemical composition assessed by XPS. The silver-free membranes presented a generalized growth of *E. coli*, which is in contrast with the inhibition patterns displayed by the membranes containing AgNP. For the surface bactericide test, the growth inhibition depends on the accessibility of *E. coli* to the silver present in the membrane; as the XPS results show, the more permeable membranes (CA30 and CA34 series) have higher silver signal detected by XPS, which is correlated with a higher growth inhibition. On the other hand, the inhibition action is independent of the membrane porous structure when the membrane is deeply immersed in an *E. coli* inoculated suspension, presenting almost complete growth inhibition.

Keywords: cellulose acetate/silver nanocomposite ultrafiltration membranes; antimicrobial properties; polyvinylpyrrolidone-coated silver nanoparticles; surface characterization

1. Introduction

Membrane pressure-driven processes are efficient, sustainable, and easy-to handle separation technologies even though membrane fouling still poses major drawbacks of flux reduction, modification of membrane selectivity, and reduction of membrane lifetime [1].

In particular, biofouling is originated by the deposition of microorganisms and extracellular polymeric substances (EPS) and causes high concern because microorganisms multiply over time and, even if the membrane can be regenerated, some cells remain and can multiply, re-forming a biofilm in a short time [2]. One of the most used solutions to overcome this limitation is the incorporation of nanoparticles in the polymeric structure of the membranes, which combines the properties of polymeric membranes with the characteristics of nanoparticles, intending to improve the mechanical properties of the membranes and enhance the permeation performance. Due to their known bactericidal potential, silver nanoparticles (AgNP) have been widely used to incorporate into membranes [3]. Qi et al. developed a one-pot method of synthesizing AgNP and immobilizing them by soaking a polysulfone ultrafiltration membrane in a mixture of silver nitrate, poly(ethylene glycol)methyl ether diol, and dopamine. The results indicated that membranes exhibited outstanding antibacterial properties with more than 90% of antibacterial efficiency against *Escherichia coli* and *Staphylococcus aureus* [4]. More recently, nanosilver stabilized with the polyhexamethylene biguanide hydrochloride (PHMB) was incorporated in situ onto the thin-film composite (TFC) NF90 membrane surface, and the results demonstrate that the system has a profound antibacterial effect against *Staphylococcus aureus* and *Escherichia coli* bacteria [5]. Peng et al. reported the deposition of silver nanoparticles (AgNP) on tunicate cellulose nanocrystals (TCNCs) by in situ hydrothermal reduction of silver nitrate, showing excellent antibacterial efficacy against *Staphylococcus aureus* and *Escherichia coli* [6].

In our earlier work, as an effort to overcome the biofilm phenomena, nanofiltration cellulose acetate/silver nanoparticle membranes with antimicrobial properties were developed [7].

Ultrafiltration (UF) cellulose acetate/silver nanoparticles (CA/Ag) membranes were prepared by phase inversion with polyvinylpyrrolidone-stabilized silver nanoparticles (AgNP) synthesized ex situ and incorporated in the casting solutions, with the solvent system having different acetone/formamide ratios [8]. The membranes casted with different acetone/formamide ratios displayed a range of asymmetric porous structures with tailored selective permeation properties. In the present work, the surfaces of these asymmetrical membranes were characterized by X-ray photoelectron spectroscopy (XPS), paying particular attention to the active layer to have a better understanding of the silver oxidation state and amount and its correlation with the antimicrobial effect against *Escherichia coli*. The interaction nanoparticles–bacteria and, therefore, the bactericide properties could be better explained by knowing the surface characteristics of the modified membranes. Therefore, it was aimed to shed light on the surface properties of the membranes using a surface driven technique such as XPS complemented with the analysis of the structural characteristics of the AgNP using X-ray diffraction.

Hypothesis:

The integral asymmetric porous structures of cellulose acetate membranes, some of them incorporating silver nanoparticles, are characterized by distinct features of the surfaces of the active and porous layers. The antibacterial activity against *Escherichia coli* in both silver-free and AgNP membranes can be correlated with the surface chemical composition, as well as with the silver accessibility, by detailed and systematic X-ray photoelectron spectroscopy (XPS) studies. Such analysis was never reported before.

2. Materials and Methods

2.1. Materials and Chemicals

Polyvinylpyrrolidone (PVP) (BDH Chemicals, Dubai, UAE, ~44,000 g/mol), sodium borohydride (Panreac, Barcelona, Spain, >96% purity), silver nitrate (Panreac, Barcelona, Spain, >99.8% purity), cellulose acetate (Sigma-Aldrich, Darmstadt, Germany, ~30,000 g/mol, 39.8 wt%. acetyl, corresponding to a degree of substitution of 2.5), formamide (Sigma-Aldrich, Darmstadt, Germany, ≥99.5% purity), acetone (Labchem, Zelienople, PA, USA, 99.9% purity), polyethylene glycol (PEG) (Merck-Schuchardt, Munich, Germany, −1000, 3000, 6000, 8000, 10,000 and 20,000 Da), dextran (Amersham Pharmacia Biotech AB, Staffanstorp, Sweden, 40,000 Da), nutrient broth (Becton, Dickinson and Company, Franklin Lakes, NJ, USA), yeast extract agar (YEA) (Biokar, Allonne, France), and *Escherichia coli* (*E. coli*) strain ATCC 700078 (WG5) were used.

2.2. Synthesis of Silver Nanoparticles

Silver nanoparticles were prepared and stabilized with PVP following a synthesis procedure adapted from Tashdjian et al. (2013) [9] and Figueiredo et al. (2015) [8].

2.3. Membranes Preparation

Cellulose acetate (CA) and cellulose acetate with AgNP (CA/Ag) flat sheet membranes were prepared by the phase inversion method [10,11].

The CA/Ag casting solutions were prepared with 0.1 and 0.4 wt% Ag, by the addition of the dispersion of AgNP in the CA casting solutions of different compositions (CA/acetone/formamide ratios used were 17/61/22, 17/53/30, and 17/49/34). The casting solutions were prepared at room temperature, according to the compositions and conditions presented in Table 1.

Table 1. Casting solutions compositions and film casting conditions of the CA membranes free of silver (CA22, CA30, and CA34), with 0.1wt% Ag (CA22Ag0.1, CA30Ag0.1, and CA34Ag0.1) and with 0.4wt% Ag (CA22Ag0.4, CA30Ag0.4, and CA34Ag0.4).

	Casting Solution (wt%)								
Membrane	CA22	CA22Ag0.1	CA22Ag0.4	CA30	CA30Ag0.1	CA30Ag0.4	CA34	CA34Ag0.1	CA34Ag0.4
Cellulose acetate	17.0	16.4	15.3	17.0	16.4	15.3	17.0	16.4	15.3
Formamide	22.0	21.2	19.8	30.0	29.0	27.0	34.0	32.8	30.6
Acetone	61.0	58.9	54.9	53.0	51.1	47.7	49.0	47.3	44.1
	AgNPs								
Dispersion	-	3.32	9.59	-	3.32	9.59	-	3.32	9.59
Silver	-	0.14	0.41	-	0.14	0.42	-	0.14	0.41
	Casting Conditions								
Temperature of solution (°C)					20–25				
Temperature of atmosphere (°C)					20–25				
Solvent evaporation time (min)					0.5				
Gelation medium					Water at temperature of 0–3 °C during 1–2 h				

The CA and CA/Ag (Table 1) casting solutions were prepared according to the procedure reported by Figueiredo et al. (2015) [8]. To prepare the CA/Ag casting solutions, the AgNP dispersions were previously introduced in the acetone. The solutions were cast on a clean glass plate using a casting knife, with the gate height fixed at 0.25 mm.

2.4. Permeation Experiments

Permeation experiments were performed for pure water to characterize the membranes in terms of hydraulic permeability and, for organic solutes to determine the molecular weight cut-off (MWCO) of the membranes. The permeation experiments were performed in flat plate units with two detachable parts separated by a porous plate (membrane support), with a membrane surface area of 13.2×10^{-4} m^2. Before the permeation experiments, the membranes were compacted for 2 h with deionized water at a transmembrane pressure of 3 bar.

2.4.1. Pure Water Permeation Experiments

The hydraulic permeability (Lp) is obtained by the slope of the straight line of pure water permeate fluxes (Jp) as a function of the transmembrane pressure (ΔP). The transmembrane pressures ranged from 1 to 3 bar with a flow rate of 180 L h^{-1}.

2.4.2. Molecular Weight Cut-Off Experiments

The MWCO parameter is defined by the molecular weight of a given macromolecule whose rejection is higher than 91% and is obtained using the rejection coefficients of organic solutes, such as polyethylene glycol (PEG), with different molecular weight (Merck Schuchardt, Munich, Germany —1000, 3000, 6000, 8000, 10,000, and 20,000 Da) and Dextran (Amersham Pharmacia Biotech AB, Staffanstorp, Sweden —40,000 Da).

The apparent rejection coefficients (f) are defined as f = ($C_f - C_p$)/C_f, where C_f and C_p are the organic solute concentrations in the bulk of the feed solution and of the permeate solution, respectively. To determine the MWCO, the curve of log(f/(1−f)) is plotted as a function of the molecular weight of the organic solutes used, and the interception of this curve with the horizontal line corresponding to a rejection of 91% gives the MWCO of the membrane.

The permeation experiments were conducted using aqueous solutions of the organic solutes at 600 ppm and with a flow rate of 180 L h^{-1} at 1 bar. The organic solute concentrations in the feed and permeate solutions were determined in terms of total organic carbon (TOC) content, using a Dohrmann Total Organic Carbon Analyzer Model DC-85A.

2.5. X-ray Diffraction Analysis

The presence of crystalline silver in the nanoparticles was analyzed using X-ray diffraction (XRD). The characterization was performed on a dried powder sample, which was obtained by precipitation of the nanoparticles, adding acetone to the aqueous dispersion and washing with water and isopropanol. XRD analysis was carried out with a Bruker D8 Advance Diffractometer (Cu Kα radiation, λ = 1.54060 Å; 40 kV, 35 mA), using a silicon single crystal ((911) orientation) as a sample holder to minimize scattering. The powder sample was analyzed in the range of 4 to 90° 2theta, with a step size of 0.02° and a counting time of 3.35 s. To estimate the average crystallite sizes and specific crystallite size anisotropy from diffraction peak broadening reflections (using the Scherrer equation), the Rietveld refinement (Le Bail method) was performed using the Bruker software TOPAS 4.2. For each Rietveld refinement, the instrumental correction was included as determined with a standard powder sample Al_2O_3 (from National Institute of Standards and Technology (NIST), as standard reference material, NIST 1976a). The diffraction pattern was analyzed qualitatively using the PDF-2 (2010) of ICDD databases.

2.6. X-ray Photoelectron Spectroscopy

Silver nanoparticle dispersions and membrane surfaces of both active and porous layers were characterized by XPS using a XSAM800 dual anode spectrometer from KRATOS. Operating conditions, data acquisition, and data treatment were described elsewhere [12]. Binding energies were corrected from the charge shift using the binding energy of aliphatic carbon as reference (285.0 eV). The following sensitivity factors were used for quantitative purposes: 0.25 for C 1s; 0.66 for O 1s; 0.42 for N 1s; and 4.05 for Ag 3d. Membranes and the AgNP dispersions were dried in a vacuum chamber before the XPS analysis. Unlike the AgNP used for XRD analysis, the particles were not purified, and the dispersion was analyzed as prepared.

2.7. Membranes Bactericide Properties

The bactericide properties of the CA and CA/Ag membranes were evaluated through tests performed under different experimental conditions, where the capability of the membranes to inhibit the generalized growth of bacteria in water, specifically of *E. coli*, was tested. The specific tests to evaluate the growth inhibition of *E. coli* in the presence of the

CA and CA/Ag membranes were conducted in three different ways designated by surface test, suspension test, and cell death test. The bactericide properties of the membranes were tested against *E. coli* strain ATCC 700078, also known as WG5. *E. coli* was grown in nutrient broth at 37 °C with orbital shaking (200 rpm) until the exponential growth phase was reached. Before incubation with the bacterial strain, each side of the membranes was sterilized for 30 min under ultraviolet (UV) radiation. The effectiveness of this sterilization was confirmed by incubating a sample of the sterilized membrane at 37 °C for 24 h in a Petri dish with YEA.

2.7.1. Surface Test

The surfaces of the active and porous layers were tested for their antibacterial properties. A sample of the membrane was placed in a Petri dish with YEA, and 100 µL of *E. coli* suspension (2.5×10^5 colony forming units (CFU)) was spread on the respective membrane surface. The plate was incubated at 37 °C for 24 h and the growth inhibition was evaluated.

2.7.2. Suspension Test

A sample of the membrane was placed in a flask with 2 mL of *E. coli* suspension (2.5×10^5 CFU) covering the membrane. The flask was capped and maintained at rest at room temperature for 24 h. On the next day, 10 mL of sterile water was added to the flask. The membrane was removed from the suspension and placed in a Petri dish with YEA. In parallel, the inoculated suspension was filtered through a Whatman membrane filter (0.2 µm), and this was placed in a Petri dish with YEA. Both were incubated at 37 °C for 24 h and the growth inhibition was evaluated.

2.7.3. Cell Death Test

A membrane sample was immersed in a solution containing 9 mL of sterilized water and 1 mL of the exponential-growth-phase culture of *E. coli* in a nutrient broth medium (8×10^7 CFU). The membrane was kept in the inoculated suspension at 37 °C with orbital shaking (200 rpm) for 18 h. To validate the bactericide effect of the membranes, samples of the suspension were collected at different contact times (143 min, 190 min, and 975 min), with the final sample taken at 18 h (1093 min) of incubation. At each time, 100 µL of the direct sample and several dilutions (10^{-2}, 10^{-4}, 10^{-5}, and 10^{-6}) were inoculated on YEA. The plates were incubated at 37 °C for 24 h, and the number of colonies was counted.

2.7.4. Growth Inhibition of Bacteria in the Water

To test the membrane's capability to inhibit the growth of bacteria in water, a membrane was immersed in water for 80 days. A sample of 10 mL of the water was filtered through a membrane filter of 0.45 µm (Whatman), the membrane filter was placed in a Petri dish with YEA and was incubated at 37 °C for 24 h. After incubation, the number of colonies was counted.

3. Results and Discussion

3.1. Membrane Synthesis and Permeation Experiments

Different asymmetric structures of cellulose acetate/AgNP (CA/Ag) membranes were synthesized with 0.1wt% Ag (CA22Ag0.1, CA30Ag0.1, and CA34Ag0.1) and 0.4wt% Ag (CA22A0.4, CA30Ag0.4, and CA34Ag0.4). Cellulose acetate (CA) membranes (CA22, CA30, and CA34) were also prepared to be used as references.

The pure water fluxes and the characteristic parameters of hydraulic permeability and MWCO of the synthesized membranes are listed in Table 2.

Table 2. CA and CA/Ag pure water fluxes (Jp), hydraulic permeabilities (Lp), and MWCO (membrane surface area: 13.2×10^{-4} m^2).

Membrane	Jp (kg m^{-2} h^{-1})			Lp (kg m^{-2} h^{-1} bar^{-1})	MWCO (kDa)
	1 bar	2 bar	3 bar		
CA22	2.62	6.62	11.03	3.50	4.17
CA22Ag0.1	5.31	13.90	21.87	7.05	6.86
CA22Ag0.4	9.75	22.40	33.89	11.16	15.35
CA30	24.43	62.73	99.59	32.05	8.32
CA30Ag0.1	55.06	136.59	202.57	66.85	17.58
CA30Ag0.4	58.28	123.51	176.93	59.72	26.52
CA34	68.55	176.68	236.81	80.88	31.43
CA34Ag0.1	76.44	187.99	243.42	84.48	41.05
CA34Ag0.4	62.60	123.17	172.80	59.10	31.96

The addition of 0.1% wt Ag introduced an increase of hydraulic permeability values in all the prepared membranes (CA22Ag0.1, CA30Ag0.1, and CA34Ag0.1), and the incorporation of 0.4% wt Ag only enhanced the permeability of the CA22 and CA30. On the other hand, the two membranes with 0.4% wt Ag, CA30Ag0.4 and CA34Ag0.4, presented lower hydraulic permeability values when compared with the CA30Ag0.1 and CA34Ag0.1 membranes. It was observed that for the tighter membranes (CA22), the hydraulic permeability was enhanced with the incorporation of silver nanoparticles in both concentrations; for the CA30 membranes, an increase of permeability was observed with the addition of 0.1% wt Ag, but the incorporation of 0.4% wt Ag induced a decrease in the hydraulic permeability value when compared with the CA30Ag0.1. The hydraulic permeability value for the more porous structure (CA34) slightly increased (5%) with the incorporation of 0.1% wt Ag and presented a strong decrease after the addition of the higher content of AgNPs (CA34Ag0.4), the results obtained being even lower than for the CA34 (with no AgNP). Taurozzi et al. (2008) [13] observed only an increase in permeability after silver nanoparticles incorporation for the denser polysulfone membrane. A higher silver nanoparticles content in casting solutions enhanced the hydraulic permeability of the denser CA22 membranes; however, for the CA30 and CA34 membranes, the increase of silver content from 0.1 to 0.4%Ag led to a decrease of the hydraulic permeability, which may be due to the pore-blocking by the AgNP [14] as a result of aggregation due to a lower compatibility between the silver dispersion and the casting solutions.

3.2. XRD Results

To obtain detailed information about the composition, crystalline structure, and mean size of nanocrystalline silver particles in the nanoparticle dispersion used to prepare the modified membranes, a quantitative phase analysis of the diffraction pattern of the precipitated particles (Figure 1) was performed using the computer program TOPAS 4.2. Precipitation of the particles was carried out by taking 1 mL of the nanoparticle dispersion, adding 1 mL of acetone, and leaving the mixture for 1 min in an ultrasound bath. After centrifugation, the supernatant was removed, and the nanoparticles were mixed thoroughly in 1 mL of water with the aid of the ultrasonicator to ensure dispersion. After that, 1 mL of acetone was added, and the entire washing procedure was repeated 6 times. The particles were dried under vacuum for several days.

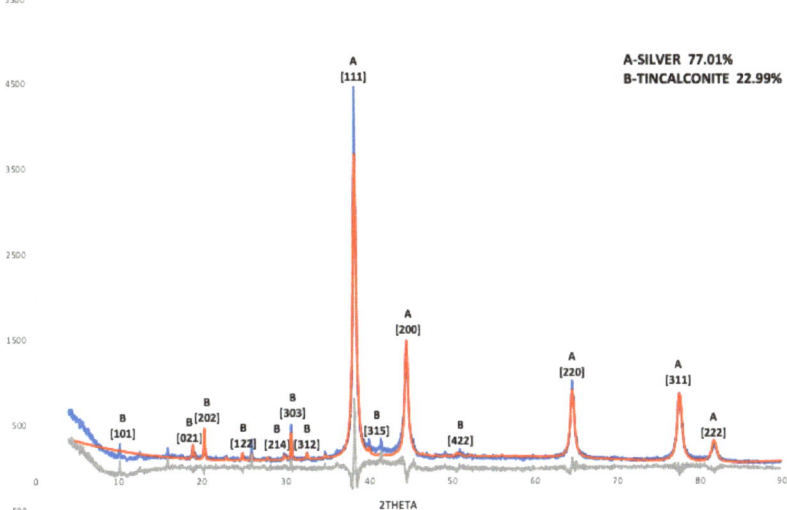

Figure 1. XRD measurements of powders obtained by precipitation from the aqueous dispersion (blue) and Rietveld refinement (red). Difference (Measured pattern−Rietveld refinement) (grey).

For the refinement, a crystal structure of silver according to PDF 01-087-0717 Silver 3C was applied. The atomic positions were determined by Rietveld refinement, obtaining a unit cell value of a = 4.08681(37) Å assuming Fm-3m symmetry and an average silver particle size of 22.6 nm (Table 3). For the fit of the sample, a Rwp value of 20.82% was obtained, Rp: 15.82, and GOF: 2.89. In addition to silver (77%), the sample contained 23% of tincalconite (PDF 00-007-0277, $Na_2B_4O_7 \cdot 5H_2O$ syn), resulting from precipitation during the washing procedure of the oxidation products of the borohydride ions. As demonstrated from XPS characterization, boron/sodium compounds were not present in the modified membranes, showing that the soluble products of silver reduction do not affect the membrane composition during preparation.

Table 3. Rietveld refinement of silver nanoparticles.

Phase Name:			Silver		Scale:	0.0007801(63)
R-Bragg:			2.028		Cell Mass:	431.470
Space Group:			Fm-3m		Cell Volume (Å^3):	68.258(19)
Wt%−Rietveld:					77.0(12)	
Crystallite Size						
Cryst. Size Lorentzian (nm):					22.26(28)	
Cryst. Linear Absorp. Coeff. (1/cm):					2247.09(61)	
Crystal Density (g/cm^3):					10.4965(29)	
Preferred Orientation (Dir 1: 1 1 1):					0.7610(77)	
Lattice parameters						
a (≈):					4.08681(37)	
Site	NP	X	Y	Z	Atom	Occ.	Beq
Ag	4	0.00000	0.00000	0.00000	Ag	1	0

3.3. XPS Results
3.3.1. Silver Nanoparticles Dispersion

XPS detailed regions of silver nanoparticles dispersion (AgNP) and qualitative XPS analysis are shown in Figures 2 and 3.

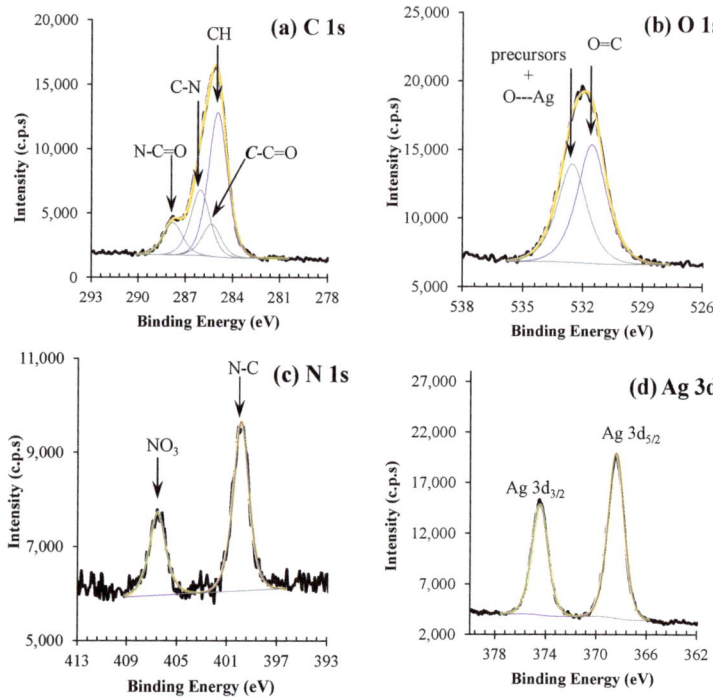

Figure 2. XPS regions of (**a**–**d**) for silver nanoparticle dispersions (AgNP).

(a) Binding Energy (eV)				
C 1s	O 1s	N 1s	Ag 3d$_{5/2}$	Ag 3d$_{3/2}$
285.0 (β)	531.6	399.9	368.4	374.4
285.4 (α)	532.5	406.6		
286.1 (σ)				
287.8 (γ)				

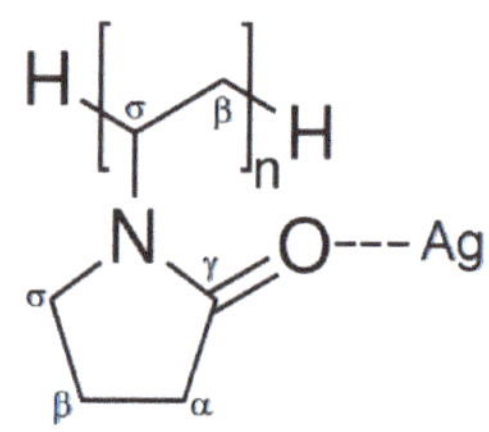

Figure 3. (**a**) Binding energy values for the PVP-stabilizing silver nanoparticle dispersion (AgNP); (**b**) Chemical structure of PVP and possible coordination with silver.

Four peaks centered at 285.0, 285.4, 286.1, and 287.8 eV, typical of PVP, were fitted into the C 1s envelope. The detailed assignment is presented in Figure 3.

The O 1s region (Figure 2b) was fitted with two peaks: one centered at 531.6 eV, which is attributed to oxygen double bound to carbon, as the one found in PVP; and another one, centered at 532.5 eV, assigned to oxygen from nitrate groups, which still remains in the colloidal system, to the interaction between oxygen and silver nanoparticles, and to some retained water [15]. The N 1s spectrum (Figure 2c) presents two peaks located at 399.9 and 406.6 eV, attributed to the N atom in PVP and NO$_3^-$ (from precursor silver nitrate), respectively. The Ag 3d region of silver nanoparticles (Figure 2d) is a simple doublet with an energy separation of 6.0 eV. The Ag $3d_{5/2}$ component is centered at 368.4 eV and, regarding the dispersion of the values reported in the literature, can be assigned to any oxidation state. However, the oxidation state can be identified from the Auger parameter, which is the sum of the experimental binding energy of the photoelectron peak Ag $3d_{5/2}$ and the experimental kinetic energy of the two most intense peaks of the Ag MNN Auger structure [16]. In this case, the computed value of 718.2 eV suggests that silver is oxidized or that the silver nanoparticles are covered with a layer of oxidized silver [16].

3.3.2. CA and CA/Ag Membranes

The results for CA and CA/Ag membranes XPS C 1s regions are displayed in Figures 4 and 5, respectively. The C 1s regions, for all the membranes, were fitted with three peaks centered at 285.0 eV (binding energy used to correct the charge shifts), attributed to aliphatic carbons, and at 286.7 ± 0.1 eV and 289.0 ± 0.2 eV, attributed to carbons from CA: C-O and carbons of ester groups, respectively [17]. Another peak at ~288 eV, assignable to carbon in the cellulose ring bonded to two oxygen atoms, exists in cellulose-derived polymers. However, given its low intensity (1/5 of C-O peak intensity), there was no need to add it in the fitting.

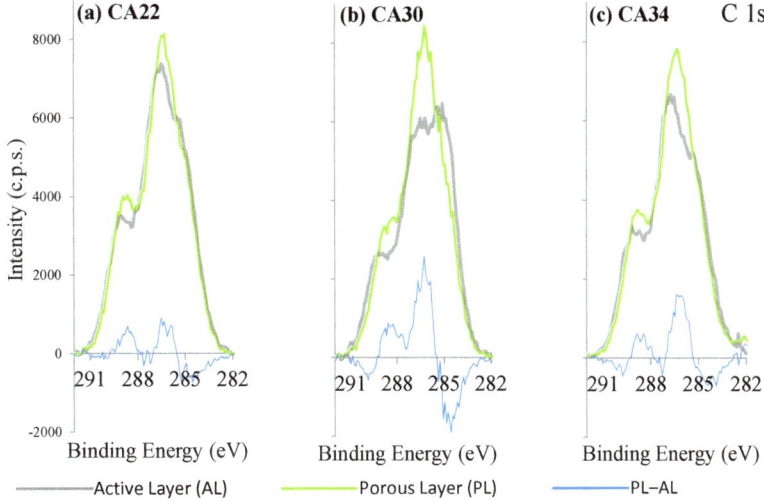

Figure 4. XPS C 1s regions of active layers (AL) and porous layers (PL) of CA membranes. The blue line is the difference between the PL and AL spectra.

The C 1s regions of the three silver-free membranes (Figure 4a–c), show some differences between the dense layer, also designated as the active layer (AL), and the porous (PL) layer. The membrane with larger difference between AL and PL is the CA30, and the one with smaller differences is the CA22, which is also the one with the smallest hydraulic permeability. The relative amount of aliphatic carbon atoms for each membrane is larger in AL than in the PL spectra. This fact may be associated with the presence of acetone residues in AL, which are expected to be larger in the denser layer (AL).

Little difference between AL and PL was observed for the silver-free CA22 membrane and for the CA22 silver-containing membranes (CA22Ag0.1 and CA22Ag0.4, Figure 5i–l), having very similar C 1s regions for AL and PL surfaces. In the other membranes, with higher hydraulic permeabilities, different C 1s regions are observed (Figure 5a–h): the relative intensity of the peak attributed to aliphatic carbon atoms (285 eV) is larger in AL than in PL, as observed for silver-free membranes.

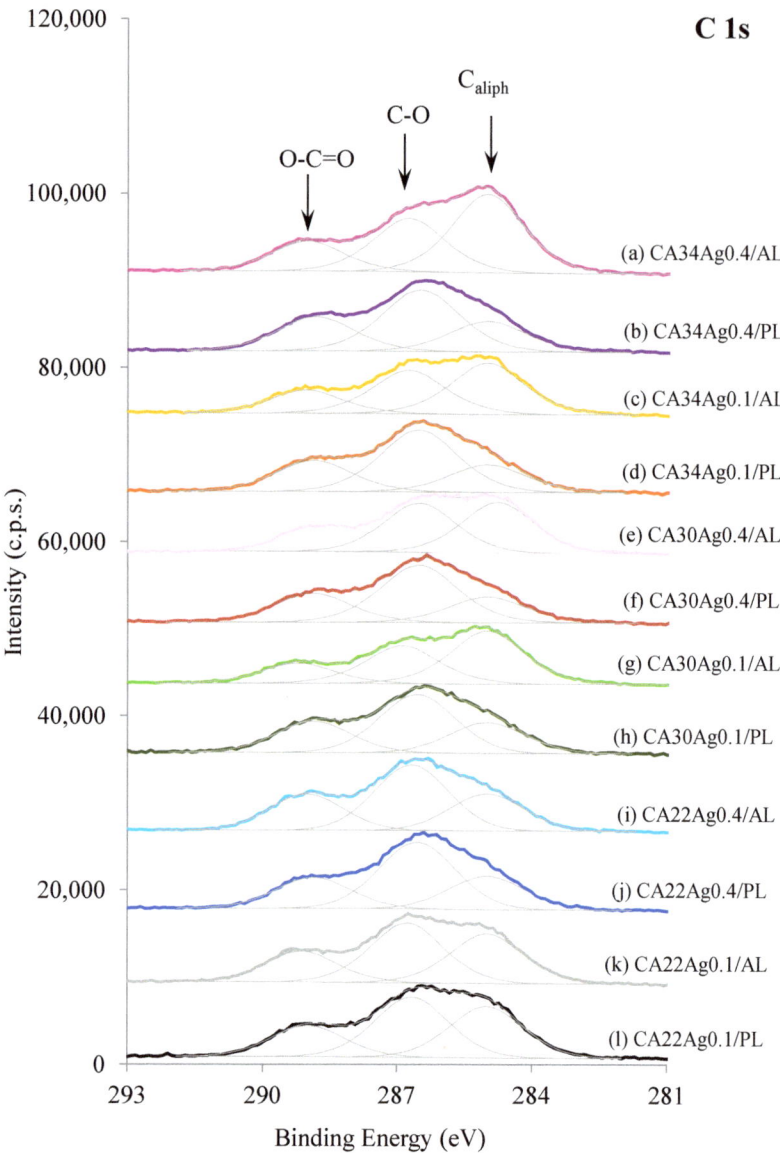

Figure 5. XPS C 1s regions of active layers (AL) and porous layers (PL) of CA/Ag membranes with 0.1% wt Ag and 0.4% wt Ag. A constant was added to each spectrum (except for l) for clarity.

The atomic ratios C/O and C3/C1 were computed from the C 1s and O 1s (not shown) regions, where C and O were computed from the total areas of C 1s and O 1s divided by the respective sensitivity factors, C3 is the amount of carbon atoms in O–C=O groups (peak at 289 eV), and C1 is the amount of aliphatic carbons (peak at 285 eV). These ratios can give a clear insight into the differences between AL and PL of CA and CA/Ag membranes. The results are gathered in Figure 6.

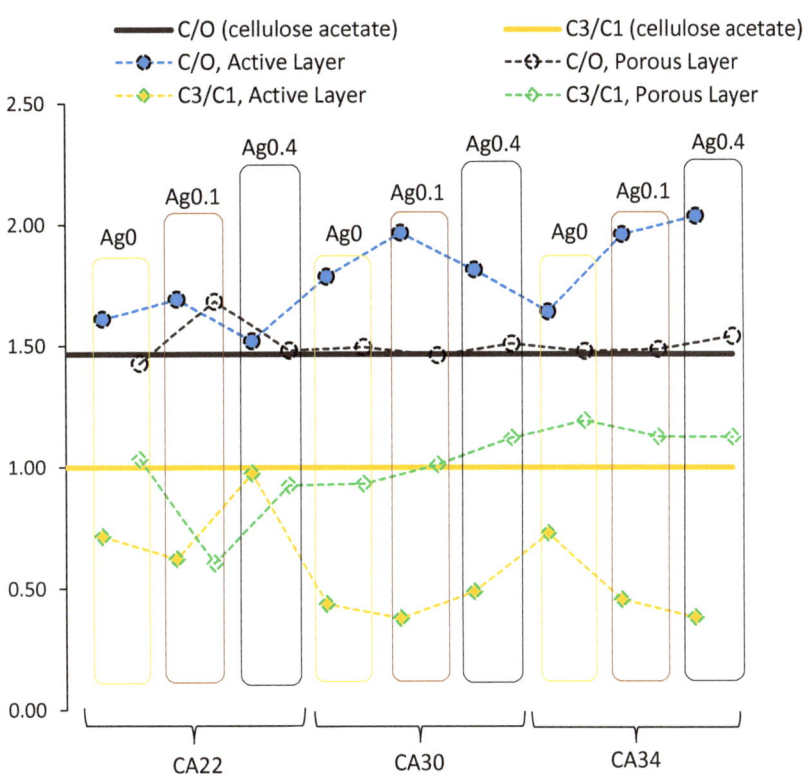

Figure 6. Atomic ratios C/O and C3/C1. The horizontal lines at 1.47 and 1 correspond, respectively, to stoichiometric ratios C/O and C3/C1 in cellulose acetate used to prepare the membranes.

For the cellulose acetate here used, the nominal ratio C/O is 1.47, and the ratio C3/C1 is 1. These stoichiometric values are represented as black and yellow horizontal lines respectively, in Figure 6. The full symbols are the experimental ratios for the membranes AL, and the empty symbols are the experimental ratios for the membranes PL. The results show, very clearly, that membrane porous layers have a composition closer to the one found in cellulose acetate than do the active layers. In fact, active layers seem to be richer in carbon than are the porous layers, particularly in aliphatic carbon, as noticed above in Figure 5a–h. Such difference may be due to the larger retention of some of the solvents used in the casting solution by the polymer-denser active layer. The exceptions are the membranes of lowest hydraulic permeabilities with 0.1 and 0.4 % wt Ag (CA22Ag0.1 and CA22Ag0.4, respectively), which present comparable C/O and C3/C1 ratios for both layers. This is in agreement with the results obtained for the C 1s region presented in Figure 5i–l.

The regions of N 1s and Si 2p were also investigated for the CA membranes (Figure 7). The presence of nitrogen in the CA membranes (silver-free membranes) was not expected at all. However, N 1s is clearly detected at the surface of active and porous layers of membranes without PVP stabilized AgNPs, as shown in Figure 7. The presence of nitrogen at the membrane surfaces may be attributed to the formamide used in the formulation of the casting solution, which could have been retained on the membrane during the phase inversion process, not diffusing completely into the aqueous phase. Silicon is clearly accumulating at the active layer surface. Silicon presence results most likely from contamination carried by the commercial CA polymer and may be washed from the porous layer by the coagulation bath.

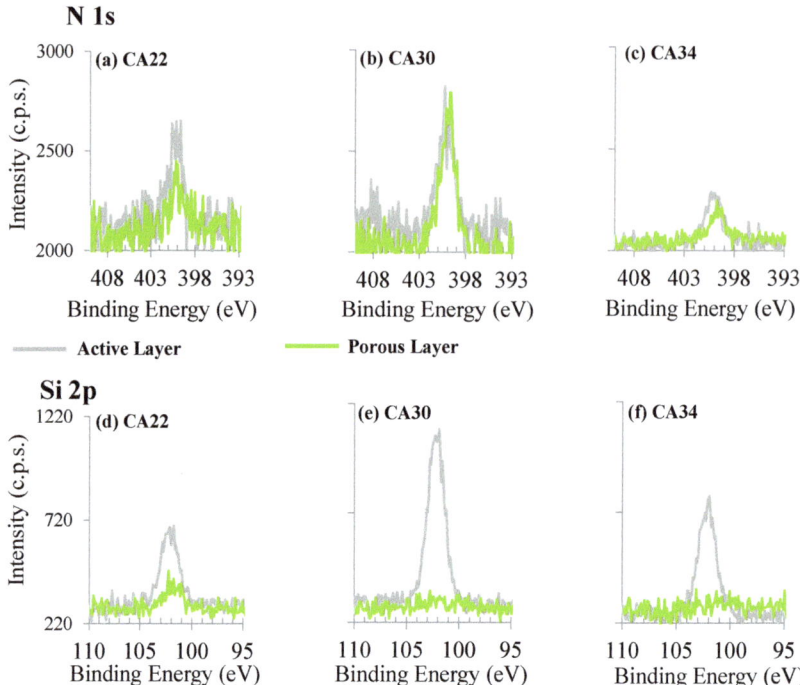

Figure 7. XPS N 1s and Si 2p regions of active layers (AL, grey spectra) and porous layers (PL, green spectra) of CA membranes (without Ag).

The source of nitrogen in CA/Ag membranes, in addition to the possible contribution of formamide used in casting solutions, can also have the input of the PVP used to stabilize the silver nanoparticles, avoiding agglomeration. High surface energy of the ultrasmall particles is normally controlled to overcome aggregation and loss of properties; PVP capability to protect the particles, and yield stable dispersions, has been previously reported [8,9,18,19]. Silicon is also detected in CA/Ag membranes. Its relative amount follows what is observed in the absence of Ag, being in a much larger amount at the AL surface than at the PL surface.

The Ag 3d XPS regions for CA/Ag membranes are displayed in Figure 8. They are doublets with energy separation of 6.0 eV. For the porous layers, two doublets were needed to fit the Ag 3d profile. This is, in principle, due to differential charge effects which are expected to occur in very rough surfaces. In some of the CA/Ag membranes, silver is not detected, which may not mean that nanoparticles are absent but that the silver nanoparticles may be located beyond the maximum depth probed by XPS, i.e., "buried" at more than 10 nm from the surface.

Membranes prepared with higher silver content (0.4% wt) are those where silver is more easily detected, particularly in the active layer of the membrane CA34. This corresponds well with the observation made by Figueiredo et al. (2015) [8] that most of the particles, particularly the smaller ones, concentrate in the active layers. In fact, for these membranes (CA34Ag0.4), a decrease was observed in the hydraulic permeability when compared with the CA34Ag0.1, which is associated with the pore-blocking caused by the AgNP in the AL. For the other two tighter membranes, CA30 and CA22, the respective increase of the membrane matrix density makes difficult the detection of AgNP leading to the decrease of the Ag 3d XPS signal of AL. On the other hand, it was observed that a decrease in silver content (0.4 to 0.1% wt) also leads to an expected decrease of the Ag 3d XPS signal of AL.

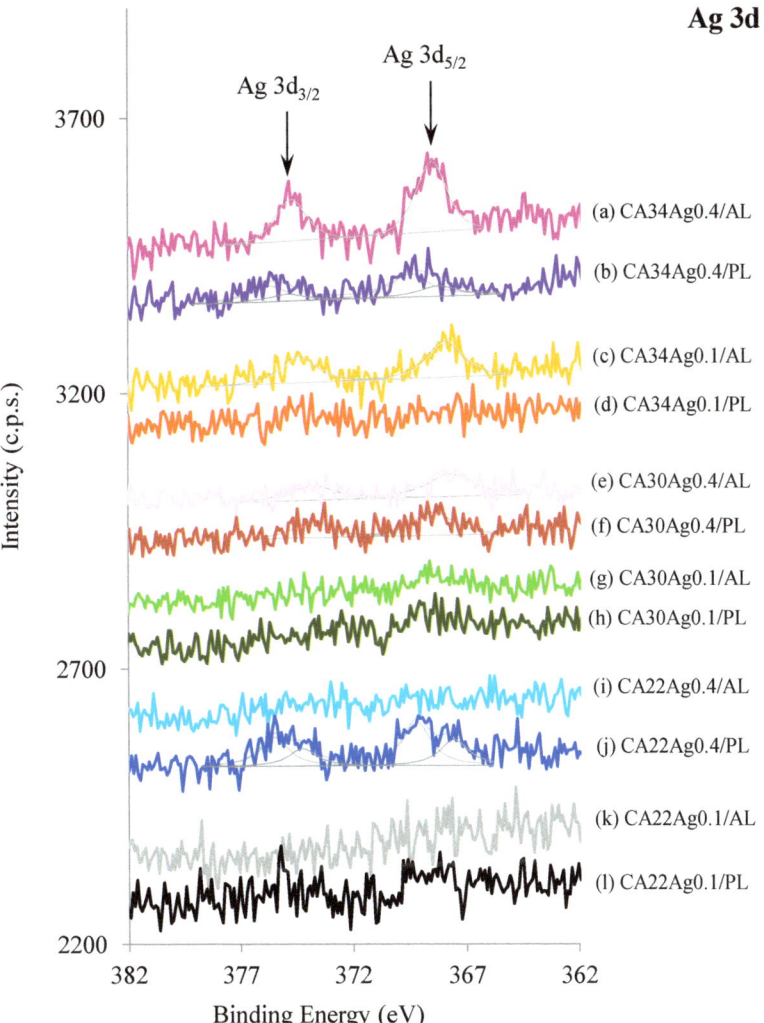

Figure 8. XPS Ag 3d regions of active layers (AL) and porous layers (PL) of CA/Ag with 0.1% wt Ag and 0.4% wt Ag membranes.

The Ag 3d XPS signals observed for the porous layers of the CA30 and CA34 are less intense than the ones for AL due to the preferential presence of AgNP in the AL [8]. In the CA22 membranes, there is a measurable Ag 3d signal from the porous layer, which is not observed for the active layer. This is because the active layers of CA22 membranes present a very dense matrix.

To identify the oxidation state of silver, one cannot rely only on the Ag 3d$_{5/2}$ binding energy; instead, the Auger parameter (AP) must be computed [16]. However, the calculation of the Auger parameter was not possible for the CA/Ag membranes because the Auger structure (Ag MNN), needed to compute this parameter, was not detected. This effect can occur when Auger electrons (which have a kinetic energy lower than the Ag

3d photoelectrons) are coming from buried silver nanoparticles and are, then, attenuated by the membrane. Nevertheless, the Auger parameter found for the silver nanoparticle suspension (AP = 718.2 eV), the same suspension which was characterized by XRD, as mentioned above, suggests that silver is oxidized or that silver nanoparticles are covered with a layer of ionized silver. Since XRD results show that silver nanoparticles present a typical metallic silver structure, one can assume the surface of the particles is either oxidized or the silver is chemically interacting with the PVP molecules adsorbed. This is an important result, as the slow release of ionized silver, as a result of the equilibrium $2Ag_2O + 4H^+ \rightarrow 4Ag^+ + 2H_2O$, favors the continuous Ag^+ release from the nanoparticles surface, particularly in aerobic media as the one used in this work, where the oxidation reaction $4Ag(0) + O_2 \rightarrow 2Ag_2O$ leads to a slow but continuous supply of silver ions [20–24]. In this sense, XPS and XRD results complement each other.

3.4. Membranes Bactericide Properties

To investigate the bactericide properties of the CA/Ag membranes, specific tests were performed against *E. coli*. The general bacteria growth in water for a long period (80 days) was also analyzed.

3.4.1. Surface Test

The surface tests were described previously in Section 2.7.1 and were conducted for the three membrane structures (CA22, CA30, and CA34) with different AgNP contents (0%, 0.1%, and 0.4%). As can be seen from Figure 9 (adapted from [8]), the active and porous layers of membranes prepared under the same experimental conditions present very different characteristics; therefore, the bacterial growth inhibition pattern of the membrane layers was evaluated to have some insights about the possibility for these membranes preventing the biofouling phenomena. Before the tests, all the membranes were sterilized in a UV camera for 30 min on each side. No bacterial growth was observed following sterilization.

Figure 9. FESEM images of (**a**) active layer, (**b**) porous layer, and (**c**) cross-section of CA30Ag0.1 membrane (adapted from Ref. [8]. 2014, John Wiley and Sons).

From the results presented in Figure 10, it was possible to observe, for the silver-free membranes (Figure 10a,d,g), a generalized growth of *E. coli* in all the membrane active layer surfaces, forming a bacteria film on the membrane surface, visible even to the naked eye.

For the membranes with AgNP, the results obtained with the CA22Ag0.1 and CA22Ag0.4 (Figure 10b,c) show different zones on the membrane active layer, indicating that the *E. coli* growth was not generalized, with existing parts of the membrane surface exhibiting *E. coli* growth inhibition. In the surface test conducted for the active layer of membrane structures CA30 and CA34 with AgNP (0.1% and 0.4%), evident growth inhibition was observed, as only a few *E. coli* colonies (white dots) were detected in the Figure 10e,f,h,i.

The results obtained from the surface tests of the membrane porous layers are presented in Figure 11. The results obtained from the silver-free membranes porous layer indicate that *E. coli* growth inhibition was not observed (Figure 11a,d,g).

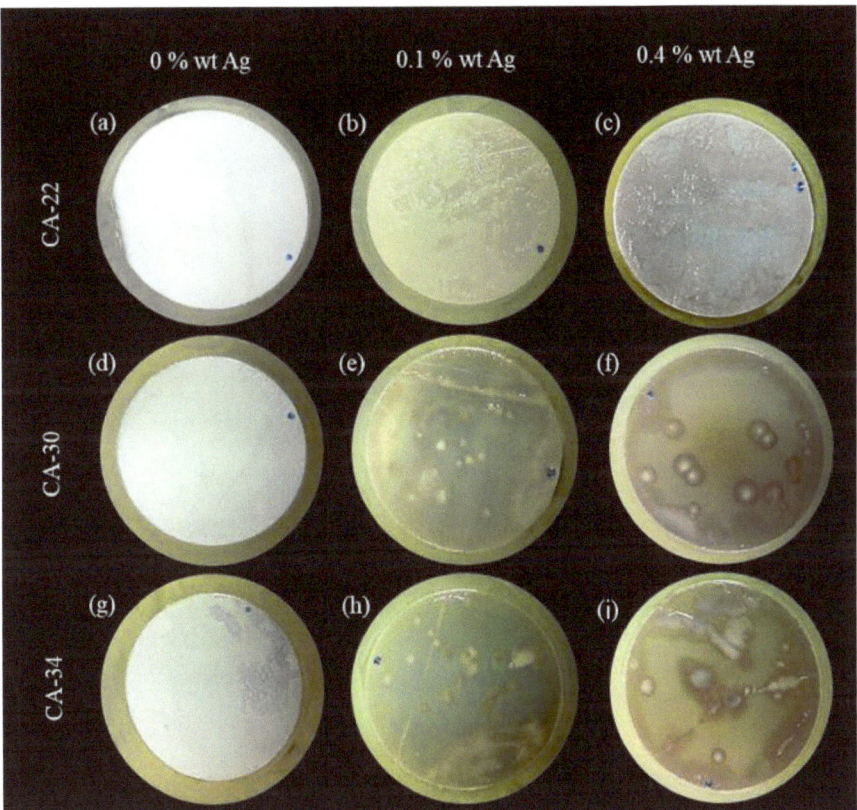

Figure 10. Results of *E. coli* growth inhibition tests on the active layers of CA membranes (**a**) CA22, (**d**) CA30, and (**g**) CA34; CA/Ag membranes with 0.1wt% Ag (**b**) CA22Ag0.1, (**e**) CA30Ag0.1, and (**h**) CA34Ag0.1; and CA/Ag membranes with 0.4wt% Ag (**c**) CA22Ag0.4, (**f**) CA30Ag0.4, and (**i**) CA34Ag0.4.

For the membranes CA22 with AgNP (0.1 and 0.4%), a slight increase in the growth inhibition for the membrane porous layer was observed, compared with the results obtained for the active layers of the CA22Ag0.1 and CA22Ag0.4 membranes. This is in line with the XPS results, which revealed a higher Ag concentration in the porous layer. For the CA30 and CA34 with AgNP, the results for the porous layer (Figure 11e,f,h,i) are comparable to those obtained for the active layers (Figure 10e,f,h,i), where significant *E. coli* growth inhibition is observed.

The difference observed in growth inhibition among the silver-containing membranes CA22 (CA22Ag0.1 and CA22Ag0.4), CA30 (CA30Ag0.1 and CA30Ag0.4), and CA34 (CA34Ag0.1 and CA34Ag0.4) may be related to the fact that the structure of the CA22 membranes is much tighter than the one of the CA30 and CA34 membranes. The presence of AgNP in the proximity of the active layer surfaces is also more difficult to detect for the CA22 silver-containing membranes (Figure 8). The differences between the surfaces of the active and porous layers of CA22Ag0.1 and CA22Ag0.4 membranes are consistent with the fact that a more porous structure promotes a better interaction between *E. coli* and the AgNP and, therefore, leads to high growth inhibition for the porous layers of CA22 silver-containing membranes. In addition, the XPS reveals that the porous layer contains a larger relative amount of silver.

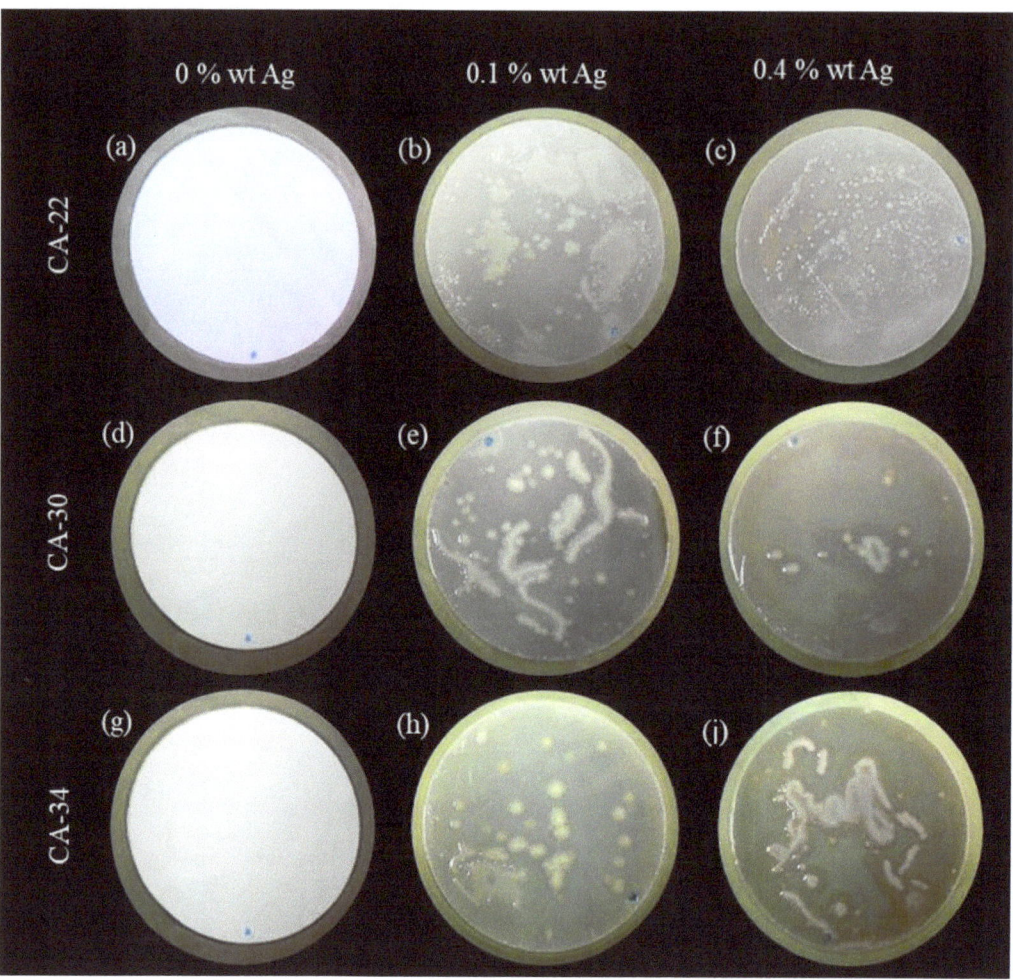

Figure 11. Results of *E. coli* growth inhibition on the porous layer of CA membranes (**a**) CA22, (**d**) CA30, and (**g**) CA34; CA/Ag membranes with 0.1wt% Ag (**b**) CA22Ag0.1, (**e**) CA30Ag0.1, and (**h**) CA34Ag0.1; and CA/Ag membranes with 0.4wt% Ag (**c**) CA22Ag0.4, (**f**) CA30Ag0.4, and (**i**) CA34Ag0.4.

As previously presented (Figure 8) and discussed, the XPS results indicate that in some membrane samples, the silver presence was not detected, which may be due to the fact that the AgNP might be located more than 10 nm away from the membrane's outermost surface. In addition to the difficulty of detecting the silver signal by XPS for the CA22 silver-containing membranes, the silver signal was perceptible for the porous layer of CA22Ag0.1 membrane (although not quantified due to the very poor signal-to-noise ratio), which is consistent with the lowest *E. coli* growth inhibition results obtained for the surface test of CA22 membranes.

3.4.2. Suspension Test

To complement the results obtained in the surface test, suspension tests were performed by immersing the membrane samples in *E. coli*-inoculated suspensions at room temperature (20 °C), different from the temperature used in the surface tests, which was considered ideal for the *E. coli* growth (37 °C).

The membranes and the suspensions were evaluated in terms of *E. coli* growth inhibition, according to the experiments previously described in the Materials and Methods section. Figure 12 presents the membrane samples after immersion in the inoculated suspension (24 h) and incubation in a Petri dish with YEA at 37 °C (24 h), and Figure 13 presents the incubated filters used to filter de-inoculated suspension.

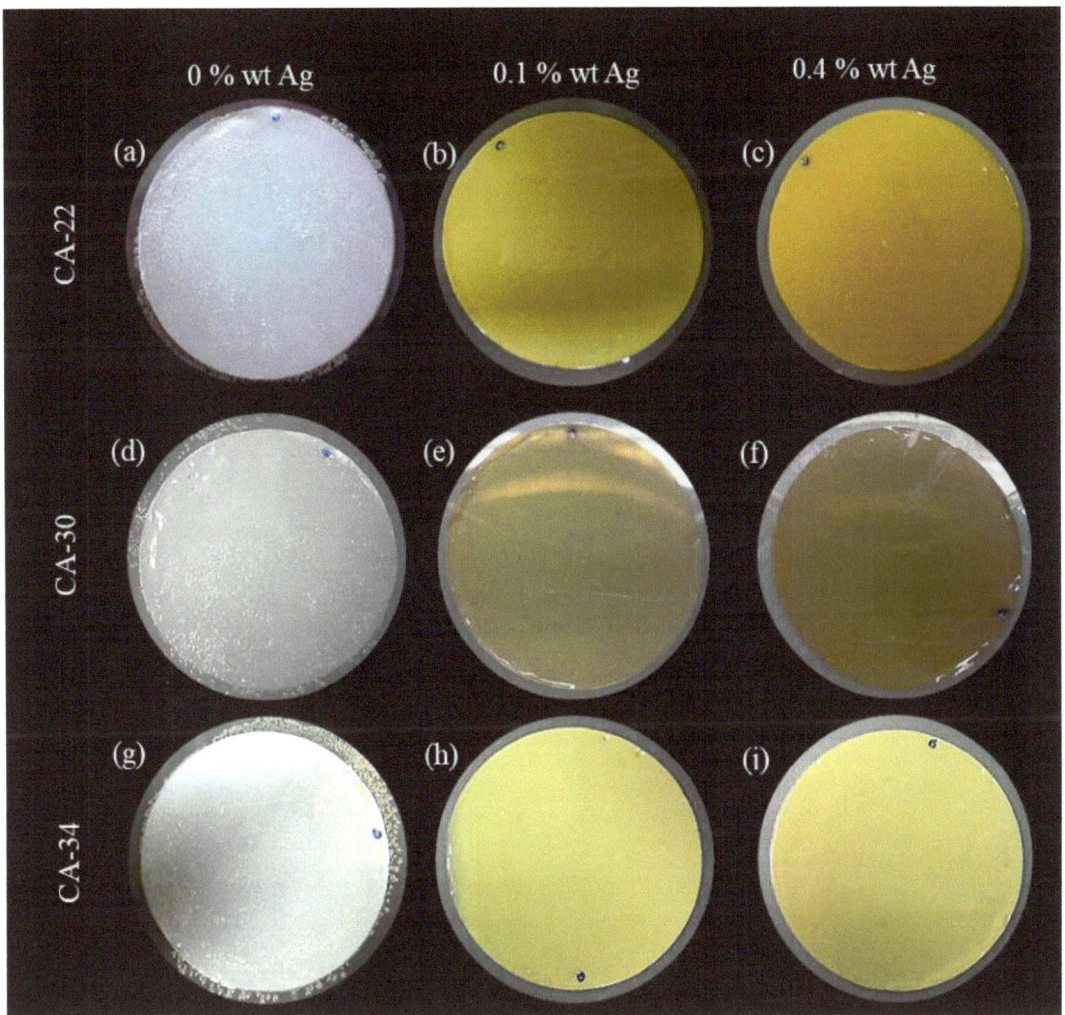

Figure 12. Results of *E. coli* growth inhibition on the membrane for the suspension test of CA membranes (**a**) CA22, (**d**) CA30, and (**g**) CA34; CA/Ag membranes with 0.1wt% Ag (**b**) CA22Ag0.1, (**e**) CA30Ag0.1, and (**h**) CA34Ag0.1; and CA/Ag membranes with 0.4wt% Ag (**c**) CA22Ag0.4, (**f**) CA30Ag0.4, and (**i**) CA34Ag0.4.

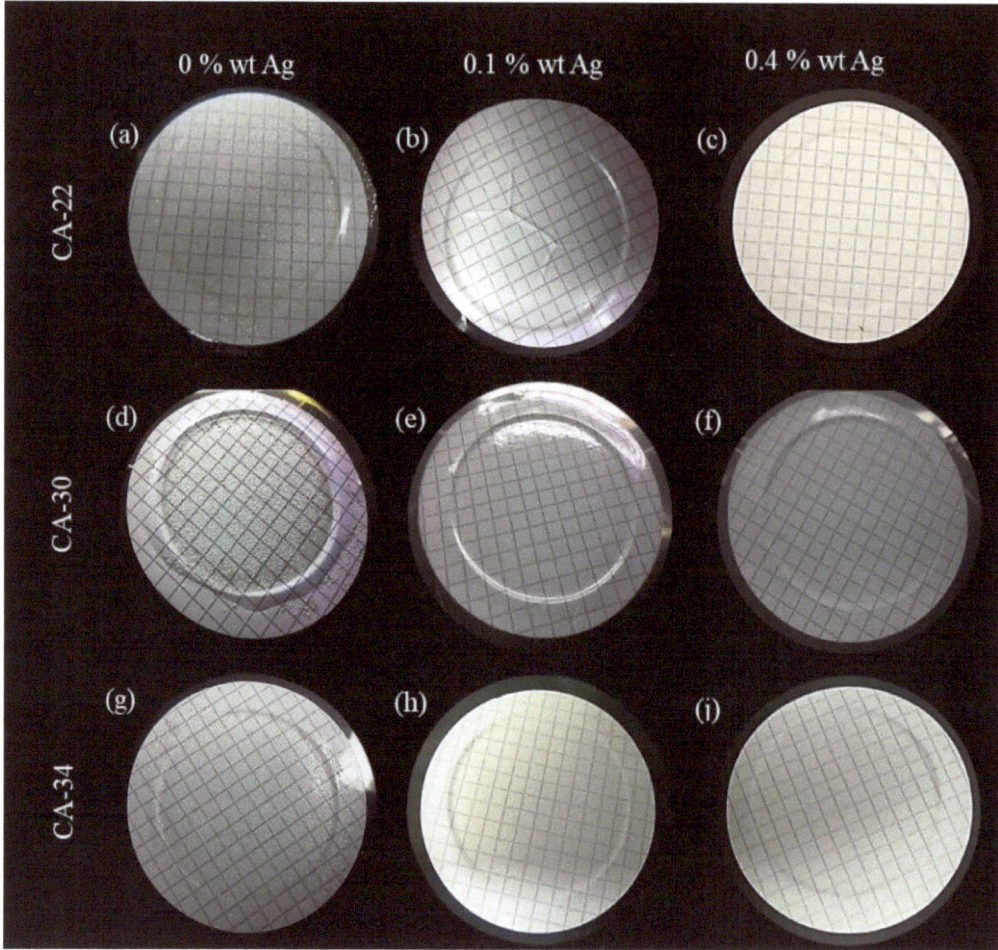

Figure 13. Results of *E. coli* growth inhibition on the suspension for the suspension test of CA and CA/Ag membranes. Silver-free membranes: (**a**) CA22, (**d**) CA30, and (**g**) CA34; CA/Ag membranes with 0.1wt% Ag: (**b**) CA22Ag0.1, (**e**) CA30Ag0.1, and (**h**) CA34Ag0.1; CA/Ag membranes with 0.4wt% Ag: (**c**) CA22Ag0.4, (**f**) CA30Ag0.4, and (**i**) CA34Ag0.4.

From the results presented in Figure 12, it can be seen very clearly that there is a total *E. coli* growth inhibition for all the membrane structures containing AgNP (Figure 12b,c,e,f,h,i). In contrast, the silver-free membranes (Figure 12a,d,g) show no *E. coli* growth inhibition.

Figure 13 presents the results for the membrane filters used to filter the *E. coli* suspensions, where the membranes were immersed for the suspension test (Figure 12). For the suspensions in contact with silver-free membranes, full development of bacterial growth was verified in the membrane filters (Figure 13a,d,g). On the other hand, the suspensions in contact with CA/Ag membranes lead to a total absence of *E. coli* colonies in the membrane filters (Figure 13b,c,e,f,h,i).

In the suspension test, with experimental conditions which are similar to the ones of the water filtration processes, there is clear evidence that the CA/Ag membranes promote total growth inhibition independently of the membrane structure and the silver content.

3.4.3. Cell Death Test

The previous results (Sections 3.4.1 and 3.4.2) demonstrate that the presence of silver nanoparticles introduces an antibacterial effect in the CA/Ag membranes, independent of the different silver content. To clarify the silver nanoparticles content effect on the antibacterial properties of silver-containing membranes, the cell death was performed for the CA34 membranes series because the membranes CA34Ag0.1 and CA34Ag0.4 are those where silver is more easily detected by XPS results. The membranes CA34, CA34Ag0.1, and CA34Ag0.4 were tested at 37 °C. The results are presented in Figures 13 and 14.

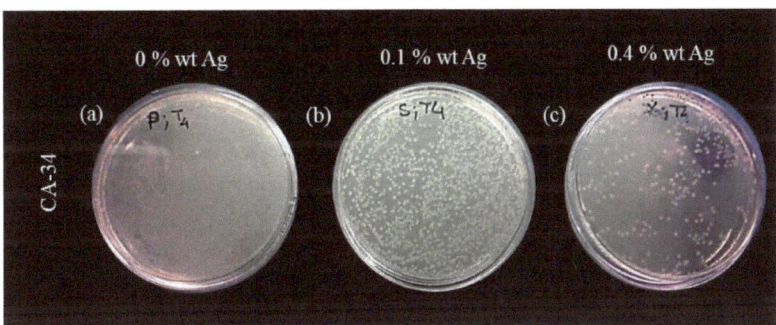

Figure 14. *E. coli* colonies after the incubation of the suspensions contacting with the membranes (1093 min). (**a**) Silver-free membranes (CA34); (**b**) CA/Ag membranes with 0.1wt% (CA34Ag0.1); and (**c**) CA/Ag membranes with 0.4wt% Ag (CA34Ag0.4).

The cell death test was carried out for 18 h and samples of the suspension were collected during this period. Figure 14 presents the image of the Petri dishes incubated with the sample collected at 1093 min of membrane/*E. coli* suspension contact. The results show a generalized growth of *E. coli* for the suspension in contact with silver-free membranes (Figure 14a). In contrast, there is growth inhibition for the tests conducted with the CA34Ag0.1 and CA34Ag0.4 membranes, resulting in a decrease in the number of *E. coli* colonies, which is even more pronounced for the CA34Ag0.4 membrane (Figure 14c).

Figure 15 displays, for the CA34, CA34Ag0.1, and CA34Ag0.4 membranes, the variation of *E. coli* concentration (CFU/mL) at different contact times of the suspensions with the membranes.

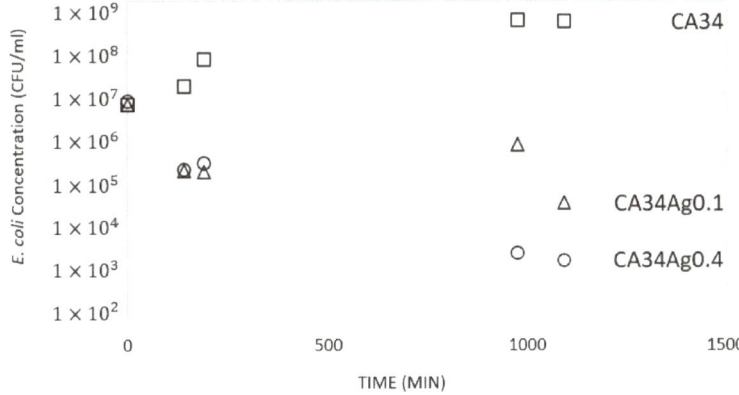

Figure 15. Variation of *E. coli* concentration (CFU/mL) at different contact times (143 min, 190 min, 975 min and 1093 min) of the suspensions with the membranes (□, CA34; △, CA34Ag0.1; and ○, CA34Ag0.4).

The concentration of E. coli decreases with the time of contact between the suspension and the CA/Ag membranes (CA34Ag0.1 and CA34Ag0.4). On the contrary, the concentration of E. coli increases with the time of contact between silver-free membranes and the E. coli suspension. For CA nanofiltration membranes incorporating silver nanoparticles or silver ion-exchanged β-zeolite, Beisl et al. (2019) [7] also verified a decrease in the concentration of E. coli along the time of contact between the suspension and the membranes.

3.4.4. Growth Inhibition of Bacteria in the Water

To consolidate the capability of silver-containing membranes to inhibit the growth of bacteria in water, the membranes CA30, CA30Ag0.1, and CA30Ag0.4 were immersed in water for a long period (80 days). The results are presented in Figure 16.

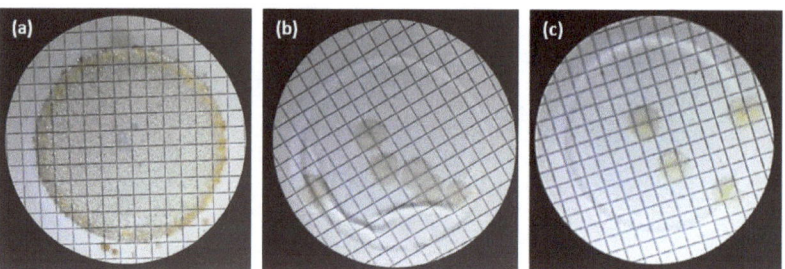

Figure 16. Inhibition growth test of microorganisms in water with membranes. (**a**) Silver-free membranes (CA30); (**b**) CA/Ag membranes with 0.1wt% Ag (CA30Ag0.1); and (**c**) CA/Ag membranes with 0.4wt% Ag (CA30Ag0.4).

The silver-free membranes showed generalized bacterial growth (Figure 16a). The silver-containing membranes, CA30Ag0.1 and CA30Ag0.4, display a strong reduction in the number of colonies in the filters. The CA30Ag0.1 membrane (Figure 16b) showed six colonies, while the CA30Ag0.4 membrane (Figure 16c) presented four colonies. These results provide strong evidence of bacterial growth inhibition in water contacting CA/Ag membranes, independently of structure, silver content, aggregation degree, and distribution of the particles in both dense and porous layers.

4. Conclusions

In this work, cellulose acetate/silver nanoparticles integral asymmetric ultrafiltration membranes were synthesized, characterized, and tested for their bactericidal properties. Combined XRD and XPS characterization showed that the silver nanoparticles are covered with ionic silver, able to solubilize and exert bactericide activity.

The active and porous layers' surfaces display different compositions, with the exception of the less permeable membranes, CA22Ag0.1 and CA22Ag0.4, that present similar C/O and O–C=O/(C-C, C-H) ratios in both surfaces. The porous layers composition of the more permeable membranes (CA30, CA30Ag0.1, CA30Ag0.4, CA34, CA34Ag0.1, and CA34Ag0.4) is closer to the composition of the CA polymer.

The silver-free membranes present a generalized growth of E. coli. This is in contrast with the inhibition pattern displayed by the membranes with AgNP. The antibacterial tests carried on the membrane surfaces of the less permeable membranes, CA22Ag0.1 and CA22Ag0.4, evidence a lower E. coli growth inhibition when compared with the more permeable membranes (CA30, CA30Ag0.1, CA30Ag0.4, CA34, CA34Ag0.1, and CA34Ag0.4) that present almost complete growth inhibition. Furthermore, the porous layers of the CA22Ag0.1 and CA22Ag0.4 show a slight increase in the E. coli growth inhibition in comparison with the corresponding active layers. This is in line with the XPS results which revealed a higher Ag concentration in the porous layer. For the more permeable membranes (CA30, CA30Ag0.1, CA30Ag0.4, CA34, CA34Ag0.1, and CA34Ag0.4), the significant E. coli growth inhibition is comparable in both active and porous layers. For those membranes, XPS reveals a higher amount of silver in the active layer than in the porous one, as already stated in [8]. However, the access to the AgNP in the porous layer is higher. Apparently, the two effects compensate for each other leading to similar bactericide properties in both layers.

Author Contributions: Conceptualization, A.S.F., M.M., M.G.S.-L. and M.N.d.P.; methodology, A.S.F., M.G.S.-L. and R.L.T.-T.; validation, A.M.F., A.M.B.d.R. and M.N.d.P.; formal analysis, A.S.F., A.M.F., A.M.B.d.R., M.G.S.-L. and M.N.d.P.; investigation, A.S.F., A.M.F., S.M., R.S., M.G.S.-L. and M.N.d.P.; data curation, A.S.F., A.M.F. and R.L.T.-T.; writing—original draft preparation, A.S.F. and A.M.F.; writing—review and editing, A.S.F., A.M.F., A.M.B.d.R., M.M., M.G.S.-L. and M.N.d.P.; supervision, M.M. and M.N.d.P. All authors have read and agreed to the published version of the manuscript.

Funding: This research was funded by FCT—Fundação para a Ciência e Tecnologia, Portugal, grant numbers: UID/CTM/04540/2019 (CeFEMA and LaPMET-Associate Laboratory of Physics for Materials and Emergent Technologies); UIDB/04565/2020 and UIDP/04565/2020 (Research Unit Institute for Bioengineering and Biosciences—iBB); and LA/P/0140/2020 (Associate Laboratory Institute for Health and Bioeconomy—i4HB). A.M.F. wishes to thank Instituto Superior Técnico for the Scientific Employment contract IST-ID/131/2018. M.N.d.P. and M.G.S.-L. thank the Cyted Network Aquamemtec for financial support.

Institutional Review Board Statement: The study did not require ethical approval.

Informed Consent Statement: Not applicable.

Data Availability Statement: Not applicable.

Conflicts of Interest: The authors declare no conflict of interest. The funders had no role in the design of the study; in the collection, analyses, or interpretation of data; in the writing of the manuscript; or in the decision to publish the results.

References

1. Jhaveri, J.H.; Murthy, Z.V.P. A comprehensive review on anti-fouling nanocomposite membranes for pressure driven membrane separation processes. *Desalination* **2016**, *379*, 137–154. [CrossRef]
2. Nguyen, T.; Roddick, F.A.; Fan, L. Biofouling of water treatment membranes: A review of the underlying causes, monitoring techniques and control measures. *Membranes* **2012**, *2*, 804–840. [CrossRef] [PubMed]
3. Makvandi, P.; Wang, C.Y.; Zare, E.N.; Borzacchiello, A.; Niu, L.N.; Tay, F.R. Metal-based nanomaterials in biomedical applications: Antimicrobial activity and cytotoxicity aspects. *Adv. Funct. Mater.* **2020**, *30*, 191002. [CrossRef]
4. Qi, L.; Liu, Z.; Wang, N.; Hu, Y. Facile and efficient in situ synthesis of silver nanoparticles on diverse filtration membrane surfaces for antimicrobial performance. *Appl. Surf. Sci.* **2018**, *456*, 95–103. [CrossRef]
5. Khaydarov, R.; Gapurova, O.; Abdukhakimov, M.; Sadikov, I.; Garipov, I.; Krishnamurthy, P.T.; Zharkov, S.M.; Zeer, G.M.; Abolentseva, P.A.; Prudnikova, S.V.; et al. Antimicrobial properties of nanofiltration membranes modified with silver nanoparticles. *Emergent Mater.* **2022**, *5*, 1477–1483. [CrossRef]
6. Peng, K.; Huang, Y.; Peng, N.; Chang, C. Antibacterial nanocellulose membranes coated with silver nanoparticles for oil/water emulsions separation. *Carbohydr. Polym.* **2022**, *278*, 118929. [CrossRef]
7. Beisl, S.; Monteiro, S.; Santos, R.; Figueiredo, A.S.; Sánchez-Loredo, M.G.; Lemos, M.A.; Lemos, F.; Minhalma, M.; de Pinho, M.N. Synthesis and bactericide activity of nanofiltration composite membranes—Cellulose acetate/silver nanoparticles and cellulose acetate/silver ion exchanged zeolites. *Water Res.* **2019**, *149*, 225–231. [CrossRef]
8. Figueiredo, A.S.; Sánchez-Loredo, M.G.; Maurício, A.; Pereira, M.F.C.; Minhalma, M.; De Pinho, M.N. Tailoring of structures and permeation properties of asymmetric nanocomposite cellulose acetate/silver membranes. *J. Appl. Polym. Sci.* **2015**, *132*, 41796. [CrossRef]
9. Tashdjian, A.; Sánchez Loredo, M.G.; González, G.A. Preparation of Silver Nanoparticles-Based Sensors for the Electrochemical Detection of Thiourea in Leaching Solutions of Waste Electrical and Electronic Equipment. *Electroanalysis* **2013**, *25*, 2124–2129. [CrossRef]
10. Kunst, B.; Sourirajan, S. An approach to the development of cellulose acetate ultrafiltration membranes. *J. Appl. Polym. Sci.* **1974**, *18*, 3423–3434. [CrossRef]
11. Kesting, R.E. *Synthetic Polymeric Membranes—A Structural Perspective*; Wiley-Interscience: New York, NY, USA, 1985.
12. Abid, M.; Bouattour, S.; Ferraria, A.M.; Conceição, D.S.; Carapeto, A.P.; Vieira Ferreira, L.F.; Botelho do Rego, A.M.; Chehimi, M.M.; Rei Vilar, M.; Boufi, S. Facile functionalization of cotton with nanostructured silver/titania for visible-light plasmonic photocatalysis. *J. Colloid. Interf. Sci.* **2017**, *507*, 83–94. [CrossRef] [PubMed]

13. Taurozzi, J.S.; Arul, H.; Bosak, V.Z.; Burban, A.F.; Voice, T.C.; Bruening, M.L.; Tarabara, V.V. Effect of filler incorporation route on the properties of polysulfone–silver nanocomposite membranes of different porosities. *J. Membr. Sci.* **2008**, *325*, 58–68. [CrossRef]
14. Maheswari, P.; Prasannadevi, D.; Mohan, D. Preparation and performance of silver nanoparticle incorporated polyetherethersulfone nanofiltration membranes. *High Perform. Polym.* **2012**, *25*, 174–187. [CrossRef]
15. Beamson, G.; Briggs, D. *High Resolution XPS of Organic Polymers, The Scienta ESCA300 Database*, 1st ed.; John Wiley & Sons Ltd.: West Sussex, UK, 1992.
16. Ferraria, A.M.; Carapeto, A.P.; Botelho do Rego, A.M. X-ray photoelectron spectroscopy: Silver salts revisited. *Vacuum* **2012**, *86*, 1988–1991. [CrossRef]
17. Boufi, S.; Ferraria, A.M.; Botelho Do Rego, A.M.; Battaglini, N.; Herbst, F.; Rei, M. Vilar Surface functionalisation of cellulose with noble metals nanoparticles through a selective nucleation. *Carbohyd. Polym.* **2011**, *86*, 1586–1594. [CrossRef]
18. Pedre, I.; Battaglini, F.; Delgado, G.J.L.; Sánchez-Loredo, M.G.; González, G.A. Detection of thiourea from electrorefining baths using silver nanoparticles-based sensors. *Sensor. Actuat. B-Chem.* **2015**, *211*, 515–522. [CrossRef]
19. Pedre, I.; Méndez DeLeo, L.P.; Sánchez-Loredo, M.G.; Battaglini, F.; González, G.A. Electrochemical sensor for thiourea focused on metallurgical applications of copper. *Sensor. Actuat. B-Chem.* **2016**, *232*, 383–389. [CrossRef]
20. Choi, O.; Deng, K.K.; Kim, N.-J.; Ross, L.; Surampalli, R.Y.; Hu, Z.; Ross Louis, J.; Surampalli, Y.; Hu, Z. The inhibitory effects of silver nanoparticles, silver ions, and silver chloride colloids on microbial growth. *Water Res.* **2008**, *42*, 3066–3074. [CrossRef]
21. World Health Organization; Fewtrell, L. Silver: Water Disinfection and Toxicity. 2014. Available online: http://www.who.int/water_sanitation_health/dwq/chemicals/Silver_water_disinfection_toxicity_2014V2.pdf?ua=1 (accessed on 19 April 2018).
22. Kim, K.J.; Sung, W.S.; Suh, B.K.; Moon, S.K.; Choi, J.S.; Kim, J.G.; Lee, D.G. Antifungal activity and mode of action of silver nano-particles on Candida albicans. *Biometals* **2009**, *22*, 235–242. [CrossRef]
23. Xiu, Z.-M.; Ma, J.; Alvarez, P.J.J. Differential Effect of Common Ligands and Molecular Oxygen on Antimicrobial Activity of Silver Nanoparticles versus Silver Ions. *Environ. Sci. Technol.* **2011**, *45*, 9003–9008. [CrossRef]
24. Xiu, Z.-M.; Zhang, Q.-B.; Puppala, H.L.; Colvin, V.L.; Alvarez, P.J.J. Negligible Particle-Specific Antibacterial Activity of Silver Nanoparticles. *Nano Lett.* **2012**, *12*, 4271–4275. [CrossRef] [PubMed]

Disclaimer/Publisher's Note: The statements, opinions and data contained in all publications are solely those of the individual author(s) and contributor(s) and not of MDPI and/or the editor(s). MDPI and/or the editor(s) disclaim responsibility for any injury to people or property resulting from any ideas, methods, instructions or products referred to in the content.

Article

Magnetic-Responsive Liposomal Hydrogel Membranes for Controlled Release of Small Bioactive Molecules—An Insight into the Release Kinetics

Luís Pereira [1], Frederico Castelo Ferreira [2,3], Filipa Pires [1,*] and Carla A. M. Portugal [1,*]

- [1] LAQV-REQUIMTE, Department of Chemistry, NOVA School of Science and Technology (FCT NOVA), Universidade Nova de Lisboa, 2829-516 Caparica, Portugal; lfc.pereira@campus.fct.unl.pt
- [2] Department of Bioengineering and iBB—Institute for Bioengineering and Biosciences, Instituto Superior Técnico, Universidade de Lisboa, 1049-001 Lisboa, Portugal; frederico.ferreira@ist.utl.pt
- [3] Associate Laboratory i4HB—Institute for Health and Bioeconomy, Instituto Superior Técnico, Universidade de Lisboa, Av. Rovisco Pais, 1049-001 Lisboa, Portugal
- * Correspondence: af.pires@campus.fct.unl.pt (F.P.); cmp@fct.unl.pt (C.A.M.P.)

Abstract: This work explores the unique features of magnetic-responsive hydrogels to obtain liposomal hydrogel delivery platforms capable of precise magnetically modulated drug release based on the mechanical responses of these hydrogels when exposed to an external magnetic field. Magnetic-responsive liposomal hydrogel delivery systems were prepared by encapsulation of 1,2-dipalmitoyl-sn-glycero-3-phosphocoline (DPPC) multilayered vesicles (MLVs) loaded with ferulic acid (FA), i.e., DPPC:FA liposomes, into gelatin hydrogel membranes containing dispersed iron oxide nanoparticles (MNPs), i.e., magnetic-responsive gelatin. The FA release mechanisms and kinetics from magnetic-responsive liposomal gelatin were studied and compared with those obtained with conventional drug delivery systems, e.g., free liposomal suspensions and hydrogel matrices, to access the effect of liposome entrapment and magnetic field on FA delivery. FA release from liposomal gelatin membranes was well described by the Korsmeyer–Peppas model, indicating that FA release occurred under a controlled diffusional regime, with or without magnetic stimulation. DPPC:FA liposomal gelatin systems provided smoother controlled FA release, relative to that obtained with the liposome suspensions and with the hydrogel platforms, suggesting the promising application of liposomal hydrogel systems in longer-term therapeutics. The magnetic field, with low intensity (0.08 T), was found to stimulate the FA release from magnetic-responsive liposomal gelatin systems, increasing the release rates while shifting the FA release to a quasi-Fickian mechanism. The magnetic-responsive liposomal hydrogels developed in this work offer the possibility to magnetically activate drug release from these liposomal platforms based on a non-thermal related delivery strategy, paving the way for the development of novel and more efficient applications of MLVs and liposomal delivery systems in biomedicine.

Keywords: drug release kinetics; magnetic field; liposomes

1. Introduction

Conventional clinical approaches, consisting of the direct administration of therapeutic compounds, have shown limited treatment efficiency, attributed to the short lifetime of these compounds, to the loss of the therapeutic molecules caused by loose guidance to the point of action and to the poor bioavailability of these bioactive compounds due to their low solubility in aqueous media (such as the case of lipophilic molecules). These problems have been circumvented by the increase of the drug doses, which may, however, potentiate drug cytotoxicity and the possibility of undesirable side effects.

Advances in this topic have been attempted through the development of biomimetic drug carriers, such as liposomes [1–8], micelles [9–13] and hydrogels [14–19]. These drug vehicles are able to accommodate and protect the drug/bioactive molecules (e.g., growth factors, peptides, genes, vaccines, and cells) from hydrolytic or enzymatic degradation in the organism. However, they lack proper delivery efficiency given their structural instability [20] and consequent incapacity to hold the drug molecules, resulting in an uncontrolled drug release, e.g., burst release [21–23]. Efforts to develop delivery systems with improved stability have been carried out through the optimization of the chemical formulation of these delivery vehicles, mainly through the addition of structural stabilizers (e.g., cholesterol and poly-L-lysine) [24–27]. Cholesterol was found to enhance the lipophilic packing in the liposome bilayer, providing improved liposome stability [24], whereas the hydrophobic molecules/polymers have been added to strength bond interactions, providing improved liposome resistance to enzymatic or immune reactions [28,29]. Thus, a fine regulation of the permeability, fluidity and temperature transition of these vesicles is possible by rigorous control of the amount of the stabilizing component in the liposome structure.

An alternative approach takes advantage of a synergetic combination of liposomal technologies and the versatility of hydrogels to obtain delivery systems- liposomal hydrogel systems, with improved stability which are able to offer an enhanced controlled drug release [30]. Liposomal hydrogels have been developed either by the immobilization of gels in the liposome core [31–34] or by the encapsulation of liposomes into the hydrogel network [35–38]. These systems provide a double advantage; on the one hand, hydrogels confer higher liposome stability and protection against pH and ionic strength [30,37]. On the other hand, the encapsulated liposomes act as drug pockets with enhanced stability, leading to a more sustained and prolonged drug release comparatively than that found for conventional liposomes [39–42] or native hydrogel matrices [25]. Liposomal hydrogel delivery systems have shown improved efficiency for the treatment of several inflammatory diseases [43] and chemotherapy [44,45] comparatively to conventional therapies, ascribed to their ability to provide longer-term and in situ release of the target therapeutic compound, minimizing the toxic side effects due to drug leakage.

The improved stability of liposomal hydrogel systems results from a symbiotic effect that combines enhanced structural liposome stability conferred by the interaction with the hydrogel matrix, but, also, liposomes were found to act as stabilizing elements providing liposomal hydrogel systems with improved viscoelastic properties, conferring them additional structural cohesion [46,47].

The release kinetics and mechanisms from liposomal hydrogels are thus dependent on the combined transport resistance of the liposomal lipidic bilayer and the polymer matrix, which creates an additional rate-limiting effect on the diffusion of the therapeutic molecules. In this respect, the drug release is regulated by the hydrogel matrix's structural and chemical properties (e.g., density, porosity and hydrophilicity) and by the permeability of the liposome membrane to the target drug. In fact, studies from other authors have also shown that the transport from liposomal hydrogels is importantly dependent on the liposome characteristics, i.e., vesicle chemical formulation and surface charge [48], and drug hydrophilicity. As reported by Mourtas et al. [36], the release of lipophilic molecules is predominantly influenced by the drug properties and loading, in contrast to that observed for hydrophilic drugs whose diffusion from liposomal hydrogels depends on the characteristics of the bilipid vesicle, i.e., chemical formulation and rigidity.

The large structural and chemical versatility of the hydrogels as well as their ability to respond to external stimuli (e. g. temperature, light, magnetic field) have been also explored to obtain liposomal hydrogels offering more precise spatiotemporal controlled drug release. Magnetic-responsive liposomal hydrogels offer the possibility of a stimuli-triggered drug release, allowing for an on-demand diffusion of the drug/bioactive molecules in the site of action while avoiding drug losses during the trajectory of the delivery vehicle to the target tissue. Magnetic liposomal systems have been developed, relying on the hyperthermic effect produced by magnetic susceptible nanoparticles (e.g., iron oxide nanoparticles)—MNPs—when exposed to a high-frequency magnetic field (AMF) [49–53]. MNPs may be present in the aqueous liposome core, embedded in the lipidic bilayer or as a solid MNP cluster forming the inner liposome core [47]. In both cases, magnetic-induced thermal release is prompted by the phase transition of the hydrogel from a solid state to a sol state at a specific temperature [51–53].

However, magnetic-responsive hydrogels exhibit unique mechanical responses upon magnetic stimulation, which have been very weakly explored in the development of magnetic-responsive liposomal systems. Magnetic-responsive hydrogels can reversibly switch their volume and shape when exposed to external magnetic field stimuli [54–56]. Such mechanical elastic distortions are due to the mobility of MNPs, imprisoned in the polymer network, in response to the attractive magnetic field forces. The mechanical distortions of hydrogels can be converted into forces and potentially interact with the entrapped liposomes, prompting the magnetically controlled release of the target bioactive molecules due to the structural destabilization of these vesicles and/or changes in the permeability of the liposome membranes. Magnetically controlled delivery may be thus obtained based on the mobility of the MNPs triggered by permanent or low-frequency magnetic field stimuli exploring the mechanical actuation of magnetic-responsive hydrogels, without thermal effects and thus extending the application of magnetic liposomal systems to compounds showing lower resistance to higher temperatures.

The present study proposes the development of magnetic-responsive liposomal hydrogels, which may be able to provide a sustainable magnetically controlled drug release prompted by the mechanical responses of magnetic-responsive hydrogels when exposed to an external magnetic field and not based on local thermal changes resulting from hyperthermia effects. The liposomal hydrogels were prepared by encapsulation of 1,2-dipalmitoyl-sn-glycero-3-phosphocoline (DPPC) liposomes loaded with ferulic acid (FA) into a gelatin matrix containing dispersed MNPs. FA was selected as a representative model of small therapeutic molecules with poor bioavailability. Furthermore, the FA's well-known anti-inflammatory and pro-angiogenic properties were also considered, envisaging the future application of these liposomal hydrogel systems in long-term therapeutics requiring sustained controlled drug release and/or as part of tissue engineering tools for enhanced tissue repair [57]. The first stage of this work is focused on the optimization of the DPPC:FA liposome formation procedures through the evaluation of the effect of different liposome preparation parameters, such as the dilution factor, sonication, chemical formulation and purification, on the DPPC:FA liposome characteristics. In a later stage, in-depth kinetic studies were carried out, aiming to provide good knowledge on the impact of the liposome encapsulation in the hydrogel matrix and on the effect of a magnetic field on the FA release kinetics and mechanisms while determining the possible use of these magnetic liposomal delivery systems in long-term treatments.

2. Materials and Methods

2.1. Materials

Gelatin from porcine skin, potassium phosphate monobasic (≥99%), sodium phosphate dibasic (≥99%), sodium chloride (≥99%), potassium chloride (≥99%), chloroform (>99%), methanol (>99%), glycerol (>99.5%) and Trans-Ferulic acid were all purchased from Sigma-Aldrich (St. Louis, MO, USA). 16:0 PC (DPPC) 1,2-dipalmitoyl-sn-glycero-3-phosphocoline was obtained from Avanti Polar Lipids, Inc. (Alabaster, AL, USA).

2.2. Methods

2.2.1. Synthesis and Characterization of the Iron Oxide Nanoparticles

MNPs were prepared following the experimental procedures described in a previous publication [58]. Briefly, the MNPs were synthesized, in alkaline media, by chemical co-precipitation of two iron salts: $FeCl_3$ and $FeCl_2$. An aqueous solution of 25% (v/v) ammonium hydroxide was added to the salt mixture, under permanent stirring at 1250 rpm, at 80 °C, in a N_2 atmosphere. The MNPs synthesized were characterized by transmission electronic microscopy (TEM) using a Hitachi H8100 TEM with a LAB6 filament and an acceleration tension of 200 kV to access the MNPs' size and size distribution.

2.2.2. Liposome Preparation, Loading and Purification

Multilamellar vesicles (MLVs) composed of 1,2-dipalmitoyl-sn-glycero-3-phosphocholine (DPPC) were prepared using the thin-film hydration method following the procedures described by Pires et al. [59]. Briefly, it consisted of the dissolution of the lipidic component (DPPC) in an organic phase, which, in this case, was a methanol/chloroform (1:4) solution, followed by the formation of a lipidic film in the recipient walls by solvent evaporation by exposure to an inert gas stream (N_2 gas stream) for three hours. Finally, the lipidic film was hydrated with a phosphate buffer solution (PBS) at 55 °C for at least 1 h under constant stirring at 320 rpms, leading to the formation of unloaded DPPC vesicle systems. The DPPC vesicles were exposed to different sonication times varying from 15 min to 45 min, aiming to avoid the presence of larger vesicle aggregates. These unloaded vesicle systems (DPPC-unloaded liposomes) were used as the control samples for comparative terms.

DPPC liposomes loaded with FA (DPPC:FA liposomes) were prepared following the methodology described above but including the FA loading step, which was carried out by adding the FA to the organic phase. The obtained vesicle systems were sonicated for different times varying between 15 min and 60 min. The effect of sonication time on liposome structural characteristics was determined based on dynamic light scattering (DLS) measurements, which allowed for the selection of the ideal sonication time based on the liposome size heterogeneity assessed by analysis of the polydispersity index (PI).

Liposomes with different chemical formulations were prepared using a constant DPPC concentration of 2 mM and different FA concentrations in a way to obtain liposomes with 3:1, 10:1 and 30:1 relative DPPC:FA mass fractions in order to evaluate the influence of chemical formulation on the liposome dimension, size heterogeneity, encapsulation efficiency (% EE) and loading capacity (% LC).

Liposomes were purified using a dialysis bag of regenerated cellulose with a molecular weight cut-off of 14 kDa (Spectra/Pro, Biotech, USA) for removal of the unloaded FA remaining in solution after liposome formation. The dialysis bag was filled with the liposomal solution and immersed in a dialysate PBS solution, at pH 7.4, with a volume 100× higher than that of the liposome solution, under constant stirring for 40 h, at room temperature. The dialysate was exchanged after 18 h. The dialysates collected at 18 h and 40 h were analyzed by UV-Vis, and the absorbance was collected in a range from 200 nm to 900 nm to verify the efficiency and completeness of the unloaded FA removal.

2.2.3. Characterization of the Liposomes

Determination of Liposome Dimension

The dimensions of the loaded (DPPC:FA) and unloaded (DPPC) liposomes over time were determined through dynamic light scattering (DLS) using a nanoparticle analyzer (Nano Partica SZ-100, Horiba Scientific (Kyoto, Japan)). DLS measurements were performed in triplicate after liposome synthesis and purification, allowing for the determination of liposome size distribution, the mean size, the polydispersity index (PI) and the respective standard deviations.

Entrapment Efficiency (EE) and Loading Capacity (LC) of Ferulic Acid-Loaded Liposomes

The FA entrapment efficiency (EE) was determined for all the liposome suspensions according to Equation (1).

$$EE\% = \frac{M_{Encap}}{M_{Total}} \times 100 \qquad (1)$$

where M_{Encap} is the amount of FA encapsulated in liposomes after dialysis (g) and M_{Total} is the initial amount of FA (g). The M_{Encap} was calculated by subtraction of the total amount of FA used to prepare the liposomal solution by the amount of FA released during dialysis. The amount of FA released during dialysis was assessed by determination of the absorbance at 311 nm (corresponding to the characteristic absorbance band of FA) in the dialysate samples collected over the dialysis time. The FA absorbance was converted to concentration using the Lambert–Beer equation and then to the absolute FA mass value by considering the dialysate volume used in liposome dialysis.

The entrapment efficiency (EE) of FA was confirmed by FA quantification upon liposome lysis. In this case, liposomes loaded with FA were added, after dialysis, to a 10% methanol solution to induce the release of FA upon liposome destruction. The released FA was quantified by determination of the absorbance at 311 nm as described above.

The drug loading capacity (LC) was also calculated using Equation (2).

$$LC\% = \frac{M_{Encap}}{M_{Lipid}} \times 100 \qquad (2)$$

where M_{Encap} is the same parameter as in the previous equation and M_{Lipid} is the total amount of lipid used to prepare the liposomal formulation.

2.2.4. Preparation of the Liposomal Hydrogel Membranes

The DPPC:FA liposomes were encapsulated into gelatin matrices/membranes by using active and passive approaches immediately after dialysis to minimize the possibility of liposomal aggregation and fusion. Active encapsulation was carried out by diffusion of unloaded (DPPC liposomes) and FA-loaded DPPC liposomes (DPPC:FA liposomes) into magnetic-responsive gelatin membranes previously prepared according to the procedure described by Manjua et al. [58]. Briefly, gelatin matrices were prepared by flat casting an 8% (w/v) porcine skin gelatin solution, containing 0.25% (w/v) of dispersed MNPs, in a silicone round mold and keeping it at 4 °C overnight for gelification. The gelatin hydrogel membranes doped with the MNPs were gently removed from the silicone round mold after gelification and crosslinked by immersion in 50 mL of an aqueous solution containing 1% (v/v) of glutaraldehyde (GA) for three hours at 4 °C. After crosslinking, the hydrogel membranes were washed with Milli-Q water for the removal of loosely bound components. Membrane washing was performed by immersion of the membranes in 10 mL of Milli-Q water under permanent stirring at 220 rpm. The washing solution was renewed every 10 min and analyzed in a UV-Vis spectrophotometer in a range from 200 nm to 900 nm. This washing procedure was repeated until the total disappearance of the bands corresponding to the release of unbound materials (loosely bound GA, gelatin molecules and MNPs). A minimum of five washing cycles were needed for complete membrane washing.

Passive encapsulation of the liposomes was performed by direct mixing of the liposomal solution into 8% gelatin solution at 40 °C. This mixture was stirred at 300 rpms for 15 min. This gelatin solution was then cast on a silicone round mold (0.2 mL/cm^2) and kept at 4 °C, overnight, for gelification.

2.2.5. Characterization of the Hydrogel Membranes

Swelling Ability

Small hydrogel membrane pieces of 1 cm^2 were cut and dried for 2 h in a fume hood. The membrane samples were immersed in 5 mL of PBS buffer (pH 7.4) and kept in an incubator at 37 °C under constant stirring at 220 rpm. The membrane weight and dimensions were determined before the immersion and frequently monitored during membrane immersion for 28 h. The swelling ratio was calculated using Equation (3).

$$\%\text{Swelling ratio} = \frac{W_s - W_d}{W_d} \times 100 \qquad (3)$$

where W_s and W_d correspond to the weight of swollen and dry hydrogel membranes, respectively.

Determination of the Hydrophilic Character and Water Uptake

The surface properties of the hydrogel membranes were assessed by determining the surface contact angle using a drop shape analyzer (KSV, CAM 100), coupled to a video camera for image acquisition. The surface contact angles were determined using glycerol as a solvent, instead of water, to minimize the penetration of the drop into the hydrogel matrix. The contact angles were determined as the average of 6 measurements.

Analysis of the Structural Integrity of the Liposomes-Encapsulated Hydrogels

The liposomal-based hydrogels were analyzed by optical microscopy to infer the impact of the encapsulation process on the structural integrity of the liposomes. Optical microscopy analyses were performed using an optical microscope (Nikon, Eclipse Ci) adapted to a digital camera (ProgRes CT3). The liposomal-based hydrogel membranes were analyzed with 10× eyepiece magnification and objective magnification ranging from 4× to 50×.

Chemical Characterization of the Liposomal Hydrogels

ATR-FTIR spectroscopy analysis was performed to evaluate the hydrogel crosslinking and for inspecting the presence of possible interactions of MNPs and DPPC liposomes with the gelatin matrix. ATR-FTIR spectra were collected using an FT-IR spectrometer (PerkinElmer (Shelton, CT, USA), Spectrum Two) in the spectral region from 400 to 4000 cm^{-1}, at a 1 cm^{-1} resolution.

2.2.6. Ferulic Acid Release Assay

FA release assays were performed, aiming at characterizing the FA release kinetics and mechanisms from liposomal hydrogel membranes while understanding the contribution of the liposome, hydrogel network and magnetic field in the regulation of the FA release. The role of the liposome and magnetic field stimuli on the controlled release of FA was determined by a comparative analysis of the FA release from liposomal hydrogels with that obtained from hydrogels doped with FA (without liposomes) and from DPPC:FA liposomal suspensions. The FA release was conducted in the presence and absence of a magnetic field to assess the effect of magnetic stimulation on the FA release profiles.

The liposomal hydrogel systems used in the release assays were prepared by passive encapsulation of 3:1 and 10:1 DPPC:FA liposomes in an 8% (w/v) porcine gelatin solution containing dispersed MNPs with concentrations of 0%, 0.25% and 1% (w/v) and crosslinked with GA, as described in Section 2.2.4.

In this regard, small pieces of the different hydrogel membranes of 1 cm^2 were immersed in 5 mL of fresh PBS and placed in an incubator with an orbital shaker (Incubating Light Duty Orbital Shakers, Ohaus, Parsippany, NJ, USA) at a controlled temperature of 37 °C (physiological temperature). The PBS solution was frequently sampled over time and analyzed in a UV-Vis spectrophotometer for absorbance spectra acquisition at the spectral region of the FA, considering the absorbance band at 311 nm to avoid superimposition with the absorbance signal from proteins at 280 nm and possible interferences from dissolved gelatin. In the experiments conducted with magnetic field stimulation, a neodymium bar allowing a magnetic field intensity of 0.08 T was placed under the beaker containing the immersed hydrogel sample.

The release assays from liposomal solutions were carried out following identical procedures. However, in this case, the liposome solution was kept inside a regenerated cellulose dialysis bag with a molecular weight cut-off of 14 kDa (Spectra/Pro, Biotech, USA), which was then immersed in a PBS dialysate solution (pH 7.4) and placed inside an incubator with controlled temperature at 37 °C under constant orbital shaking. The FA release was monitored by periodic measurement of the dialysate absorbance at 311 nm. In this case, the resistance of the dialysis bag to the FA transport was considered negligible, considering the differences between the MWCO of the dialysis bag (14 kDa) and the FA molecular size (194.18 g/mol).

The FA release kinetics and mechanisms were studied by adjustment of the experimental data points to release kinetic models, such as zero-order, 1st-order, Higuchi and Korsmeyer–Peppas models expressed by Equations (4)–(7) and their respective linearized forms [60,61].

$$\frac{dM_R}{dt} = k, \text{ zero-order model} \tag{4}$$

$$M_R = kt, \text{ linearized zero-order model}$$

$$\frac{dM_R}{dt} = k \times M_R, \text{ 1st-order model} \tag{5}$$

$$Ln(M_R) = Ln(M_{R0}) \times kt, \text{ linearized 1st-order model}$$

$$M_R = k_H t^{0.5}, \text{ Higuchi model} \tag{6}$$

$$Log(M_R) = Log(k_H) + 0.5 Log(t), \text{ linearized Higuchi model}$$

$$\frac{M_R}{M_T} = k_{KP} \times t^n, \text{ Korsmeyer–Peppas model} \tag{7}$$

$$Log\left(\frac{M_R}{M_T}\right) = Log(k_{KP}) + nLog(t), \text{ linearized Korsmeyer–Peppas model}$$

where M_R is the amount of the FA released to dissolution media at time t, M_T is the total amount of FA in the delivery platform (i.e., the hydrogel membrane, the liposome and the liposomal hydrogel) at the beginning of the experiment (t = 0 min), n is the exponential diffusion, and k_0, k_1, k_H and k_{KP} are the zero-order, 1st-order, Higuchi and Korsmeyer–Peppas release constants, corresponding to the constant release rates.

The FA release kinetics and mechanism were determined by a comparative analysis of the values obtained for the different release parameters.

3. Results and Discussion

3.1. Synthesis and Characterization of FA-Loaded Liposomes

The first stage of this work was focused on the synthesis of liposomes with fine-tuned characteristics, i.e., low-size heterogeneity, aiming to obtain monodispersed liposomal suspensions and thus avoiding the presence of vesicle aggregates. In this regard, the effect of different liposome preparation variables, such as the dilution factor of the liposome suspensions, sonication time, chemical formulation and purification, on liposome dimension and size heterogeneity was studied to ensure the optimal adjustment of the parameters, which may lead to the preparation of liposomes with desirable characteristics. The liposome dimension and size heterogeneity were evaluated based on the vesicle mean size and the polydispersity index (PI) values obtained by DLS analysis.

Table 1 shows the mean size and PI values obtained for unloaded liposomes with different dilution factors and prepared by exposure to different sonication times.

Table 1. Mean sizes and the respective PI values obtained for DPPC liposomes dispersed in PBS solution with different dilution factors and exposed to different sonication times.

Liposome	Dilution Factor	Sonication Time (min)	Mean Size (nm)	PI
DPPC_5_30	5×	30	705.3 ± 155.8	0.52 ± 0.03
DPPC_10_30	10×	30	559.5 ± 27.6	0.76 ± 0.21
DPPC_5_45	5×	45	270.6 ± 139.2	0.53 ± 0.09
DPPC_10_45	10×	45	237.7 ± 65.4	0.51 ± 0.12

As shown in Table 1, a remarkable decrease of ca. 60% in the liposome mean sizes was observed with the increase in the sonication time from 30 min to 45 min. The observed decrease in liposome sizes from 500–700 nm to the 200–300 nm range suggested an increase in the content of small unilamellar vesicles (SUVs) to the detriment of the multilamellar vesicles (MLVs). The increase in the dilution factor also led to a decrease in the liposomal size, but, in this case, the effect was much less expressive than that obtained by the increase in the sonication time. The effect of the sonication time and the dilution factor on the PI values was not totally evident. However, PI values varying from 0.51 ± 0.12 and 0.53 ± 0.09 were consistently obtained when increasing the sonication time to 45 min. The DLS correlograms respective to the liposomes described in Table 1 were added in Figure A1A in Appendix A for a better perception of the effect of the sonication time on the liposome polydispersity (PI). A comparative analysis of the DLS correlograms in Figure A1A shows that the increase in the sonication time is followed by a narrowing of the correlation function, compatible with the increase in liposome size homogeneity (e.g., higher size monodisperse character) in suspension.

The impact of FA loading in the liposome and the liposome purification step on the liposome morphology was then inspected. DPPC liposomes loaded with FA (DPPC:FA liposomes) were prepared using different DPPC:FA ratios of 3:1, 10:1 and 30:1. Table 2 shows the mean size and PI values obtained for a non-diluted solution of the DPPC:FA liposomes, upon 60 min sonication, before and after purification by dialysis.

Table 2. Mean size and PI values obtained for non-diluted solutions of unloaded DPPC liposomes and DPPC liposomes loaded with 3:1, 10:1 and 30:1 DPPC:FA ratios, upon 60 min of sonication, before and after purification by dialysis.

Liposome	Purification	Mean Size (nm)	PI
DPPC	No	106.6 ± 45.9	0.36 ± 0.34
DPPC:FA_30:1		716.5 ± 103.7	1.49 ± 0.44
DPPC:FA_10:1		1385.7 ± 443.3	0.12 ± 0.01
DPPC:FA_3:1		1200.0 ± 257.8	0.12 ± 0.11
DPPC:FA_30:1	Yes	714.1 ± 152.9	0.71 ± 0.01
DPPC:FA_10:1		992.6 ± 121.0	0.15 ± 0.06
DPPC:FA_3:1		775.9 ± 39.4	0.17 ± 0.02

The sonication time was further increased from 45 min (Table 1) to 60 min (Table 2), allowing for an additional reduction in the liposome size to 106.6 ± 45.9 and size polydispersity, expressed by a reduction in PI values of the unloaded DPPC, to 0.36 ± 0.34. The impact of sonication time on the dimension of FA-loaded liposomes was also studied (Table A1 in Appendix A), and it was observed to produce a significant decrease in the PI values from the value of 0.63 ± 0.05 obtained for DPPC:FA 10:1 liposomes (Table A1 in Appendix A) to 0.12 ± 0.01 for DPPC:FA 10:1 liposomes (Table 2). No significant changes were noticed in DPPC:FA liposomes with a lower FA fraction, i.e., DPPC:FA 30:1 liposomes, suggesting a possible role of FA in the structural characteristics of these liposomes. In fact, Table 2 also shows that the encapsulation of FA led to an increase in the liposomal size, more evident in liposomes with higher FA fractions, i.e., 3:1 and 10:1 DPPC:FA liposomes, followed by a significant decrease in the PI to half of its value, thus evidencing the positive contribution of FA for the size homogeneity of the liposomal suspension.

Also, liposome purification by dialysis was found to have a remarkable effect on the liposome size, leading to a significant decrease in the average size of the liposomes from the 1200–1400 nm range to values below the microscale dimensions (<1000 nm), as illustrated in Figure 1 and Table 2. Minor changes were observed in the mean size of liposomes formed by the lowest FA proportion of 30:1 DPPC:FA. The liposome mean sizes obtained before purification suggested the predominance of MLVs with the potential presence of vesicle aggregates. Yet, despite the decrease in the average liposome sizes, MLVs were still predominant after dialysis, with the 3:1 DPPC:FA sizes varying from 750 nm to 1000 nm, whereas the 10:1 DPPC:FA sizes were mostly contained in the 750 nm–1750 nm range, as illustrated in Figure 1. The effect of dialysis was also noticed in the DLS correlograms of purified and non-purified liposomes shown in Figure A1B, confirming the efficiency of dialysis in removing possible forming aggregates and large vesicles present in suspension. However, it is noteworthy that these changes in the liposome size distribution did not have a pronounced effect on the size heterogeneity of liposomes with higher FA content (3:1 and 10:1 DPPC:FA liposomes), as reflected by the insignificant changes in PI values (Table 2). Figure A1C also illustrates the DLS correlograms obtained for DPPC:FA liposomes after dialysis, prepared by an optimized selection of the liposome synthesis variables considered in the present work.

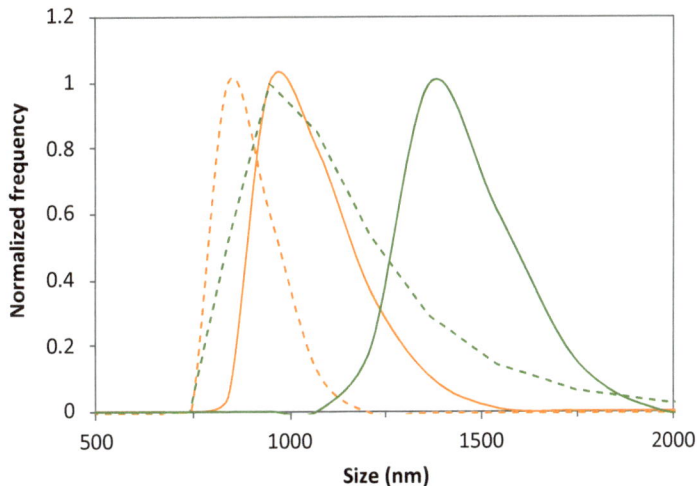

Figure 1. Illustration of the effect of FA loading, purification by dialysis and DPPC:FA formulation on the liposome size distribution. Size distribution of 3:1 (orange lines) and 10:1 (green lines) DPPC:FA liposomes before (solid lines) and after (dashed lines) dialysis.

The entrapment efficiency (EE) and loading capacity (LC) obtained for FA-loaded DPPC liposomes prepared using different DPPC:FA ratios were determined using Equations (1) and (2), respectively. The results obtained are listed in Table 3.

Table 3. Encapsulation efficiency (EE) and loading capacity (LC) obtained for liposomes prepared with different DPPC:FA ratios, upon purification by dialysis.

Liposome	EE (%)	LC (%)
DPPC:FA_30:1	62.1	1.28
DPPC:FA_10:1	73.53	6.73
DPPC:FA_3:1	88.06	14.01

The results depicted in Table 3 show that the increase in the FA fraction in the liposomal formulations led to the increase in entrapment efficiency, EE, and loading capacity, LC. This result was already expectable as it results from the higher availability of FA for encapsulation in the lipidic bilayer of liposomes per amount of the lipidic component.

Considering the characteristics of the prepared DPPC:FA liposomes described above, the MLVs with the highest fractions of FA, i.e., 10:1 and 3:1 DPPC:FA, were selected for the development of the liposomal hydrogels described in the following sections of this paper.

3.2. Design and Characterization of the Liposomal Gelatin Membranes

The main goal of this work was to develop liposomal hydrogel delivery systems allowing for a magnetically modulated and long-term drug release while providing a deeper understanding of the impact of liposomal encapsulation and the effect of the magnetic field on the release mechanisms and kinetics of small bioactive molecules with poor bioavailability. In this regard, FA, a small organic bioactive molecule with anti-inflammatory and pro-angiogenic properties, was used as a representative model molecule in this work.

The magnetic-responsive liposomal hydrogels were prepared by encapsulation of the MLV DPPC liposomes loaded with FA (3:1 and 10:1), previously described and listed in Table 2, into gelatin membranes containing dispersed iron oxide nanoparticles (MNPs)—magnetic-responsive gelatin. On the one hand, the liposomes were used as drug reservoirs offering a more efficient imprisonment of small therapeutic molecules than that achieved by direct drug immobilization in hydrogel matrices. On the other hand, the unique mechanical behavior of magnetic-responsive hydrogels (magnetic-responsive gelatin in the case of the present study) in the presence of magnetic stimuli was explored for improved in situ and on-demand regulation of FA delivery.

3.2.1. Encapsulation of Liposomes into the Gelatin Membranes

MLVs are vesicles consisting of multiple concentric lipid bilayers, which are considered less effective drug delivery systems than SUVs as they are more prone to fusion and aggregation, which may result in the loss of their enclosed payload. Furthermore, the possible immunogenicity and toxicity of MLV liposomes can also present difficulties; hence, SUVs tend to be preferred for more accurate targeting and effective cellular uptake. However, MLVs present specific characteristics that make them more attractive considering the primary objective of the present work, which is the development of liposomal hydrogel systems allowing for long-term drug storage and transportation. MLVs are larger than SUVs, allowing for a higher capacity for drug loading, and their multilayered lipid bilayers enable higher tolerance to harsh conditions such as pH changes or temperature variations while working as additional barriers to drug transport, delaying drug release and thus providing a controlled and prolonged drug release profile. Furthermore, the additional premises of using MLVs were that (i) MLVs largely remain intact during the hydrogel preparation process (crosslinking method), contrary to SUVs (higher tendency to rupture); (ii) the mechanical stability of SUVs, even in a hydrogel matrix, can be lower compared to the MLVs; and (iii) the lower size of SUVs can enable their easier diffusion through the gelatin hydrogel, and their unilamellarity tends to destabilize and immediately releases high levels of ferulic acid (burst release).

The encapsulation of DPPC:FA liposomes into gelatin membranes was attempted by active and passive immobilization procedures. Active immobilization involved the diffusion of the liposomes from the liposomal solution by swelling of the pre-formed magnetic-responsive gelatin membranes crosslinked with glutaraldehyde. Passive immobilization consisted of the direct addition of the liposome solution to the gelatin solution containing dispersed MNPs with an average size of 12.7 nm (Figure 2A) before casting and gelification. The liposomal gelatin membranes were then immersed in a glutaraldehyde solution for crosslinking, rendering magnetic-responsive gelatin membranes resembling the one shown in Figure 2B.

The liposomal gelatin membranes obtained by active and passive liposome encapsulation were inspected by using optical microscopy. Microscopy analysis of the liposomal hydrogels was performed for evaluation of the efficiency of the different experimental approaches used for encapsulation based on the presence of the encapsulated liposomes, the liposome dispersion quality and structural integrity. Figure 3 shows the images obtained through optical microscopy of the liposome hydrogel membranes with different crosslinking degrees, prepared by active and passive immobilization of DPPC:FA liposomes, before washing (on the left). The images shown on the right side correspond to the same respective gelatin membranes after washing by immersion for different durations (from 0 to 6 h) in PBS at pH 7.4 and 37 °C.

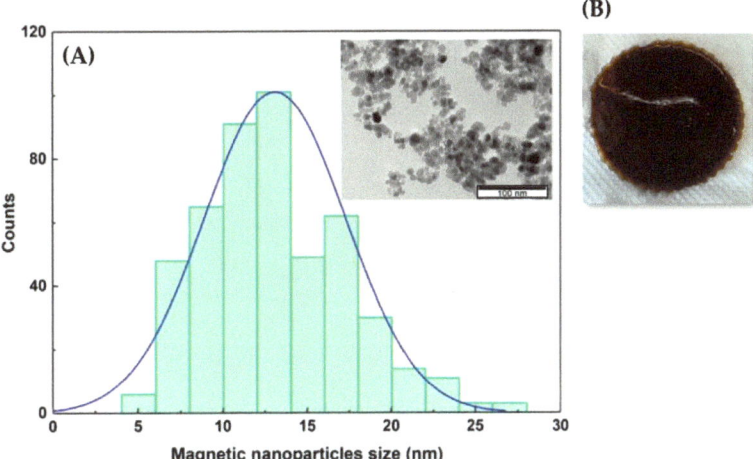

Figure 2. (**A**) Size distribution of the MNPs adjusted to a Gaussian function. Inset. Image of the MNPs obtained by transmission electron microscopy (TEM). (**B**) Photo of the magnetic-responsive gelatin doped with 1% of MNPs.

Several randomly dispersed vesicles resembling the typical structure of liposomes and/or liposome aggregates were identified before hydrogel washing in hydrogels prepared by active immobilization. A comparative analysis of Figure 3A,C,E seems to indicate that the number of immobilized vesicles depended on the hydrogel crosslinking degree, decreasing with the increase in the hydrogel crosslinking reaction time from 3 h to 6 h. The decrease in the number of immobilized vesicles was possibly explained by the limited diffusion of the liposomes into the hydrogel network, which was expected to increase with the increase in the crosslinking degree due to the reduction in the dimension of the hydrogel voids. An inspection of the microscopy images obtained for liposomal hydrogel membranes after washing, as shown in Figure 3B,D and F, revealed the total loss of the liposomal vesicles from the membranes during the washing process, suggesting that active immobilization led to a non-efficient internalization of liposomes into the gelatin matrix.

The removal of liposomes from the hydrogel membranes was possibly due to the limited access of liposomes to hydrogel cavities with reduced dimensions or located in less accessible regions of the crosslinked hydrogel, which would assure more efficient imprisonment of the liposomes. The limited diffusion of liposomes into the hydrogel network forced their location at the surface of the hydrogel and in the larger cavities in the hydrogel matrix, allowing for easier liposome removal by back diffusion to the clean washing solution, upon hydrogel swelling at 37 °C, due to the osmotic difference effect.

Finally, the passive immobilization of liposomes was tested, aiming at reducing the diffusional limited encapsulation of the liposomes. The microscopy images obtained for liposomal hydrogels prepared by passive immobilization in Figure 3G,H seem to indicate a much lower number of entrapped vesicles than that observed for liposomal hydrogels prepared by active immobilization, before washing (Figure 3G). Despite the reduced number, the immobilized vesicles showed good structural integrity and a liposome-like structure, and, thus, they were also ascribed to the presence of liposomes (non-discarding the possible presence of liposomal aggregates).

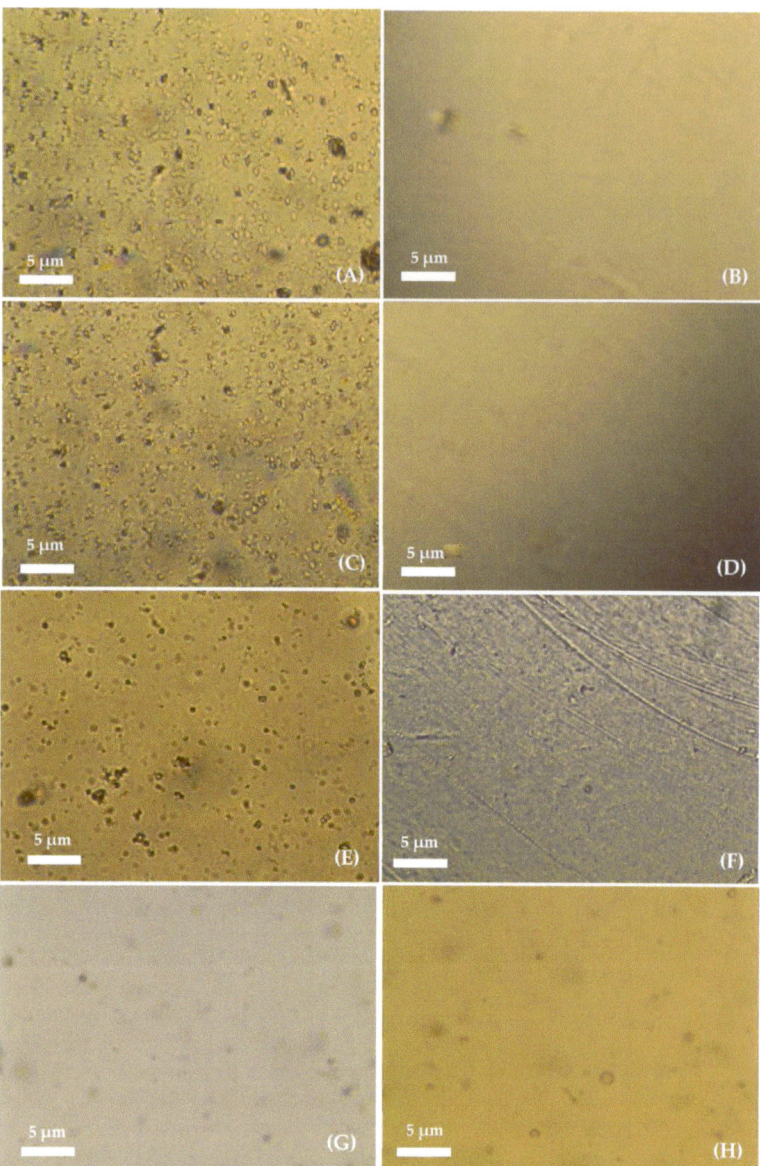

Figure 3. Images of the non-crosslinked liposomal-based hydrogel prepared by active immobilization of DPPC:FA liposomes acquired (**A**) before and (**B**) after washing for 6 h; liposomal-based hydrogel prepared by active immobilization of DPPC:FA liposomes in a gelatin membrane crosslinked for 3 h acquired (**C**) before and (**D**) after washing for 6 h; liposomal-based hydrogel prepared by active immobilization of DPPC:FA liposomes in gelatin membranes crosslinked for 6 h acquired (**E**) before (**F**) and after washing for 6 h; and liposomal-based hydrogel prepared by passive immobilization of DPPC:FA liposomes crosslinked for 3 h acquired (**G**) before and after (**H**) washing for 6 h. Hydrogel washing was performed through immersion of the hydrogels in PBS solution with pH 7.4 at 37 °C. All images were taken with 10× magnification with 10× eyepiece.

The comparative analysis of the hydrogel images before (Figure 3G) and after washing (Figure 3H) revealed a minimal loss of liposomes during the washing step, evidencing that passive immobilization allowed for more efficient development of liposomal hydrogel delivery systems with improved stability (with respect to the capacity to retain the encapsulated liposomes).

The ability of the gelatin membranes to retain and stabilize the encapsulated liposomes and to regulate FA release is very much dependent on their swelling capacity. The swelling ability is influenced by the temperature, the capacity of the crosslinking agent to covalently bind two polymer chains and the potential contribution of the embedded liposomes. Liposomes might possibly produce two opposite effects: 1. liposomes may be able to interact with the gelatin molecules, increasing the structural cohesion of the hydrogel matrix and contributing positively to the hydrogel stability but negatively to the hydrogel swelling capacity, and 2. liposomes may potentially perturb the access of the crosslinking agent to the gelatin molecules through a stereochemical effect, thus reducing the crosslinking efficiency. In this last case, the liposome would exert a negative contribution in terms of hydrogel stability and a much possible positive contribution in terms of the hydrogel swelling capacity.

3.2.2. Determination of the Liposomal Gelatin Membranes Swelling Ability

Studies were performed to evaluate the effect of liposomes and MNPs on the swelling ability of the gelatin membranes under specific physiological conditions, i.e., at pH 7.4 and 37 °C.

The swelling ratios of the gelatin membranes with embedded MNPs (magnetic-responsive gelatin membranes), gelatin membranes with encapsulated liposomes (without MNPs) and magnetic-responsive gelatin membranes containing 0.25% and 1% MNPs, with encapsulated liposomes, were determined using Equation (3) and compared with that obtained for native gelatin membranes, i.e., without any encapsulated component. The swelling ability of each liposomal gelatin membrane was determined based on an analysis of the swelling ratio profiles over time, represented in Figure 4.

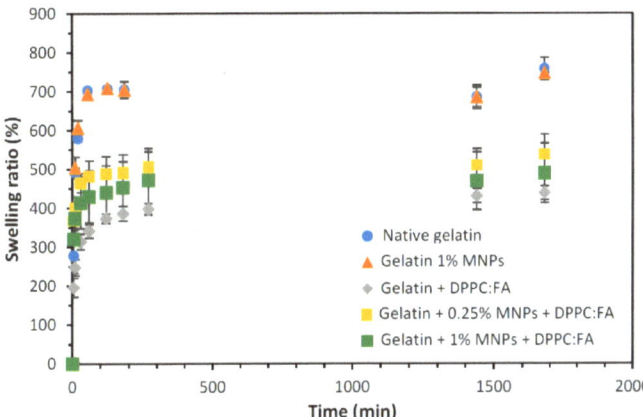

Figure 4. Swelling ratio obtained for gelatin membranes without (dark blue circles) and with 1% of MNPs (orange triangles), gelatin membranes with internalized DPPC:FA liposomes (grey diamonds) and magnetic-responsive gelatin membranes doped with 0.25% (yellow squares) and 1% MNPs (green squares) and with encapsulated 3:1 DPPC:FA liposomes, by immersion in PBS solution at 37 °C.

The analysis of the swelling ratio profiles revealed a more accentuated increase in the gelatin membrane swelling during the first hour, reaching a plateau after 120 min. Native gelatin membranes and gelatin membranes with embedded MNPs showed identical swelling ratios, evidencing that the presence of MNPs (in the concentrations used in this study) does not significantly affect the swelling and water uptake capacity of these membranes. Contrastingly, a remarkable decrease in the maximum swelling was observed for gelatin membranes with immobilized DPPC:FA liposomes, suggesting the contribution of these liposomal structures to an additional cohesion of the gelatin matrix. A similar effect has been reported by other authors. Wu et al. [46] and Lee et al. [62] showed that the encapsulation of liposomes in hydrogel matrices improved the hydrogel strength, increasing the compressive modulus of the liposomal gel and thus justifying a higher resistance to deformation and lower swelling capacity. In a previous work, Lee et al. [62] explained the increase in the strength of gelatin-DPPC liposomal systems due to the interaction of liposomes with the hydrophobic moieties in a gelatin network, which acted as additional crosslinkers. However, given the diversity of the functional groups present in a protein chain and the high hydrophilicity of the hydrogel, the crosslinking effect of liposomes may also be plausibly explained by the interaction of the polar phospholipidic heads of the DPPC located at the external surface of liposomes with the hydrophilic moieties present in the gelatin chain.

The active contribution of the liposomes in hydrogel crosslinking may explain the higher retention of liposomes in liposomal hydrogel systems prepared by passive immobilization, as the addition of liposomes to the polymer solution before gelification may render the imprisonment of liposomes in more internal hydrogel regions, followed by a decreased liposome back diffusion ascribed to the lower swelling ability of these liposomal gelatins.

3.2.3. Chemical Characterization of the Liposomal Gelatin Membranes

The ATR-FTIR spectra of the liposomal gelatin membranes were determined and analyzed to obtain information on the chemical characteristics of the gelatin membranes and to confirm the presence of FA, MNPs and DPPC:FA liposomes in magnetic-responsive liposomal gelatin membranes. Figure 5 shows the transmittance spectra obtained for the different gelatin membranes.

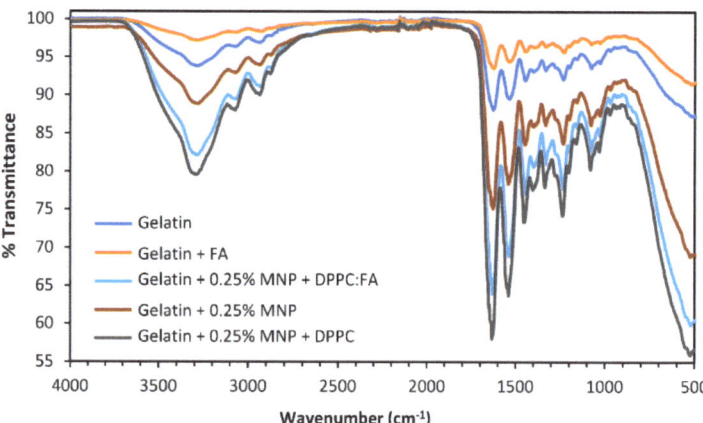

Figure 5. ATR-FTIR spectra for the native gelatin membranes with (orange line) and without (dark blue line) immobilized FA (without liposomes) and magnetic-responsive gelatin doped with 0.25% MNPs without liposomes (brown line), with unloaded liposomes (dark grey line) and with liposomes loaded with FA (light blue).

Similar FTIR spectra were obtained for all membranes analyzed, mostly characterized by the presence of the bands located in the same spectral region, indicating that the presence of FA, MNPs and DPPC:FA liposomes, possibly due to their small amounts in the membrane, have little influence on the FTIR spectra. A broad band identified between

3600 cm^{-1} and 3100 cm^{-1} was attributed to the -NH and -OH stretching vibrations [63], visible for all samples. The spectral region from 1700 cm^{-1} to 800 cm^{-1} shows major spectral bands at 1630 cm^{-1}, 1539 cm^{-1} and 1239 cm^{-1}, characteristic of gelatin spectra [63] and attributed, respectively, to the C=O and C-N stretching vibrations of the amide carbonyl group in Amide I, the N-H and C-N stretching vibrations of groups in Amide II and C-N and N-H stretching vibrations in Amide III.

A small spectral band at 523 cm^{-1} only observed for gelatin membranes embedded with MNPs was attributed to the presence of the MNPs. As reported in further literature, MNP's characteristic band is located from 580 cm^{-1} to 400 cm^{-1} [64], denoting a spectral shift for MNPs embedded in the gelatin membranes. This spectral shift from 580 cm^{-1} to 523 cm^{-1} might be explained by the low MNP concentrations, which make their detection difficult.

FA is the most difficult element to identify in these spectra, which may result from its low solubility in aqueous phases, justifying a low FA content outside liposome structures. Actually, the quantification of FA performed through UV-Vis analysis, after the destruction of a membrane containing 0.25% MNPs and embedded DPPC:FA liposomes, showed that the total amount of FA in the membranes was as low as 2.5×10^{-5} g/cm^2. Furthermore, most of the characteristic spectra bands of FA are located between 1668 cm^{-1} and 685 cm^{-1} and are coincident with many of the characteristic spectral bands of gelatin, which also perturbs the detection of FA spectral signals.

3.3. Ferulic Acid Release Assays

The FA release profiles obtained from liposomal gelatin membranes embedded with 3:1 and 10:1 DPPC:FA liposomes were determined and compared with those obtained with native gelatin membranes with immobilized FA and with a free MLVs DPPC:FA liposome suspension in order to understand the impact of the hydrogel matrix on FA release from liposome delivery systems.

The liposomal hydrogels used for FA release assays were prepared via passive encapsulation of DPPC:FA liposomes, as previously described, and all FA release experiments were conducted at pH 7.4 and 37 °C to mimetic the physiological conditions of the human organism.

As shown in Figure 6, the FA release profiles obtained for all delivery systems studied were characterized by a faster FA release in the initial process stage, followed by a decrease in the FA release rates over the process time, suggesting the dependence of the release rate on the FA concentration. This trend was, however, less notorious for liposomal gelatins with encapsulated 3:1 DPPC:FA liposomes, which showed a more linear and smoother FA release over time. The release of FA directly immobilized into the gelatin matrices was complete after 2880 min (48 h) in contrast to that observed for liposome suspensions and liposomal gelatin membranes, which showed delayed FA delivery, with maximum cumulative FA release reaching values of 36.8%, 32.9%, 22.1% and 6.2% for liposome suspensions and liposomal hydrogels with 10:1 and 3:1 DPPC:FA liposomes, respectively, over an identical experimental period of 48 h.

For a deeper comparative analysis, the FA release profiles were adjusted to different mathematical models, such as zero-order (Equation (4)) and first-order (Equation (5)) kinetic models and to the empirical Higuchi (Equation (6)) and Korsmeyer–Peppas (Equation (7)) models, aiming at obtaining a good understanding of the FA release mechanisms and kinetics from the different delivering systems.

The FA release obtained for the different delivery systems was found not to be properly described by zero-order kinetic models, since the cumulative FA release does not depend linearly on time. Also, FA release profiles were not best described by a 1st-order kinetic function (Figure A2, in Appendix A), as evidenced by R^2 values < 0.94 (Table A2, in Appendix A), but they finely fitted to Higuchi models (Figure A3, in Appendix A), as expressed by R^2 values of 0.968 and 0.983 (Table A3, in Appendix A). However, despite the good fitting quality, it is important to notice that diffusional exponent values of 0.641,

0.807 and 0.141 were obtained for native hydrogels with immobilized FA and for 3:1 and 10:1 DPPC:FA liposome suspensions, which are significantly different from the exponential diffusion value of 0.5, characteristic of a Higuchian release. For this reason, it is not possible to conclude that the FA release mechanisms in these delivery systems followed a Higuchian release mechanism.

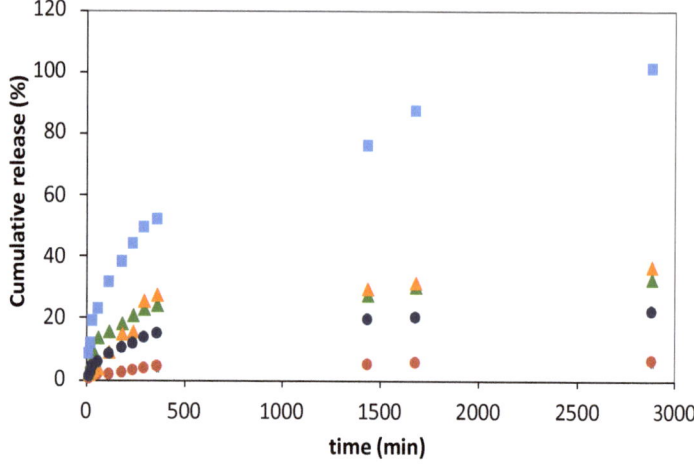

Figure 6. Ferulic acid (FA) release profiles from gelatin membranes with immobilized FA (light blue squares), from liposome suspensions prepared using 3:1 (green triangles) and 10:1 (orange triangles) DPPC:FA fractions and from magnetic−responsive liposomal gelatin membranes doped with 0.25% MNPs containing encapsulated 3:1 (red circles) and 10:1 (blue circles) DPPC:FA liposomes at 37 °C. The cumulative release of FA was determined by $M_R/M_T \times 100$.

FA release profiles from gelatin membranes with immobilized FA and gelatin membranes with encapsulated 3:1 and 10:1 DPPC:FA liposomes were acceptably described by Korsmeyer–Peppas models as shown in Figure A4 in Appendix A, showing R^2 values of 0.980, 0.934 and 0.915 (Table 4).

Table 4. Release constant (k_{KP}) and diffusion exponential (n) obtained for the FA release from gelatin membranes doped with free FA, free liposomes prepared using 3:1 and 10:1 DPPC:FA fractions and 3:1 and 10:1 DPPC:FA liposomes encapsulated in gelatin membranes over 48 h, at 37 °C.

	Hydrogel + FA	DPPC:FA Hydrogel		DPPC:FA Suspension	
		3:1	10:1	3:1	10:1
k_{KP}	3.818 ± 0.022	0.264 ± 0.068	1.104 ± 0.090	0.352 ± 0.197 5.810 ± 0.059	0.238 ± 0.067 12.460 ± 0.080
n	0.428 ± 0.051	0.426 ± 0.028	0.405 ± 0.037	0.920 ± 0.129 0.223 ± 0.022	0.794 ± 0.033 0.129 ± 0.027
R^2	0.980	0.934	0.915	0.926 0.940	0.970 0.875

The FA release mechanism of each delivery system was determined by analysis of the releasing parameters, i.e., the release constant k_{KP} and the diffusion exponent, n, obtained by adjustment of the linearized Korsmeyer–Peppas models to the experimental data points as shown in Figure 7.

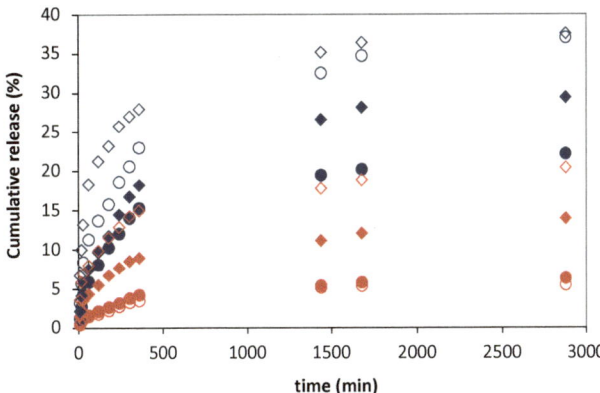

Figure 7. Ferulic acid (FA) release profiles from 3:1 (red symbols) and 10:1 (blue symbols) DPPC:FA liposomes embedded in magnetic-responsive gelatin membranes containing 0.25% MNPs (filled symbols) and 1% MNPs (empty symbols) in the absence (circles) and presence (diamonds) of magnetic field (MF) with an intensity of 0.08 T at 37 °C. The cumulative release of FA was determined by $M_R/M_T \times 100$.

The values obtained for the releasing parameters for each delivery system are listed in Table 4. The release constant, k, expresses the molecular releasing rates, whereas the diffusion exponent, n, indicates the transport mechanism underlying the FA release in each delivery system [60].

The diffusion exponent n was found to be closer to 0.5 for FA delivery from gelatin membranes with dispersed FA and from the DPPC:FA liposomes encapsulated within gelatin membranes, which indicates that the FA release in these systems occurs in a diffusional controlled regime [60], similarly to that previously reported [35]. A comparative analysis of the release constant k_{KP} revealed that the FA release process is significantly faster in gelatin membranes with dispersed FA, showing a release constant k_{KP} of 3.818, than that observed from DPPC:FA liposomes immobilized in gelatin membranes, characterized by k_{KP} values of 0.264 and 1.104 for liposomal hydrogel membranes containing 3:1 and 10:1 DPPC:FA liposomes, respectively, resulting in a remarkably delayed FA release, reaching maximum releasing values of ca. 20% in the same period of 48 h, as shown in Figure 6. The decrease in the FA release rates observed for liposomal hydrogels agrees with previous reports from other authors [35,38], being attributable to the imprisonment of FA into the liposomal vesicles, the higher resistance offered by liposomal hydrogels and the FA transport to the external media. Furthermore, it evidences the important regulatory effect of the liposomal structure in the FA release and its higher capacity to retain FA in comparison to that achieved with native gelatin membranes.

The results also unveiled that FA release from liposomal gelatin membranes is also dependent on the chemical formulation of DPPC:FA liposomes. DPPC:FA liposomes formed by a lower FA proportion, i.e., 10:1 DPPC:FA liposomes, allowed for a faster FA release than that observed for 3:1 DPPC:FA liposomes. Liposomal hydrogels containing dispersed 10:1 DPPC:FA liposomes allowed for 22.1% release of FA in 48 h, leading to a k_{KP} value of 1.104, which contrasts with the 6.2% FA release obtained with the liposomal hydrogels doped with 3:1 DPPC:FA liposomes in the same experimental period, which resulted in a k_{KP} value of 0.264. These results seem to be partially contradictory to those observed by Mourtas et al. [36]. These authors identically reported the existence of a dependence of the release rates of lipophilic molecules, in liposomal hydrogels, on the liposome payloads. However, in contrast with that observed in this work, they concluded that it increased with the increase in the drug load. The delayed FA release observed in the present work from liposomal hydrogels containing liposomes with higher FA loads

(3:1 DPPC:FA liposomes) may hypothetically be explained by the higher stability of these liposomes due to a strong interaction between the FA and the aliphatic chains of the phospholipids in the lipidic bilayer. Due to its hydrophobic character, FA is more likely allocated within the lipidic bilayer than in the hydrophilic liposome core. FA will possibly act as a stabilizer of the lipidic bilayer, creating a more cohesive vesicle and thus enhancing the liposome stability with a consequent decrease in FA release. However, changes in the FA release profiles associated with differences in the structural morphology and mechanical behavior of these liposomal gelatins caused by the different dimensions of the encapsulated liposomes cannot be excluded.

The FA release mechanisms obtained for the hydrogel-based delivery systems contrast with those obtained with DPPC:FA liposome suspensions. The FA-releasing profiles obtained for DPPC:FA liposome suspensions showed FA cumulative releases of 32.9% and 36.8% after 48 h for 3:1 and 10:1 DPPC:FA liposomes, respectively, corresponding to a delayed release of FA comparatively to that obtained with hydrogels with immobilized FA (Figure 6). These results reveal the higher ability of liposomes to regulate the release of small molecules than that obtained with the native gelatin hydrogels. The poorer drug delivery control of gelatin hydrogel membranes is attributed to the large dimension of the gelatin voids, characteristic of the swollen hydrogel networks, allied in this case to a low interaction of FA with the gelatin matrix, considering the different polarities of these two molecules.

It is interesting to note that the analysis of the FA release profiles with the Korsmeyer–Peppas model seemed to indicate that the release of FA from liposome suspensions obeys a dual-release regime (Figure A4 in Appendix A). As observed in this figure, FA delivery consists of two different releasing stages, with each one described by a different Korsmeyer–Peppas function (Table 4), evidencing a change in the FA release mechanism at the mid-term process. As shown in Table 4, in the initial stage, the FA release was characterized by n values of 0.794 and 0.920, respectively, for 10:1 and 3:1 DPPC:FA liposomes, indicating that the FA release occurs through an anomalous Fickian diffusion. The second stage was characterized by an abrupt decrease in the exponential diffusion n to values <0.3, followed by a strong increase in the release rate constant, k_{KP}, from values lower than 0.5 in the first stage to values of 5.81 and 12.46 in the second process stage for 3:1 and 10:1 DPPC:FA liposomes, respectively. This change in the diffusion exponential value, n, evidences a change in the FA release mechanism from an anomalous Fickian diffusion, in the first stage, to a quasi-Fickian diffusion, in the second stage, corresponding to a limited diffusional transport of FA. This dual-release regime may be potentially explained by the lower stability of MLVs in suspension, which justifies a FA release associated with the destabilization/erosion of the outer layers of these vesicles. The FA release observed in the second stage might be explained by the release of FA from the more stable vesicle inner layers/vesicle core, leading to FA release governed by controlled diffusion mechanisms. Yet, this does not clearly explain the rise in the k_{KP} values in the second stage of the regime. The absence of these dual FA release regimes in liposomal hydrogels may thus be once more interpreted as the additional structural stabilization provided by liposome confinement in the hydrogel network, in agreement with that reported in further literature [25].

3.4. Magnetically Controlled Release of Ferulic Acid

Magnetic-responsive hydrogels, i.e., hydrogel matrices doped with magnetic susceptible components, such as the iron oxide nanoparticles (Fe_3O_4), are able to switch their volume and shape when exposed to a magnetic field [54,55]. Magnetic-responsive liposomal hydrogels reported in further literature have been developed for magnetic modulation of drug release based on their ability to produce hypothermia effects when exposed to an alternate magnetic field (AMF), resulting in the thermally induced release of the target drug [49,53]. In contrast, the magnetic-responsive gelatin membranes used in this work for encapsulation of DPPC:FA liposomes may also be used as mechanical actuators capable of producing mechanical forces under magnetic stimulation, which are expected to activate the release of FA from the encapsulated liposomes.

Hence, studies were conducted to evaluate the ability of the magnetic field to control the FA release from magnetic-responsive liposomal hydrogels prepared in this work. In this case, FA release assays from magnetic-responsive gelatin membranes doped with 0.25% and 1% MNPs and embedded 3:1 and 10:1 DPPC:FA liposomes were performed in the absence (reference condition) and presence of an external permanent magnetic field with an intensity of 0.08 T, produced by a neodymium magnet.

Figure 7 shows the FA release profiles obtained for the magnetic-responsive liposomal gelatin membranes exposed and non-exposed to a magnetic field. A comparative analysis of the FA release profiles immediately evidences the effect of the liposome formulation and the %MNP present in the hydrogel membrane on the FA release. The results showed a consistently higher cumulative release of FA, varying between 22.1% and 36.8%, after 48 h for liposomal hydrogels embedded with 10:1 DPPC:FA liposomes (blue symbols), confirming the lower ability of 10:1 DPPC:FA liposomal hydrogel systems to retain the FA molecules.

Furthermore, the results show a higher FA release when the %MNP was increased from 0.25% to 1%, easily perceptible by comparing the filled (0.25 % MNPs) and empty (1% MNPs) symbols in Figure 7, evidencing that MNPs influence the delivery of FA even in the absence of a magnetic field. This effect cannot be explained by differences in the hydrogel swelling capacity, since, as discussed before, the swelling ratio obtained with gelatins with 0.25% MNPs was higher than that obtained with gelatin with a higher MNP concentration (Figure 4). However, it might be due to an additional instability of the FA molecules in the hydrogel matrix resulting from the increased polarity of the hydrogel matrix after increasing the %MNP, associated with the high hygroscopic properties of the MNPs.

The effect of hydrogel formulation on the FA release kinetics and mechanisms was accessed by comparative analysis of the releasing parameters obtained by adjustment of different release kinetic models. The experimental FA release profiles are clearly not described by a zero-order kinetic model since there is not a linear increase in the cumulative mass of the released FA over time. The FA release profiles are also not properly described by first-order kinetic functions, as expressed by the low R^2 values < 0.85 (Table A4) obtained for all delivery systems, denoting a poor fitting quality in each case (Figure A5, in Appendix A). In contrast, the FA release from these magnetic-responsive liposomal hydrogels was perfectly adjusted to Higuchi models (Figure A6A in Appendix A), with R^2 > 0.980 in most cases (Table 5), as well as to Korsmeyer–Peppas models, as evidenced by fittings to the Korsmeyer–Peppas function shown in Figure A6B and confirmed by the resultant R^2 values > 0.960 (Table 5).

Table 5. Release constants (k_H and k_{KP}) and diffusion exponential (n) obtained for 3:1 and 10:1 DPPC:FA liposomal gelatin systems for a period of 360 min in the presence and absence of a 0.08 T magnetic field, at 37 °C.

Liposomal Hydrogel		0.25% MNPs				1% MNPs			
		3:1	3:1_MF	10:1	10:1_MF	3:1	3:1_MF	10:1	10:1_MF
Higuchi	k_H (×10^{-7})	0.225 ± 0.107	0.567 ± 0.042	0.428 ± 0.044	0.464 ± 0.064	0.331 ± 0.080	4.594 ± 0.043	0.837 ± 0.073	2.193 ± 0.057
	n = 0.5	0.651 ± 0.054	0.494 ± 0.021	0.531 ± 0.022	0.551 ± 0.032	0.562 ± 0.040	0.363 ± 0.022	0.501 ± 0.036	0.381 ± 0.029
	R^2	0.983	0.988	0.989	0.991	0.988	0.990	0.984	0.966
Korsmeyer–Peppas	k_{KP}	0.174 ± 0.044	0.550 ± 0.055	0.333 ± 0.103	0.715 ± 0.064	0.129 ± 0.079	1.784 ± 0.043	1.235 ± 0.073	3.237 ± 0.057
	n	0.531 ± 0.022	0.481 ± 0.028	0.663 ± 0.052	0.551 ± 0.032	0.562 ± 0.040	0.363 ± 0.022	0.501 ± 0.036	0.381 ± 0.029
	R^2	0.989	0.990	0.990	0.991	0.988	0.990	0.984	0.966

As shown in Table 5, according to both Higuchi and Korsmeyer–Peppas models, for liposomal hydrogel delivery systems a decrease in the FA fraction from 3:1 to 10:1 DPPC:FA and an increase in the MNP content from 0.25% to 1% led a significant increase in the FA release rates, with K_H and K_{KP} release constants registering an increase >50% (with the only exception of the liposomal hydrogels formed with 3:1 DPPC:FA liposomes). Besides the changes in the FA release rates, the FA release showed a good adjustment to the Higuchi model, leading to exponent values in good agreement with the characteristic Higuchi model exponent, i.e., 0.5 (Table 5), which evidences a diffusional controlled release of FA in the absence of magnetic field. An identical conclusion was taken from the analysis with Korsmeyer–Peppas models. This model assumes variable n values dependent on the FA release mechanism, but as shown in Table 5, Korsmeyer—Peppas led to n values ~0.5, quite similar to those estimated with Higuchi models, confirming a Fickian diffusion delivery of FA in the absence of a magnetic field.

It was interesting to see the effect of the magnetic field on the FA release mechanisms. In this case, the magnetically induced FA release profiles were also perfectly fit to the Higuchi model but with a significant decrease in the diffusion exponential parameter to values lower than 0.5, which suggests that despite the good fitting quality, the magnetic field stimuli shift the FA from a Higuchian model. For this reason, the effect of the magnetic field on the FA release was only evaluated based on the releasing parameters estimated by the Korsmeyer–Peppas model.

In fact, the presence of a magnetic field resulted in a remarkable increase in the FA release rates, with k_{KP} registering values more than two-fold higher than that obtained for the same liposomal hydrogel delivery system in the absence of a magnetic field. It is important to note that the increase in FA release due to the thermal effect due to the magnetic field was excluded. The presence of thermal effects is mainly associated with the use of an alternate magnetic field (AMF) and not with permanent magnetic field conditions, such as those used in the release assays described in this work. Despite this, the absence of thermal effects on magnetic-responsive polymeric systems when exposed to a magnetic field up to 1.5 T was investigated and confirmed in previous works from the same authors [56,65]. The transport of FA from the liposomal hydrogel membrane under magnetic stimulation was kept in the controlled diffusion regime, with n < 0.5. However, the magnetic stimulation induced a decrease in the n values (n < 0.4 for liposomal hydrogel systems doped with 1% MNPs), suggesting a change from a Fickian diffusional to a quasi-diffusion mechanism. This change was clearer in liposomal hydrogels containing higher %MNPs, revealing, as expected, a more accentuated effect of the magnetic stimuli in hydrogels with higher MNP concentrations.

In analogy to the dual-release regime observed for the FA release profiles from free MLV suspensions, it might be plausible to think that the low FA release observed in the 48 h experiment in the absence of a magnetic field corresponds mainly to the release of FA located on the outer layers of the encapsulated MLVs. When the magnetic field is applied, it might force the release of FA from the internal layers or the core of MLVs by mechanical compression of the liposomes triggered by a magnetic-induced contraction of the gelatin matrix resulting from the mobility of MNPs imposed by the attraction magnetic field forces, as reported in further literature [54–59]. The magnetic-induced release may thus be associated with changes in the permeability of the liposome membrane, with or without rupture of the vesicle. Nevertheless, it might be also hypothesized that it may benefit from a lower accumulation of FA in the hydrogel matrix (due to a higher diffusion of FA from the hydrogel matrix upon magnetic hydrogel contraction), thus minimizing the potential decrease in the driving force for the FA liposomal release during the process.

Overall, these results prove that the magnetic behavior of hydrogels doped with MNPs may effectively be used for the activation of drug release, suggesting the potential use of magnetic-responsive liposomal hydrogels as delivery platforms for an enhanced controlled and on-demand delivery of therapeutic compounds/bioactive molecules. However, further work is still needed to better clarify the impact of this magnetic stimulatory strategy on the structural integrity of the encapsulated liposomes and the reproducibility of this drug release process. This work shows that magnetic-induced release can be accomplished using low magnetic field intensities (0.08 T), which is minimally invasive to the human organism, contributing to the development of magnetic-responsive drug delivery for long-term treatments and simulating its integration into tissue engineering systems which may be further explored as novel and more efficient magnetic-responsive therapeutics for tissue repair [58].

4. Conclusions

This work was focused on the development of magnetic-responsive liposomal hydrogel membranes while proving their ability to provide enhanced magnetically controlled release of small molecules with therapeutic relevance and poor water solubility, such as ferulic acid (FA), based on the mechanical responsive behavior (mechanical distortions) of hydrogels doped with iron oxide nanoparticles (MNPs) upon magnetic stimulation. Magnetic-responsive liposomal hydrogels were found to delay the release of FA comparatively to that observed with conventional drug release systems, e.g., MLV suspensions or native hydrogel matrices. The delayed FA release was attributable to the additional stability of entrapped multilayered DPPC vesicles loaded with FA as well as to some resistance offered by the hydrogel matrix to the FA diffusion, resulting in a more sustained and longer-term release of this therapeutic molecule. The FA release from liposomal hydrogel systems was well described by the Korsmeyer–Peppas model, which showed that FA release followed a Fickian diffusion mechanism identical to that found for native hydrogels in the absence of a magnetic field. However, the encapsulation of MLVs into the hydrogel matrix renders a remarkable decrease in the constant release rates, confirming the delayed FA release from liposomal hydrogels. These results primarily suggest the promising use of these liposomal (MLVs) hydrogel platforms in therapeutics requiring a longer-term drug administration as an alternative to the less efficient conventional drug delivery systems while evidencing the valuable utilization of MLVs in biomedicine following their inclusion into hydrogel matrices.

The liposomal hydrogel delivery platforms were shown to be able to magnetically stimulate drug release, owing to the presence of dispersed MNPs. The magnetic-induced release of FA from these liposomal gelatin membranes was expressed by a significant increase in the release constant rates while keeping a diffusional controlled FA release mechanism but changing the FA release from Fickian diffusion to quasi-Fickian diffusion.

Magnetic-responsive liposomal hydrogels may be used as an independent non-cellular drug delivery strategy offering an on-demand and efficient magnetically controlled drug release into the target tissue, thus minimizing the drug losses caused by burst release. Furthermore, it may be easily combined with tissue engineering approaches, e.g., by integration of liposomes into magnetic-responsive tissue scaffolds. The impact of these magnetic-responsive liposomal hydrogels on the metabolic behavior of mammalian cells will be the focus of the following studies aiming to further explore and evaluate their potential for the development of novel and more efficient tissue repair strategies. On the one hand, they will take advantage of the magnetic responsiveness of these hydrogels to obtain in situ fine-tuned delivery of therapeutic agents; on the other hand, they will be used for the magnetic stimulation of key cell mechanotransduction processes for triggering improved cell development and faster tissue regeneration.

Author Contributions: Conceptualization: C.A.M.P. and F.P.; Methodology: C.A.M.P., F.P. and L.P.; Data analysis: L.P.; Validation: C.A.M.P. and F.P.; Original draft preparation: L.P. and C.A.M.P.; Writing, Review, and Editing: All the authors; Supervision: C.A.M.P. and F.P.; Funding acquisition: C.A.M.P. and F.C.F. All authors have read and agreed to the published version of the manuscript.

Funding: The authors acknowledge the financial support from Fundação para a Ciência e a Tecnologia (FCT-MEC), Portugal, through the dedicated project (PTDC/EDM-EDM/30828/2017) (Be-Live) and through the research units REQUIMTE (UIDB/50006/2020 and UIDP/50006/2020), iBB (UIDB/04565/2020 and UIDP/04565/2020) and i4HB (LA/P/0140/2020).

Institutional Review Board Statement: Not applicable.

Data Availability Statement: Experimental data are available upon request.

Acknowledgments: The authors acknowledge the collaboration of Clara Biechy through a scholar internship in the frame of a mobility grant from ERASMUS+Agence (France) and of César Laia (LAQV-REQUIMTE, FCT-NOVA) for the access to the Dynamic Light Scattering (DLS) equipment.

Conflicts of Interest: The authors declare no conflict of interest.

Appendix A

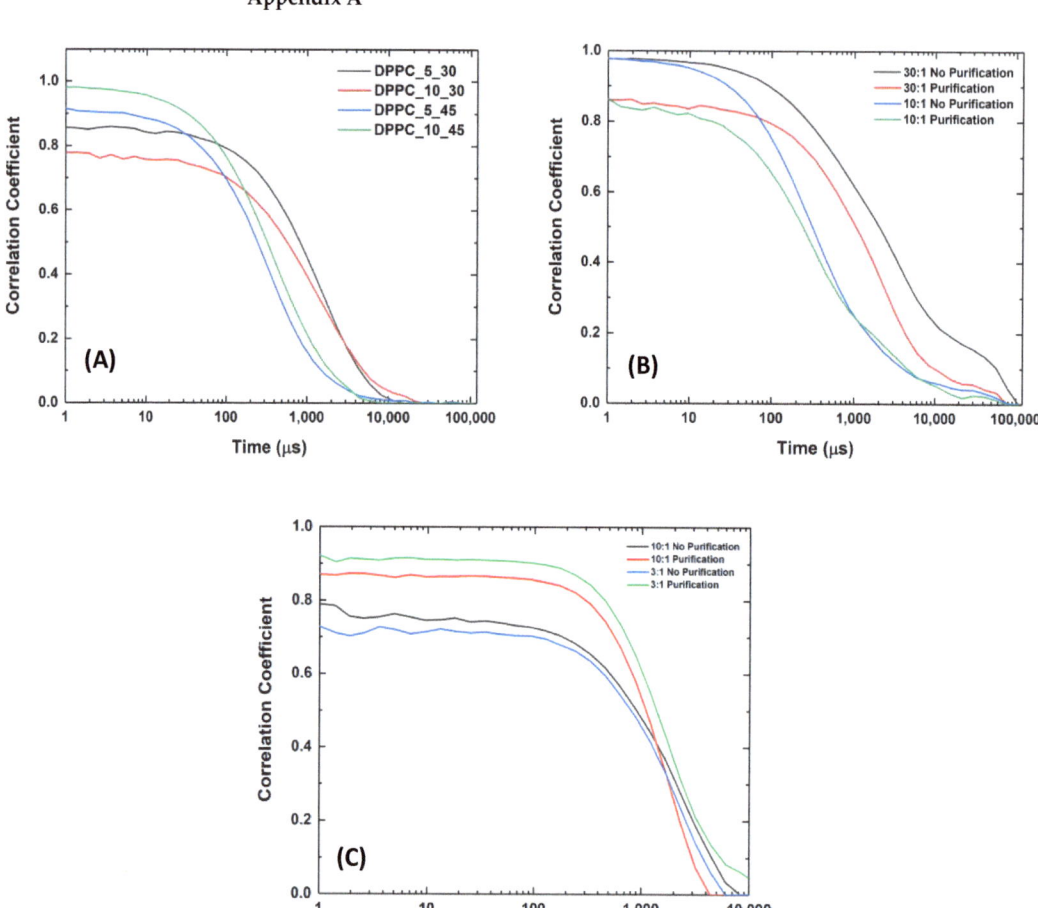

Figure A1. DLS correlograms obtained for (**A**) empty DPPC liposomes dispersed in PBS solution with different dilution factors (5× and 10×) and exposed to different sonication times (30 and 45 min); (**B**) 30:1 and 10:1 DPPC:FA liposomes, upon 45 min of sonication, before and after purification by dialysis; and (**C**) 3:1 and 10:1 DPPC:FA liposomes upon 60 min of sonication, before and after purification by dialysis.

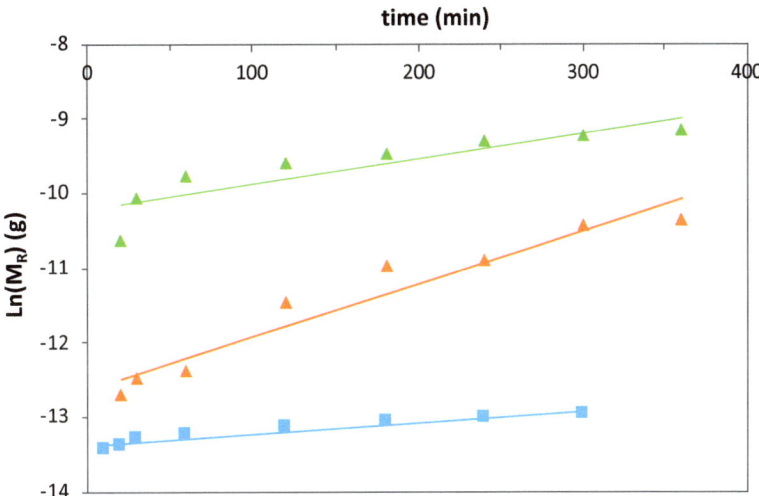

Figure A2. Logarithmic function of the ferulic acid (FA) mass release profiles obtained with magnetic−responsive gelatin membranes with embedded FA (blue squares) and liposome suspension with 10:1 (orange triangles) and 3:1 (green triangles) DPPC:FA fractions, at 37 °C, fit to 1st-order kinetic functions.

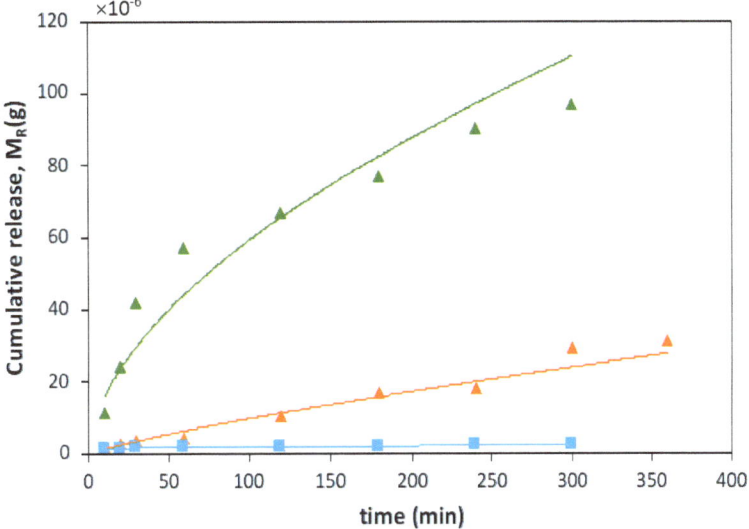

Figure A3. Ferulic acid (FA) mass release profiles obtained with magnetic−responsive gelatin membranes with embedded FA (blue squares) and liposome suspension with 10:1 (orange triangles) and 3:1 (green triangles) DPPC:FA fractions, at 37 °C, fit to Higuchi model.

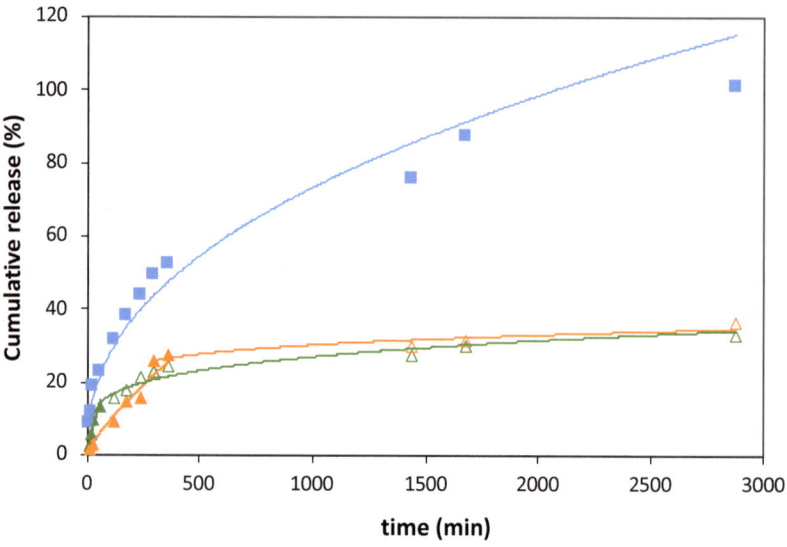

Figure A4. Ferulic acid (FA) release profiles obtained with magnetic–responsive gelatin membranes with embedded FA (blue squares) and liposome suspensions with 10:1 (orange triangles) and 3:1 (green triangles) DPPC:FA fractions, at 37 °C, fit to Korsmeyer–Peppas model. The cumulative release of FA was determined by $M_R/M_T \times 100$.

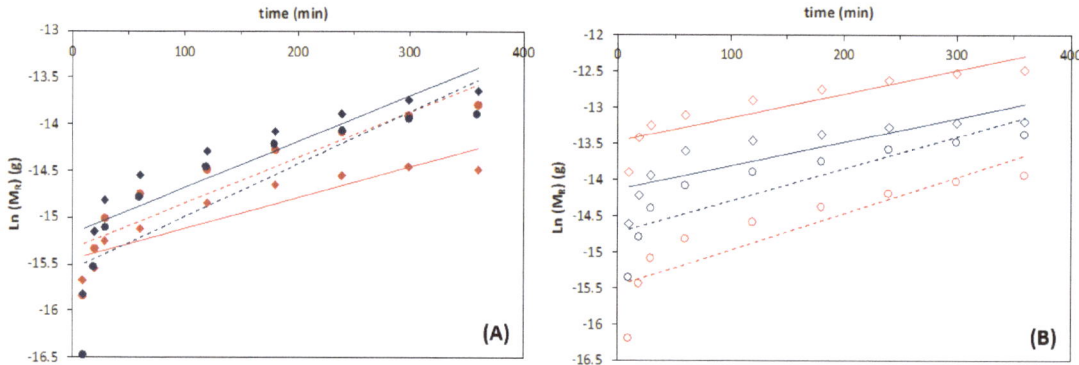

Figure A5. Logarithmic function of the ferulic acid (FA) mass release profiles obtained with magnetic–responsive liposomal gelatins containing (**A**) 0.25% and (**B**) 1% MNPs, with encapsulated 3:1 (red symbols) and 10:1 (blue symbols) DPPC:FA liposomes in the absence (circles) and presence (diamonds) of a magnetic field with an intensity of 0.08 T, at 37 °C, fit to a 1st-order kinetic model.

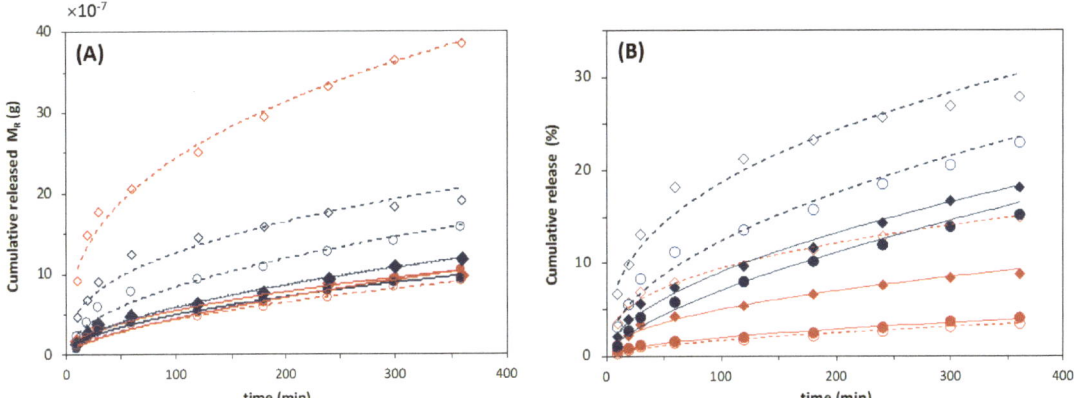

Figure A6. Ferulic acid (FA) release profiles obtained with a 3:1 (red symbols) and 10:1 (blue symbols) DPPC:FA liposomes encapsulated in magnetic–responsive gelatin membranes containing 0.25% MNPs (filled symbols) and 1% MNPs (empty symbols), in the absence (circles) and presence (diamonds) of magnetic field (MF) with an intensity of 0.08 T, for a period of 360 min, at 37 °C, fit to (**A**) Higuchi model, which considers the FA mass release, M_R, and (**B**) Korsmeyer–Peppas model which considers the FA release profiles determined by $M_R/M_T \times 100$.

Table A1. Mean size and PI values obtained for DPPC liposomes loaded with 30:1 and 10:1 DPPC:FA ratios, upon 45 min of sonication, before and after purification by dialysis.

Liposome	Purification	Mean Size (nm)	PI
DPPC:FA_30:1	No	716.5 ± 103.7	1.49 ± 0.44
DPPC:FA_10:1		1930.1 ± 101.7	0.63 ± 0.09
DPPC:FA_30:1	Yes	714.1 ± 152.9	0.71 ± 0.01
DPPC:FA_10:1		998.5 ± 131.0	0.63 ± 0.05

Table A2. Release constant k_1 (min^{-1}) obtained by fitting of the 1st-order kinetic function to the FA release from magnetic-responsive gelatin membranes with embedded FA and from liposome suspensions with 10:1 and 3:1 DPPC:FA fractions at 37 °C.

	Hydrogel + FA	DPPC:FA Suspension	
		3:1	10:1
k_1 (min^{-1})	1.6×10^{-3}	3.4×10^{-3}	7.1×10^{-3}
R^2	0.9205	0.767	0.933

Table A3. Higuchi constant, k_H, and diffusion exponential, obtained by fitting of Higuchi model to the FA release from magnetic-responsive gelatin membranes with embedded FA and from liposome suspensions with 10:1 and 3:1 DPPC:FA fractions at 37 °C.

	Hydrogel + FA	DPPC:FA Suspension	
		3:1	10:1
k_H	44.1×10^{-7}	2.4×10^{-7}	10.2×10^{-7}
n = 0.5	0.641	0.807	0.141
R^2	0.945	0.968	0.983

Table A4. First-order rate constant, k_1 (min^{-1}), and R^2 values obtained by fitting of first-order kinetic function to the FA release profiles from magnetic-responsive liposomal gelatins containing 0.25% and 1% MNPs and encapsulated 3:1 and 10:1 DPPC:FA liposomes in the absence and presence of a magnetic field with an intensity of 0.08 T, at 37 °C.

	0.25% MNPs				1% MNPs			
	3:1	10:1	3:1 MF	10:1 MF	3:1	10:1	3:1 MF	10:1 MF
k_1 (min^{-1})	0.0049	0.0056	0.0033	0.0049	0.0050	0.0044	0.0033	0.0032
R^2	0.719	0.847	0.797	0.844	0.777	0.753	0.8035	0.7128

References

1. Dana, P.; Bunthot, S.; Suktham, K.; Surassmo, S.; Yata, T.; Namdee, K. Active Targeting Liposome-PLGA Composite for Cisplatin Delivery against Cervical Cancer. *Colloids Surf. B Biointerfaces* **2020**, *196*, 111270. [CrossRef] [PubMed]
2. Jeon, M.; Kim, G.; Lee, W.; Im, H.-J. Development of Theranostic PEGylated Liposomal Au-Liposome for Effective Tumor Passive Targeting and Photothermal Therapy. *Soc. Nucl. Med.* **2020**, *61*, 1076.
3. Pires, F.; Geraldo, V.P.; Antunes, A.; Marletta, A.; Oliveira, O.N., Jr.; Raposo, M. On the role of epigallocatechin-3-gallate in protecting phospholipid molecules against UV irradiation. *Colloids Surf. B Biointerfaces* **2019**, *173*, 312–319. [CrossRef]
4. Pires, F.; Magalhaes-Mota, G.; Geraldo, V.P.; Ribeiro, P.A.; Oliveira, O.N., Jr.; Raposo, M. The impact of blue light in monolayers representing tumorigenic and nontumorigenic cell membranes containing epigallocatechin-3-gallate. *Colloids Surf. B Biointerfaces* **2020**, *193*, 111129. [CrossRef]
5. Zhang, J.; Chen, Y.; Li, X.; Liang, X.; Luo, X. The Influence of Different Long-Circulating Materials on the Pharmacokinetics of Liposomal Vincristine Sulfate. *Int. J. Nanomed.* **2016**, *11*, 4187–4197. [CrossRef]
6. Portilla, S.; Fernández, L.; Gutiérrez, D.; Rodríguez, A.; García, P. Encapsulation of the Antistaphylococcal Endolysin LysRODI in pH-Sensitive Liposomes. *Antibiotics* **2020**, *9*, 242. [CrossRef]
7. Figueroa-Robles, A.; Antunes-Ricardo, M.; Guajardo-Flores, D. Encapsulation of Phenolic Compounds with Liposomal Improvement in the Cosmetic Industry. *Int. J. Pharm.* **2020**, *593*, 120125. [CrossRef]
8. Matole, V.; Shegaonkar, A.; Kumbhar, S.; Thorat, Y.; Hosmani, A. Need of Liposomes as a Novel Drug Delivery System. *Res. J. Pharm. Dos. Forms Technol.* **2020**, *12*, 285–294. [CrossRef]
9. Fan, Q.-Q.; Zhang, C.-L.; Qiao, J.-B.; Cui, P.-F.; Xing, L.; Oh, Y.-K.; Jiang, H.-L. Extracellular Matrix-Penetrating Nanodrill Micelles for Liver Fibrosis Therapy. *Biomaterials* **2020**, *230*, 119616. [CrossRef]
10. Wei, X.; Liu, L.; Li, X.; Wang, Y.; Guo, X.; Zhao, J.; Zhou, S. Selectively Targeting Tumor-Associated Macrophages and Tumor Cells with Polymeric Micelles for Enhanced Cancer Chemo-Immunotherapy. *J. Control. Release* **2019**, *313*, 42–53. [CrossRef]
11. Wang, Y.; van Steenbergen, M.J.; Beztsinn, N.; Shi, Y.; Lammers, T.; Van Nostrum, C.F.; Hennink, W.E. Biotin-decorated all-HPMA polymeric micelles for paclitaxel delivery. *J. Control. Release* **2020**, *328*, 970–984. [CrossRef]
12. Hussein, Y.H.A.; Youssry, M. Polymeric Micelles of Biodegradable Diblock Copolymers: Enhanced Encapsulation of Hydrophobic Drugs. *Materials* **2018**, *11*, 688. [CrossRef] [PubMed]
13. Kapse, A.; Anup, N.; Patel, V.; Saraogi, G.K.; Mishra, D.K.; Tekade, R.K. Chapter 6—Polymeric micelles: A ray of hope among new drug delivery systems. In *Advances in Pharmaceutical Product Development and Research, Drug Delivery Systems*; Tekade, R.K., Ed.; Academic Press: Cambridge, MA, USA, 2020; pp. 235–289. ISBN 9780128144879. [CrossRef]
14. Salehi, S.; Naghib, S.M.; Garshasbi, H.R.; Ghorbanzadeh, S.; Zhang, W. Smart stimuli-responsive injectable gels and hydrogels for drug delivery and tissue engineering applications: A review. *Front. Bioeng. Biotechnol.* **2023**, *11*, 1104126. [CrossRef] [PubMed]
15. Jiang, B.; Liu, X.; Yang, C.; Yang, Z.; Luo, J.; Kou, S.; Liu, K.; Sun, F. Injectable, photoresponsive hydrogels for delivering neuroprotective proteins enabled by metal directed protein assembly. *Sci. Adv.* **2022**, *6*, eabc4824. [CrossRef] [PubMed]
16. Kurian, A.G.; Singh, R.K.; Patel, K.D.; Lee, J.-H.; Kim, H.-W. Multifunctional GelMA platforms with nanomaterials for advanced tissue therapeutics. *Bioact. Mat.* **2022**, *8*, 267–295. [CrossRef] [PubMed]
17. Liu, Y.; Zhong, D.; He, Y.; Jiang, J.; Xie, W.; Tang, Z.; Qiu, J.; Luo, J.; Wang, X. Photoresponsive hydrogel-coated upconversion cyanobacteria nanocapsules for myocardial infarction prevention and treatment. *Adv. Sci.* **2022**, *9*, 2202920. [CrossRef]
18. Sabourian, P.; Tavakolian, M.; Yazdani, H.; Frounchi, M.; van de Ven, T.G.M.; Maysinger, D.; Kakkar, A. Stimuli-responsive chitosan as an advantageous platform for efficient delivery of bioactive agents. *J. Control. Release* **2020**, *317*, 216–231. [CrossRef]
19. Shen, X.; Li, S.; Zhao, X.; Han, J.; Chen, J.; Rao, Z.; Zhang, K.; Quan, D.; Yuan, J.; Bao, Y. Dual-crosslinked regenerative hydrogel for sutureless long-term repair of corneal defect. *Bioact. Mat.* **2023**, *20*, 434–448. [CrossRef]
20. Wang, X.; Swing, C.J.; Feng, T.; Xia, S.; Yu, J.; Zhang, X. Effects of environmental pH and ionic strength on the physical stability of cinnamaldehyde-loaded liposomes. *J. Dispers. Sci. Technol.* **2020**, *41*, 1568–1575. [CrossRef]
21. Huang, X.; Brazel, C.S. On the importance and mechanisms of burst release in matrix-controlled drug delivery systems. *J. Control. Release* **2001**, *73*, 121–136. [CrossRef]
22. Sankaranarayanan, J.; Mahmoud, A.E.; Kim, G.; Morachis, J.M.; Almutairi, A. Multiresponse strategies to modulate burst degradation and release from nano-particles. *ACS Nano* **2010**, *4*, 5930–5936. [CrossRef]
23. Martínez, A.W.; Caves, J.M.; Swathi, L.W.; Chaikof, E.L. Effects of crosslinking on the mechanical properties, drug release and cytocompatibility of protein polymers. *Acta Biomater.* **2014**, *10*, 26–33. [CrossRef]

24. Nakhaei, P.; Margiana, R.; Bokov, D.O.; Abdelbasset, W.K.; Jadidi, K.M.A.; Varma, R.S.; Marofi, F.; Jarahian, M.; Beheshtkhoo, N. Liposomes: Structure, Biomedical Applications, and Stability Parameters with Emphasis on Cholesterol. *Front. Bioeng. Biotechnol.* **2021**, *9*, 705886. [CrossRef]
25. Panwar, P.; Pandey, B.; Lakhera, P.C.; Singh, K.P. Preparation, characterization, and in-vitro release study of albendazole encapsulated nanosize liposomes. *Int. J. Nanomed.* **2010**, *5*, 101–108. [CrossRef]
26. Trucillo, P.; Campardelli, R.; Reverchon, E. Supercritical CO_2 Assisted Liposomes Formation: Optimization of the Lipidic Layer for an Efficient Hydrophilic Drug Loading. *J. Co2 Util.* **2017**, *18*, 181–188. [CrossRef]
27. Ricci, M.; Oliva, R.; Del Vecchio, P.; Paolantoni, M.; Morresi, A.; Sassi, P. DMSO-induced Perturbation of Thermotropic Properties of Cholesterol-Containing DPPC Liposomes. *Biochim. Biophys. Acta (Bba)—Biomembr.* **2016**, *1858*, 3024–3031. [CrossRef] [PubMed]
28. Samuni, A.M.; Lipman, A.; Barenholz, Y. Damage to liposomal lipids: Protection by antioxidants and cholesterol-mediated dehydration. *Chem. Phys. Lipids* **2000**, *105*, 121–134. [CrossRef]
29. Popova, A.V.; Hincha, D.K. Effects of cholesterol on dry bilayers: Interactions between phosphatidylcholine unsaturation and glycolipid or free sugar. *Biophys. J.* **2007**, *93*, 1204–1214. [CrossRef]
30. Mou, Y.; Zhang, P.; Lai, W.-F.; Zhang, D. Design and applications of liposome-in-gel as carriers for cancer therapy. *Drug Deliv.* **2022**, *29*, 3245–3255. [CrossRef]
31. Ivashkov, O.; Yakimova, T.; Evtushenko, E.; Gelissen, A.; Plamper, F.; Richtering, W.; Yaroslavov, A. On the mechanism of payload release from liposomes bound to temperature-sensitive microgel particles. *Colloids Surf. A Physicochem. Eng. Asp.* **2019**, *570*, 396–402. [CrossRef]
32. Qin, C.; Lv, Y.; Xu, C.; Li, J.; Yin, L.; He, W. Lipid-bilayer-coated nanogels allow for sustained release and enhanced internalization. *Int. J. Pharm.* **2018**, *551*, 8–13. [CrossRef]
33. Moustafa, M.A.; El-Refaie, W.M.; Elnaggar, Y.S.R.; Abdallah, O.Y. Gel in core carbosomes as novel ophthalmic vehicles with enhanced corneal permeation and residence. *Int. J. Pharm.* **2018**, *546*, 166–175. [CrossRef]
34. Petralito, S.; Spera, R.; Pacelli, S.; Relucenti, M.; Familiarim, G.; Vitalone, A.; Paolicelli, P.; Casadei, M.A. Design and development of PEG-DMA gel-in-liposomes as a new tool for drug delivery. *React. Funct. Polym.* **2014**, *77*, 30–38. [CrossRef]
35. Yang, J.; Zhu, Y.; Wang, F.; Deng, L.; Xu, X.; Cui, W. Microfluidic liposomes-anchored microgels as extended delivery platform T for treatment of osteoarthritis. *Chem. Eng. J.* **2020**, *400*, 126004. [CrossRef]
36. Mourtas, S.; Fotopoulou, S.; Duraj, S.; Sfika, V.; Tsakiroglou, C.; Antimisiaris, S. Liposomal drugs dispersed in hydrogels. *Colloids Surf. B Biointerfaces* **2007**, *55*, 212–221. [CrossRef] [PubMed]
37. GuhaSarkar, S.; More, P.; Banerjee, R. Urothelium-adherent, ion-triggered liposome-in-gel system as a platform for intravesical drug delivery. *J. Control. Release* **2017**, *10*, 147–156. [CrossRef]
38. O'Neill, H.S.; Herron, C.C.; Hastings, C.L.; Deckers, R.; Noriega, A.L.; Kelly, H.M.; Hennink, W.E.; McDonnell, C.O.; O'Brien, F.J.; Ruiz-Hernández, E.; et al. A stimuli responsive liposome loaded hydrogel provides flexible on-demand release of therapeutic agents. *Acta Biomater.* **2017**, *48*, 110–119. [CrossRef]
39. Bilard, L.; Pourchet, S.; Malaise, P.; Alcouffe, A.; Montembault, C.; Ladavière, C. Liposome-loaded chitosan physical hydrogel: Toward a promising delayed-release biosystem. *Carbohydr. Polym.* **2015**, *115*, 651–657. [CrossRef]
40. Ruel-Gariepy, E.; Leclair, G.; Hildgen, P.; Gupta, A.; Leroux, J.-C. Thermosensitive chitosan-based hydrogel containing liposomes for the delivery of hydrophilic molecules. *J. Control. Release* **2002**, *82*, 373–383. [CrossRef]
41. Klotz, B.J.; Gawlitta, D.; Rosenberg, A.J.E.P.; Malda, J.; Melchels, F.P.W. Gelatin-methacryloyl hydrogels: Towards biofabrication-based tissue repair. *Trends Biotechnol.* **2016**, *34*, 394–407. [CrossRef]
42. Ciobanu, B.C.; Cadinoiu, A.N.; Popa, M.; Desbrières, J.; Peptu, C.A. Modulated release from liposomes entrapped in chitosan/gelatin hydrogels. *Mater. Sci. Eng. C* **2014**, *43*, 383–391. [CrossRef] [PubMed]
43. Lajavardi, L.; Carmelo, S.; Agnely, F.; Luo, W.; Goldenberg, B.; Naud, M.-C.; Behar-Cohen, F.; de Kozak, Y.; Bochot, A. New formulations of vasoactive intestinal peptide using liposomes in hyaluronic acid gel for uveitis. *J. Control. Release* **2009**, *139*, 22–30. [CrossRef] [PubMed]
44. Lee, J.-H.; Hyuntaek, O.; Baxa, U.; Raghavan, S.R.; Blumenthal, R. Biopolymer-Connected Liposome Networks as Injectable Biomaterials Capable of Sustained Local Drug Delivery. *Biomacromolecules* **2012**, *13*, 3388–3394. [CrossRef]
45. GuhaSarkar, S.; Pathak, K.; Sudhalkar, N.; More, P.; Goda, J.S.; Gota, V.; Banerjee, R. Synergistic locoregional chemoradiotherapy using a composite liposome-in-gel system as an injectable drug depot. *Int. J. Nanomed.* **2016**, *11*, 6435–6448. [CrossRef]
46. Wu, W.; Dai, Y.; Liu, H.; Cheng, R.; Ni, Q.; Ye, T.; Cui, W. Local release of gemcitabine via in situ UV-crosslinked lipid-strengthened hydrogel for inhibiting osteosarcoma. *Drug Deliv.* **2018**, *25*, 1642–1651. [CrossRef] [PubMed]
47. Veloso, S.R.S.; Andrade, R.G.D.; Castanheira, E.M.S. Review on the advancements of magnetic gels: Towards multifunctional magnetic liposome-hydrogel composites for biomedical applications. *Adv. Colloid Interface Sci.* **2021**, *288*, 102351. [CrossRef]
48. Hurler, J.; Żakelj, S.; Mravljak, J.; Pajk, S.; Kristl, A.; Schubert, R.; Škalko-Basnet, N. The effect of lipid composition and liposome size on the release properties of liposomes-in-hydrogel. *Int. J. Pharm.* **2013**, *456*, 49–57. [CrossRef]
49. TS, A.; Lu, Y.J.; Chen, J.P. Optimization of the Preparation of Magnetic Liposomes for the Combined Use of Magnetic Hyperthermia and Photothermia in Dual Magneto-Photothermal Cancer Therapy. *Int. J. Mol. Sci.* **2020**, *21*, 5187. [CrossRef]
50. Amiri, M.; Gholami, T.; Amiri, O.; Pardakhti, A.; Ahmadi, M.; Akbari, A.; Amanatfard, A.; Salavati-Niasari, M. The magnetic inorganic-organic nanocomposite based on $ZnFe_2O_4$-Imatinib-liposome for biomedical applications, in vivo and in vitro study. *J. Alloys Compd.* **2020**, *849*, 156604. [CrossRef]

51. Kong, Y.; Dai, Y.; Qi, D.; Du, W.; Ni, H.; Zhang, F.; Zhao, H.; Shen, Q.; Li, M.; Fan, Q. Injectable and thermosensitive liposomal hydrogels for NIR-II light-triggered photothermal-chemo therapy of pancreatic cancer. *ACS Appl. Bio Mater.* **2021**, *4*, 7595–7604. [CrossRef]
52. Mao, Y.; Li, X.; Chen, G.; Wang, S. Thermosensitive hydrogel system with paclitaxel liposomes used in localized drug delivery system for in situ treatment of tumor: Better antitumor efficacy and lower toxicity. *J. Pharm. Sci.* **2016**, *105*, 194–204. [CrossRef]
53. Hosny, K.M. Preparation and Evaluation of Thermosensitive Liposomal Hydrogel for Enhanced Transcorneal Permeation of Ofloxacin. *Aaps Pharmscitech* **2009**, *10*, 1336–1342. [CrossRef]
54. Filipcsei, G.; Csetneki, I.; Szilágyi, A.; Zrínyi, M. Magnetic responsive-field smart polymer composites. *Adv. Polym. Sci.* **2007**, *206*, 137–189.
55. Szabo, D.; Szeghy, G.; Zrínyi, M. Shape Transition of Magnetic Field Sensitive Polymer Gels. *Macromolecules* **1998**, *31*, 6541–6548. [CrossRef]
56. Manjua, A.C.; Alves, V.D.; Crespo, J.G.; Portugal, C.A.M. Magnetic responsive PVA hydrogels for remote modulation of protein sorption. *ACS Appl. Mater. Interfaces* **2019**, *11*, 21239–21249. [CrossRef] [PubMed]
57. Kumar, N.; Pruthi, V. Potential applications of ferulic acid from natural sources. *Biotechnol. Rep.* **2014**, *4*, 86–93. [CrossRef] [PubMed]
58. Manjua, A.C.; Cabral, J.M.S.; Portugal, C.A.M.; Ferreira, F.C. Magnetic stimulation of the angiogenic potential of mesenchymal stromal cells in vascular tissue engineering. *Sci. Technol. Adv. Mater.* **2021**, *22*, 461–480. [CrossRef] [PubMed]
59. Pires, F.; Santos, J.F.; Bitoque, D.; Silva, G.A.; Marletta, A.; Nunes, V.A.; Raposo, M. Polycaprolactone/gelatin nanofiber membranes containing EGCG-loaded liposomes and their potential use for skin regeneration. *ACS Appl. Bio Mater.* **2019**, *2*, 4790–4800. [CrossRef]
60. Peppas, N.A.; Hoffman, A.S. *Biomaterials Science: An Introduction to Materials in Medicine*, 4th ed.; Academic Press: Cambridge, MA, USA, 2020; Volume 1, pp. 153–166. [CrossRef]
61. Pal, K.; Singh, V.K.; Anis, A.; Thakur, G.; Bhattacharya, M.K. Hydrogel-Based Controlled Release Formulations: Designing Considerations, Characterization Techniques and Applications. *Polym.-Plast. Technol. Eng.* **2013**, *52*, 1391–1422. [CrossRef]
62. Lee, J.; Gustin, J.; Chen, T.; Payne, G.; Raghavan, S. Vesicle−biopolymer gels: Networks of surfactant vesicles connected by associating biopolymers. *Langmuir* **2005**, *21*, 26–33. [CrossRef]
63. Derkach, S.R.; Voronko, N.G.; Sokolan, N.I.; Kolotova, D.S.; Kuchina, Y.A. Interactions between gelatin and sodium alginate: UV and FTIR studies. *J. Dispers. Sci. Technol.* **2020**, *41*, 690–698. [CrossRef]
64. Gaihre, B.; Khil, M.S.; Lee, D.R.; Kim, H.Y. Gelatin-coated magnetic iron oxide nanoparticles as carrier system: Drug loading and in vitro drug release study. *Int. J. Pharm.* **2009**, *365*, 180–189. [CrossRef] [PubMed]
65. Upadhyaya, L.; Semsarilar, M.; Quémener, D.; Fernández-Pacheco, R.; Martinez, G.; Mallada, R.; Coelhoso, I.M.; Portugal, C.A.M.; Crespo, J.G. Block copolymer based novel magnetic mixed matrix membranes-magnetic modulation of water permeation by irreversible structural changes. *J. Membr. Sci.* **2018**, *551*, 273–282. [CrossRef]

Disclaimer/Publisher's Note: The statements, opinions and data contained in all publications are solely those of the individual author(s) and contributor(s) and not of MDPI and/or the editor(s). MDPI and/or the editor(s) disclaim responsibility for any injury to people or property resulting from any ideas, methods, instructions or products referred to in the content.

Article

Functionalized GO Membranes for Efficient Separation of Acid Gases from Natural Gas: A Computational Mechanistic Understanding

Quan Liu [1], Zhonglian Yang [1,*], Gongping Liu [2], Longlong Sun [1], Rong Xu [3,*] and Jing Zhong [3]

[1] Analytical and Testing Center, School of Chemical Engineering, Anhui University of Science and Technology, Huainan 232001, China
[2] State Key Laboratory of Materials-Oriented Chemical Engineering, College of Chemical Engineering, Nanjing Tech University, 30 Puzhu Road (S), Nanjing 211816, China
[3] Key Laboratory of Advanced Catalytic Materials and Technology, School of Petrochemical Engineering, Changzhou University, Gehu Road, Changzhou 213164, China
* Correspondence: zhlyang@aust.edu.cn (Z.Y.); xurong@cczu.edu.cn (R.X.)

Abstract: Membrane separation technology is applied in natural gas processing, while a high-performance membrane is highly in demand. This paper considers the bright future of functionalized graphene oxide (GO) membranes in acid gas removal from natural gas. By molecular simulations, the adsorption and diffusion behaviors of several unary gases (N_2, CH_4, CO_2, H_2S, and SO_2) are explored in the 1,4-phenylenediamine-2-sulfonate (PDASA)-doped GO channels. Molecular insights show that the multilayer adsorption of acid gases evaluates well by the Redlich-Peterson model. A tiny amount of PDASA promotes the solubility coefficient of CO_2 and H_2S, respectively, up to 4.5 and 5.3 mmol·g^{-1}·kPa^{-1}, nearly 2.5 times higher than those of a pure GO membrane, which is due to the improved binding affinity, great isosteric heat, and hydrogen bonds, while N_2 and CH_4 only show single-layer adsorption with solubility coefficients lower than 0.002 mmol·g^{-1}·kPa^{-1}, and their weak adsorption is insusceptible to PDASA. Although acid gas diffusivity in GO channels is inhibited below 20×10^{-6} cm^2·s^{-1} by PDASA, the solubility coefficient of acid gases is certainly high enough to ensure their separation efficiency. As a result, the permeabilities (P) of acid gases and their selectivities (α) over CH_4 are simultaneously improved (P_{CO2} = 7265.5 Barrer, $\alpha_{CO2/CH4}$ = 95.7; $P_{(H2S+CO2)}$ = 42075.1 Barrer, $\alpha_{H2S/CH4}$ = 243.8), which outperforms most of the ever-reported membranes. This theoretical study gives a mechanistic understanding of acid gas separation and provides a unique design strategy to develop high-performance GO membranes toward efficient natural gas processing.

Keywords: acid gas removal; graphene oxide; membrane separation; molecular simulation; natural gas

Citation: Liu, Q.; Yang, Z.; Liu, G.; Sun, L.; Xu, R.; Zhong, J. Functionalized GO Membranes for Efficient Separation of Acid Gases from Natural Gas: A Computational Mechanistic Understanding. *Membranes* **2022**, *12*, 1155. https://doi.org/10.3390/membranes12111155

Academic Editor: Alexander Toikka

Received: 19 October 2022
Accepted: 15 November 2022
Published: 16 November 2022

Publisher's Note: MDPI stays neutral with regard to jurisdictional claims in published maps and institutional affiliations.

Copyright: © 2022 by the authors. Licensee MDPI, Basel, Switzerland. This article is an open access article distributed under the terms and conditions of the Creative Commons Attribution (CC BY) license (https:// creativecommons.org/licenses/by/ 4.0/).

1. Introduction

Methane (CH_4), as the main constituent of natural gas, is one kind of renewable energy source [1]. The raw natural gas coming from crude oil wells always exists in the form of mixtures, containing other light hydrocarbons, nitrogen (N_2), carbon dioxide (CO_2), hydrogen sulfide (H_2S), and sulfur dioxide (SO_2). Among these impurities, significant amounts of CO_2, H_2S, and SO_2 commonly called acid gases are the most harmful components in raw natural gas, which not only lowers the calorific value of CH_4 but also causes internal corrosion in gas pipelines [2,3]. Therefore, to meet the requirements of end users and the specifications of transportation pipelines, the removal of acid gases is an essential process in natural gas processing [3,4]. Several processes can be adopted to remove acid gases, including pressure swing adsorption, supersonic separation, and membrane separation. In addition, natural gas can also be purified by forming CO_2 hydrates from the gas mixtures [5–7]. The commercialized technology is amine scrubbing [8], which uses plenty of alkanolamine solutions in absorption columns to dissolve acid gases. However, it requires the use of large equipment, rapidly increasing the operating cost [9], and lots of undesirable liquid wastes produced in this process pose a threat to the environment. Alternatively, with low energy consumption, low pollution and high separation efficiency, membrane gas separation technology is regarded as a potential candidate for acid gas removal [10]. Especially under ordinary operation conditions (i.e., room temperature and low operating pressure), it will achieve better economic benefits in natural gas processing.

Various membrane materials have been developed to address these challenging separations, such as polymer [11,12], metal-organic framework (MOF) [13,14] and graphene [15]. Among them, the polymeric membrane is the most large-scale development for commercial, while its performance is somewhat low primarily due to the trade-off effect. Fortunately, two-dimensional (2D) graphene oxide (GO) membranes with tailorable channels and abundant active sites are emerging candidates for boosting molecular separation performance [15,16]. It is reported that their inherent transport channels can be regulated for selective permeation at the sub-nanometer scale [17]. For instance, by adjusting ultraviolet irradiation, the interlayer spacing of GO membrane was precisely controlled by Zheng et al. to improve the separation efficiency of these two species with a very low molecular weight difference [18]. Our previous work also showed that the 1,4-phenylenediamine-2-sulfonate (PDASA)-functionalized GO channels facilitated the adsorption of the polar molecule (i.e., water), and then largely promoted its permeation [19]. For acid gas removal, the CO_2 permeability was successfully enhanced by incorporating GO nanosheets as the filler to create additional gas transport channels in polymers of intrinsic microporosity [20]. Additionally, using the strong affinity between GO and CO_2 was a brilliant strategy to enhance the CO_2 solubility in polyimide hybrid membranes [21]. After doping GO nanosheets, the CO_2/CH_4 separation performance of various polymeric membranes was promoted to outperform the 2008 Robeson upper bound [15,22].

However, as mentioned above, the GO nanosheet is mostly dispersed as a filler into mixed matrix membranes or prepared as hybrid membranes to separate CO_2/CH_4 [15,20,21,23], thus lack of exploration on pure GO membrane especially on its separation mechanism for acid gas removal. Fortunately, a few molecular simulations attempted to explore the CO_2/CH_4 separation process through pure GO membranes [24,25]. Whereas, for other 2D membranes, most previous simulations demonstrated that there were two main dominated separation mechanisms (i.e., the size-sieving effect and preferential adsorption) in natural gas processing [26–28]. A suitable aperture is key to the high separation performance of CO_2/CH_4 [26,27]. While in order to further improve the removal efficiency of CO_2, the separation mechanism should be governed by preferential adsorption, which helps to improve CO_2 separation selectivity [28]. However, until now, there has been no theoretical model established for acid gas separation through GO membranes. Therefore, in order to establish this theoretical model, it is necessary to study the acid gas permeation behavior in GO channels from the perspectives of adsorption and diffusion. Moreover, CO_2 and other acid gases (i.e., H_2S and SO_2) need to be studied at the same time. Furthermore, to improve the removal efficiency, a rational design of a GO membrane at the molecular level is highly in demand. This study aims to theoretically design a high-performance GO membrane toward acid gas removal and explore the separation models.

In this work, GO membranes are functionalized by PDASA (this selection is inspired by our previous experimental work [19]) to examine how it performs in removing acid gases (CO_2, H_2S, and SO_2) from CH_4 and N_2. By Grand Canonical Monte Carlo (GCMC) simulations, unary isotherms of different gases in GO membranes with variable doping amounts of PDASA are first studied by several adsorption models. To accurately describe the adsorption characteristics of different gases and provide molecular insights, structural and energetic analyses are conducted in GO channels via molecular distribution probability, radial distribution function (RDF), isosteric heat, and hydrogen bonds. The solubility coefficient is calculated to characterize the adsorption ability of different gases. Then gas diffusion behavior is explored by molecular dynamical (MD) simulations. After that, the acid gas separation performance is predicated on the basis of the solution-diffusion mechanism. Finally, a performance comparison with previous reports is enclosed to demonstrate the potential of the PDASA-doped GO membranes in natural gas processing.

2. Models and Methods

Figure 1 shows the simulation models. First of all, GO nanosheets with the format of $C_{312}(O)_{65}(OH)_{79}(COOH)_4$ were constructed by the Material studio in amorphous cell as per our previous works [16,29–33]. Functional groups were randomly distributed on the sp^2-conjugated surface of which the dimensions were 3×3 nm^2, as shown in Figure 1a. The numbers of epoxy, hydroxyl and carboxyl groups were 65, 79, and 4, respectively, similar to our previous experimental reports [19]. As a result, the oxidized ratio that was defined by the total number of oxygen atoms to carbon atoms was about 0.48, which is feasible in membrane process simulation for both gas and liquid separations [16,32,33]. Five gases with variable electronegativities and kinetic diameters were investigated, as shown in Figure 1b. Electrostatic potentials show that the acid gases of CO_2, H_2S and SO_2 exhibit higher electronegativity compared to CH_4 and N_2. To reveal gas sorption and diffusion behaviors in the lamellar structure of GO membranes, two GO nanosheets were parallelly aligned with interlayer spacing initially set as 0.8 nm (Figure 1c). To increase the affinity between GO membrane and acid gases, interlayer channel was functionalized with PDASA groups (Figure 1f) that have a great affinity to polar molecules [19]. The number of doped PDASA molecules increased from 1 to 5, correspondingly to the doping amounts varying from 1.5 to 7.5 wt%. The atomic positions of GO nanosheets were flexible during simulations. After being loaded with PDASA groups, GO membranes were relaxed well, and then interlayer spacing was slightly enlarged, as shown in Figure 1d,e where the doping amounts are 4.5 wt% and 7.5 wt%, respectively.

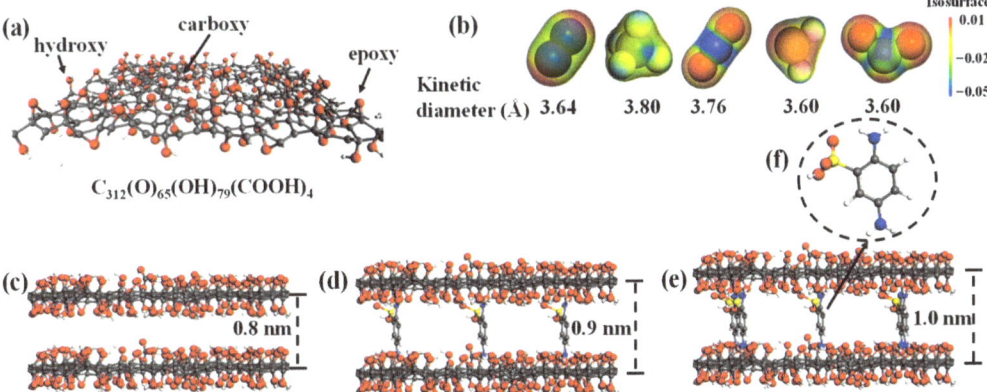

Figure 1. Simulation models. (**a**) GO nanosheet with the format of $C_{312}(O)_{65}(OH)_{79}(COOH)_4$. (**b**) Electrostatic potentials and kinetic diameters of gases. Configurations of GO membranes with variable PDASA-doping amounts: (**c**) 0.0%; (**d**) 4.5 wt%; (**e**) 7.5 wt%. (**f**) Molecular model of PDASA.

Before GCMC simulations, GO membranes and gases were performed with geometry optimization to search for a minimum energy structure. In this process, the convergence thresholds of energy, force and displacement were specified as 10^{-5} kcal/mol, 10^{-3} kcal/mol/Å and 10^{-5} Å, respectively. To calculate adsorption isotherms of gases in flexible GO membranes, the Configurational bias method [34] was performed with 10^7 equilibration and production steps. The temperature was maintained at 298 K by the algorithm of Nosé-Hoover thermostat [35]. Production frame was output every 10,000 steps. Partial charges were taken from the Compass force field [36], which was also used to describe interatomic interactions among membrane and variable gases. Here, nonbonded interactions were summarized by electrostatic and van der Waals potentials. Long-range electrostatic interactions were handled with the Ewald method [37] with an accuracy of 10^{-5} kcal/mol, whereas van der Waals interaction potentials were predicated by the atom-based method with a 9.8 Å cut-off distance. Periodic boundary conditions are applied in all three directions. After adsorption simulations, the lowest energy configuration returned from the GCMC calculation was used as the initial frame to explore gas diffusion properties. In MD simulations, there were a total of 50 gas molecules inserted in GO membranes and they could freely roam in GO interlayers. The system reached temperature (298 K) equilibrium first in an isothermal-isobaric ensemble for 1 ns. The pressure was controlled at 1 bar by the Berendsen barostat [38] with a decay constant of 0.1 ps. Subsequently, the production runs were performed in a canonical ensemble. The time step was set as 0.5 fs and trajectories were recorded every 2 ps, and the total simulation time was 2 ns. The final results were averaged over three independent trials.

3. Results and Discussion

3.1. Adsorption Evaluation

To calculate the adsorption isotherms of different gases in GO membranes, GCMC simulations were performed under low pressures (0.01 KPa~1000 Kpa). The fugacity coefficients of unary gases (N_2, CH_4, CO_2, H_2S and SO_2) are close to 1.0 under these pressures by physical property estimation in Aspen using the Peng-Robinson equation-of-state [39], indicating that the gas behavior approximates the ideal gas model. Therefore, the fugacity and pressure are approximately equal. Figure 2 shows the absolute adsorption isotherms of five gases are dependent on the relative pressures in GO membranes with variable doping amounts of PDASA. The adsorption capacities of CH_4 and N_2 slowly rise with increasing pressure. While for acid gases (CO_2, H_2S and SO_2), their isotherms grow rapidly, especially a sudden increase at relatively low pressures, behaving in a different adsorption mode. As a result, the adsorption capacities of acid gases in GO membranes are obviously larger than those of CH_4 and N_2. In addition, the maximum absorption capacity increases in the order of $N_2 < CH_4 < CO_2 < SO_2 < H_2S$. With increasing the doping amounts of PDASA from 0.0 to 7.5 wt%, the adsorption capacities of three acid gases increase at first and then decrease, as shown in Figure 1a–f. In view of the low density of adsorbed gases at low pressure and low temperature, the absolute adsorption capacity (Q_{ab}) obtained in our simulations is close to the excess adsorption capacity (Q_{ex}) that is determined in the experiment according to Equation (1) [40] where ρ_g is the gas density at simulated pressure and V_f is the free volume in GO membranes. Therefore, the absolute adsorption isotherms in Figure 1 without further conversion can be directly described by adsorption models.

$$Q_{ex} = Q_{ab} - \rho_g V_f \tag{1}$$

$$S_0 = \lim_{p \to 0} \frac{Q_e}{P} \tag{2}$$

$$Q_{ex} = \delta P + \frac{\beta P}{1+\gamma P^n} = \begin{cases} \frac{Q_L K_L P}{1+K_L P} & \delta = 0; n = 1 \ (Langmuir, for\ CH_4\ and\ N_2) & (3) \\ \frac{\beta P}{1+\gamma P^n} & \delta = 0; 0 < n < 1 \ (Redlich - Peterson, for\ H_2S\ and\ CO_2) & (4) \\ \delta P + \frac{\beta P}{1+\gamma P^n} & \delta \neq 0; 0 < n < 1 \ (Dual - mode, for\ SO_2) & (5) \end{cases}$$

$$S_0 = \begin{cases} Q_L K_L & (for\ CH_4\ and\ N_2) & (6) \\ \beta & (for\ H_2S\ and\ CO_2) & (7) \\ \delta + \beta & (for\ SO_2) & (8) \end{cases}$$

The solubility coefficient (S_0) of infinite dilution is an important factor in characterizing membrane separation properties, which is defined as the slope of isotherm at infinite dilution (Equation (2)) [41–43]. When gas concentration is extremely low, several theoretical models (Equations (3)–(5)) are applied to fit isotherms to obtain the S_0 of gases in GO membranes, where P is the sorbate pressure, and δ, β and γ are fitting parameters. After curve fitting, it shows that the adsorption of CH_4 and N_2 obey the Langmuir model [44]

(Equation (3)) where Q_L is the maximal adsorption capacity and K_L is the adsorption equilibrium constant, indicating a simple adsorption process. While simulation results suggest a three-parameter model (i.e., Redlich-Peterson [45], Equation (4)) for CO_2 and H_2S, where n is the empirical constant. The adsorption behavior for SO_2 is a little complex as it needs more variables to fit the isotherm based on the dual-mode sorption model [46] as Equation (5). All fitting parameters are presented in Table S1. A high correlation coefficient (R^2) above 0.992 for most systems indicates the reliability of these adopted adsorption models [45]. These different theoretical models are ascribed to the variable adsorption mechanism of gases in GO membranes, which will be discussed below. Thereafter, the S_0 of different gases in GO membranes is accordingly calculated by Equations (6)–(8) [41–43].

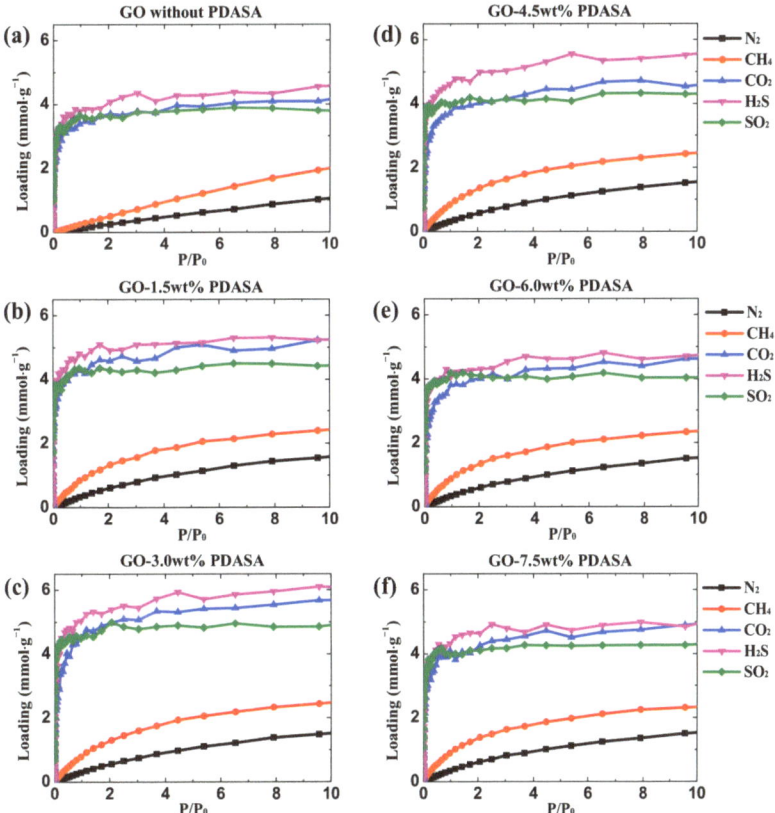

Figure 2. Unary isotherms of different gases in GO membranes with variable doping amounts of PDASA. (**a**) 0.0 wt%. (**b**) 1.5 wt%. (**c**) 3.0 wt%; (**d**) 4.5 wt%; (**e**) 6.0 wt%; (**f**) 7.5 wt%.

3.2. Adsorption Insight

To quantitatively evaluate the adsorption ability of different gases in GO membranes and understand the variable adsorption models, Figure 3 presents the calculated S_0 and the corresponding adsorption behaviors. The S_0 as a function of variable doping amounts of PDASA is shown in Figure 3a. For CH_4 and N_2, the S_0 values in different GO membranes are less than 0.002 mmol·g^{-1}·kPa^{-1}, almost invariable with the doped PDASA. The distribution probability in Figure 3b reveals that the particles of CH_4 and N_2 are highly concentrated, forming single-layer adsorption. Snapshots in Figure 3c,d provide a visual perspective for these single-adsorbate cases, where CH_4 and N_2 deposit in the center of GO channels, indicating a weak adsorption ability. That is the reason their adsorption behaviors in GO membranes can be accurately represented by Langmuir model [44]. On the contrary, CO_2 and H_2S exhibit a strong adsorption ability with the S_0 all above 3.4 mmol·g^{-1}·kPa^{-1}. As seen in Figure 3a, when the PDASA-doping amount is 3.0 wt%, CO_2 and H_2S exhibit the maximum S_0 values of 4.5 and 5.3 mmol·g^{-1}·kPa^{-1}, respectively, almost 2.5 times higher than those values of GO membranes without doping PDASA. Continuously increasing the doping amounts, the S_0 shows a downward trend. The adsorption ability of SO_2 in GO membranes is extremely strong as there is an almost vertical ascent motion at the start point of isotherms (Figure 2). Therefore, the S_0 of SO_2 are all above 80 mmol·g^{-1}·kPa^{-1} and not compared in Figure 3a. Compared to CH_4 and N_2, for acid gases, their maximum distribution probability is not in the center of channels but on either side of the center. By visual of Figure 3e–g, CO_2, H_2S and SO_2 present multilayer adsorption in GO channels. In addition, they also have a probability to distribute "outside" channels due to periodic boundary conditions. The above complex adsorption behavior of CO_2 and H_2S indicates a strong adsorption ability, thus deserving the Redlich-Peterson model [45,47].

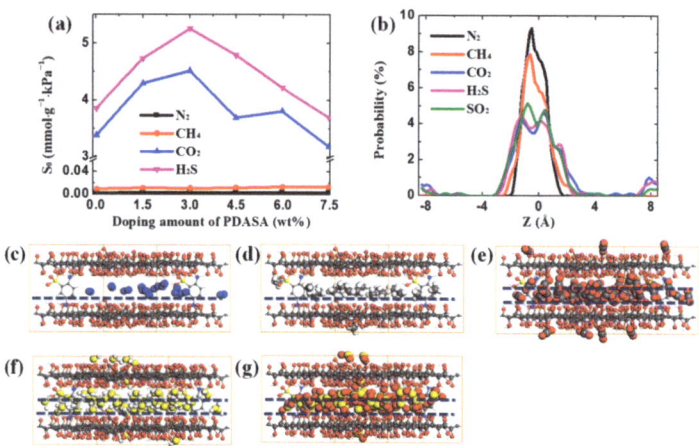

Figure 3. Adsorption behavior. (**a**) Solubility coefficient of different gases in GO membranes. (**b**) Distribution probability of gases in GO channels. Snapshots of variable gases adsorbed in GO channels. (**c**) N_2; (**d**) CH_4; (**e**) CO_2; (**f**) H_2S and (**g**) SO_2.

To reveal the positive effect of PDASA on acid gas adsorption in GO membranes, RDF, isosteric heat and hydrogen bonds are analyzed in Figure 4 to provide molecular insight into the adsorption process. The dynamic binding process between gases and PDASA is evaluated with RDF graph g(r) based on Equation (9) [33], where r is the distance from species i to j, N_i represents the number of species i, $N_{ij}(r, r + \Delta r)$ is the number of i around j within a shell and V is the volume. The RDF value is a measure of binding affinity, whereas a high RDF value means a strong affinity of PDASA to gases. As seen in Figure 4a, the affinity increases following the sequence of $N_2 \approx CH_4 < CO_2 < H_2S \approx SO_2$. The high affinity of PDASA to acid gases is the primary reason for its positive effect on acid gas adsorption, while the weak guest-membrane affinities lead to the weak adsorption of CH_4 and N_2 in GO channels. Isosteric heat, a decisive factor of adsorption strength, is analyzed in Figure 4b. Obviously, the isosteric heats of five gases in GO membranes increase in the order of $N_2 < CH_4 < CO_2 \approx H_2S < SO_2$, confirming the strong adsorption strength of acid gases in GO membranes, especially for SO_2. Besides the binding affinity and isosteric heat, the strong adsorption of acid gases is also related to hydrogen bonds. Based on these two geometrical criteria [16], (1) r(H···O) ≤ 0.35 nm; (2) α(O-H···O) ≤ 30°, hydrogen bonds in acid gases adsorption process are pictured in Figure 4c–e. A great number of hydrogen bonds are formed between GO membranes and acid gases. In addition, the doped PDASA also contributes to the formation of hydrogen bonds, as shown in Figure 4f, which further helps GO membranes to capture H_2S. The above effects synergistically promote acid gas adsorption, while large doping amounts will decrease the effective adsorption sites and reduce the packing efficiency of acid gases in GO channels due to the narrowing of the passage, which will be discussed below.

$$g_{ij}(r) = \frac{N_{ij}(r, r+\Delta r)V}{4\pi r^2 \Delta r N_i N_j} \quad (9)$$

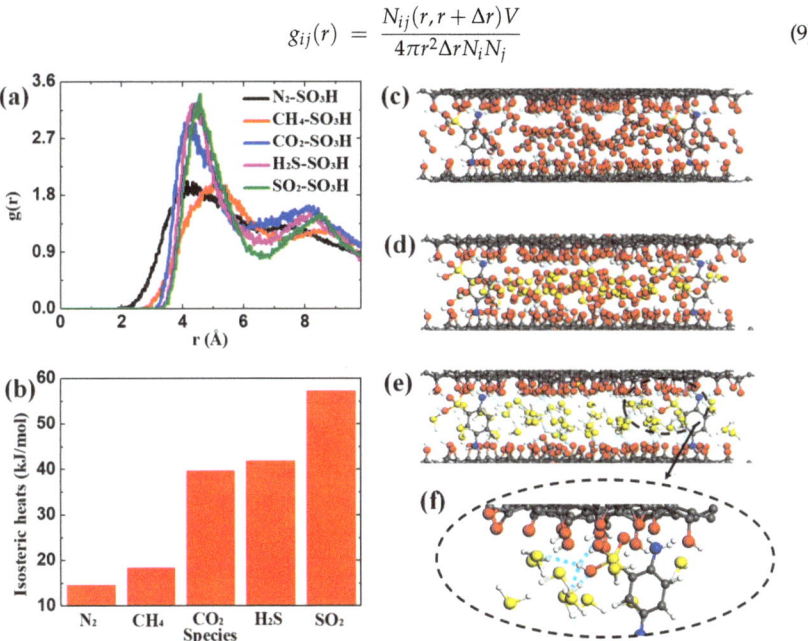

Figure 4. Adsorption Insights. (**a**) RDF of the doped PDASA to various gases. (**b**) Isosteric heats. Hydrogen bonds formed in the adsorption process of acid gases. (**c**) CO_2. (**d**) SO_2. (**e**) H_2S. (**f**) Hydrogen bonds around PDASA.

3.3. Diffusion Evaluation

Dynamical properties of gases in GO channels are evaluated by mean square displacement (MSD) according to Equation (10) [32,33] in which the N refers to the total number of particles and $r_i(t) - r_i(t_0)$ is the displacement distance of particle i from the initial state t_0 to the final state t. As shown in Figure 5, the gas mobility in GO channels with variable doping amounts of PDASA follows the sequence of $N_2 \approx CH_4 > H_2S > CO_2 > SO_2$, which means the diffusion process is not governed by the size-sieving effect. The large mobilities of CH_4 and N_2 in GO channels are attributed to their weak interactions with GO membranes, thus resulting in low mass-transfer resistance. Although with smaller molecular size, acid gases exhibit slow mobility in that the strong interactions generate a large transport resistance [16]. After doping the PDASA into GO channels, the mobilities of all gases slow down. Diffusion coefficient (D) is another key role in determining separation performance, which is calculated by the linear slope of MSD based on Equation (11) [32,33]. Taking the cases in pure GO membrane as examples, the logarithmic form shown in Figure S1 can be fitted linearly from 100 to 1000 ps with slopes larger than 0.94, indicating that the gas diffusion tends to stabilize and approach to a normal diffusion state [48]. Then the D can be obtained from this region in MSD curves. To uncover the diffusion mechanism of gases in GO channels, the quantitative diffusivity, accessible free volume (AFV) [49] and effective transport channels are analyzed in Figure 6. Figure 6a illustrates that the diffusion coefficient generally shows a decreasing trend with the increase in the PDASA-doping amount. For N_2 and CH_4, both have diffusion coefficients larger than 240×10^{-7} $cm^2 \cdot s^{-1}$ due to the low transfer resistance, which agrees well with previous work [25], demonstrating the reliability of our calculations. In contrast, for acid gases, their diffusivities in GO channels are relatively low. Especially for SO_2, its dynamic motion is severely restricted with diffusion coefficients lower than 80×10^{-7} cm$^{-2}\cdot$s^{-1}. The AFV in variable GO membranes as a function of probe radius is shown in Figure 6b based on Equation (12) where V_f and V_o denote the free and occupied volumes, respectively. It shows that the AFV is sensitive to the probe radius. In addition, when the probe radius is larger than the molecular sizes of acid gases, the AFV nearly declines with the increase in the PDASA-doping amounts (Figure 6c). Figure 6d–i show the visualization of free volume. Apparently, the PDASA severed as barriers in GO channels to block the passage of gases (green region). With increasing the doping amounts, the effective passage is narrowed especially in GO-7.5 wt% PDASA (Figure 6i). That is the reason molecular diffusion is severely inhibited by doping PDASA in GO channels. This confirms that doping PDASA into GO channels brings a change not only in their adsorption but also in their diffusion. However, in this condition, diffusion is not supposed to govern the separation process of acid gases through the PDASA-doped GO membranes.

$$MSD(t) = \frac{1}{N}\left\langle \sum_{i=1}^{N}[r_i(t) - r_i(t_0)]^2 \right\rangle \tag{10}$$

$$D = \frac{1}{6}\lim_{t\to\infty}\frac{dMSD}{dt} \tag{11}$$

$$AFV = \frac{V_f}{V_f + V_o} \times 100\% \tag{12}$$

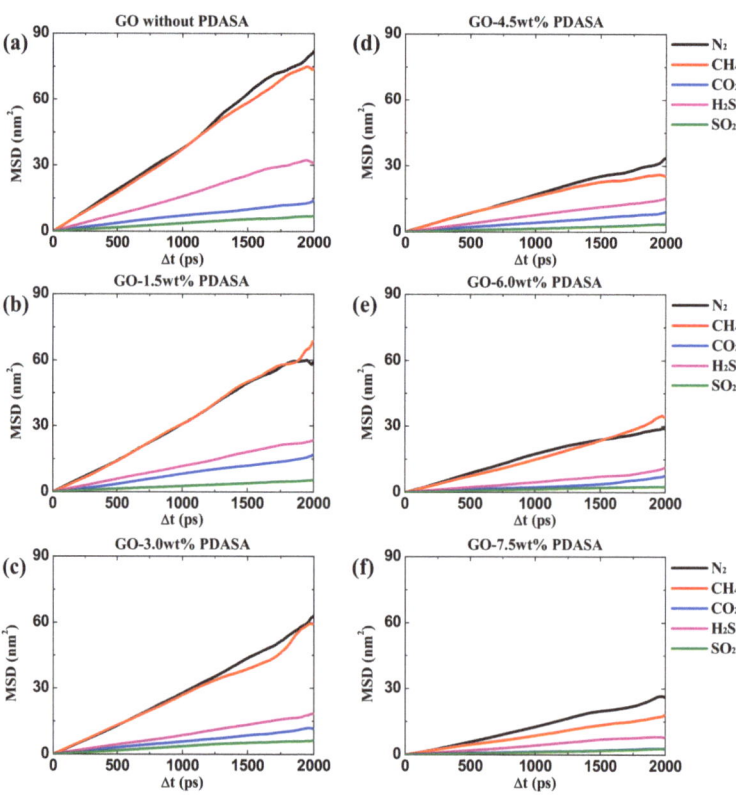

Figure 5. Mobility of gases in GO channels with variable doping amount of PDASA. (**a**) 0.0 wt%. (**b**) 1.5 wt%. (**c**) 3.0 wt%; (**d**) 4.5 wt%; (**e**) 6.0 wt%; (**f**) 7.5 wt%.

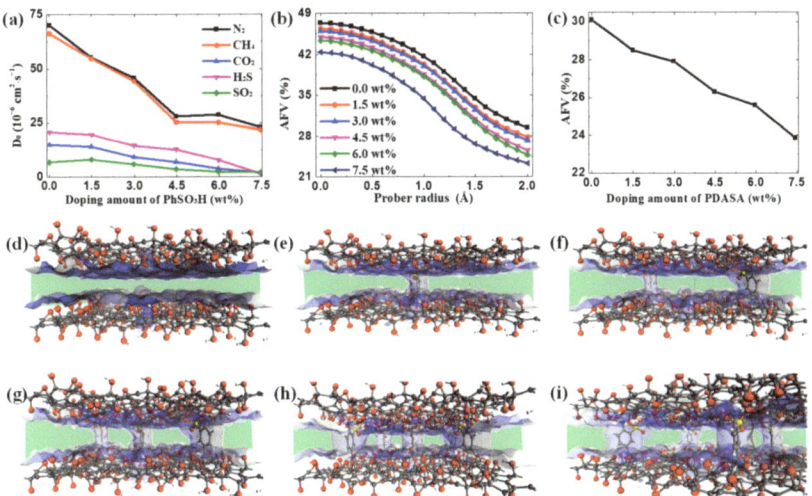

Figure 6. Diffusion insights. (**a**) Diffusion coefficient of different gases. (**b**) The free accessible volume of variable GO membranes. (**c**) The detected AFV with a 1.9 Å-sized prober is dependent on the doping amount of PDASA. Visualization of passage in variable GO channels. (**d**) 0.0 wt%. (**e**) 1.5 wt%. (**f**) 3.0 wt%; (**g**) 4.5 wt%; (**h**) 6.0 wt%; (**i**) 7.5 wt%.

3.4. Separation Performance Prediction

The permeability coefficient, P_i, with a typically reported unit of Barrer is determined on the basis of the solution-diffusion model in Equation (13), where the corresponding S_i and D_i have a unit of $cm^3(STP) \cdot cm^{-3} \cdot mmHg$ and $10^{-7} \ cm^2 \cdot s^{-1}$, respectively, which are included in Table S2. The ideal gas selectivity, $\alpha_{i/j}$, is defined as the ratio of permeabilities of i and j by Equation (14). The separation performance of acid gases (CO_2 and H_2S) through PDASA-doped GO membranes is predicated in Figure 7. For CH_4 and N_2, their permeabilities are relatively low, as shown in Figure 7a; in contrast, acid gases exhibit high permeabilities thanks to their extraordinarily high S_0 in GO membranes, which indicates that this permeation process is governed by preferential adsorption. Doping a tiny amount of PDASA into GO channels helps to promote the permeability of CO_2 and H_2S by 21% and 18%, respectively. Figure 7b shows the ideal selectivities of CO_2/CH_4, CO_2/N_2, H_2S/CH_4 and H_2S/N_2. Apparently, the selectivities of the above four gas pairs also increase first and then decrease with the increase of PDASA-doping amounts, and their highest selectivities can be up to 95.7, 290.3, 200.8, and 608.2, respectively. The predicted separation performance is compared with experimental results. As shown in Figure 7c,d, the separation performance for both CO_2/CH_4 and $(CO_2 + H_2S)/CH_4$ of the PDASA-doped GO membranes were several orders of magnitude greater than most of the ever-reported membranes (Table S3) and far exceed the 2008 Robeson upper bound [22], suggesting the promising potential of the adsorption-dominated separation in acid gas treatment.

$$P_i = S_i D_i \tag{13}$$

$$\alpha_{i/j} = \frac{P_i}{P_j} = \frac{S_i D_i}{S_j D_j} \tag{14}$$

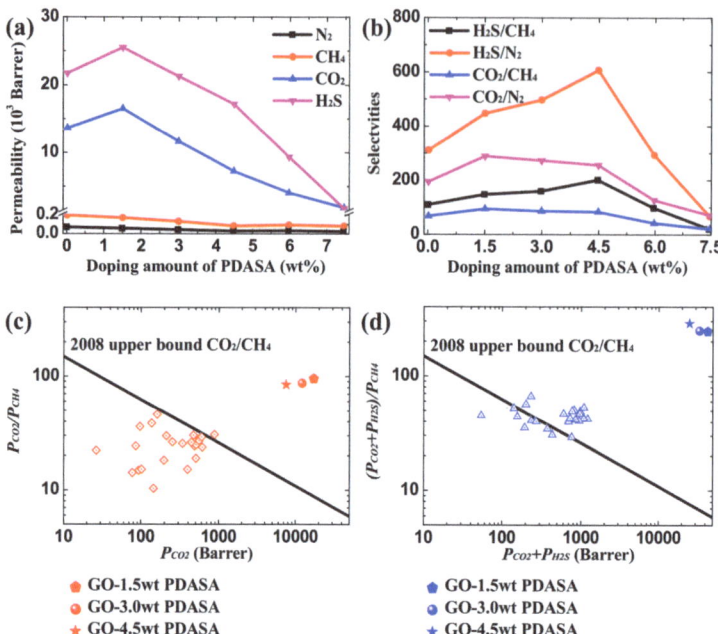

Figure 7. Separation performance. (**a**) Gas permeability. (**b**) Ideal selectivities of H_2S/CH_4, H_2S/N_2, CO_2/CH_4 and CO_2/N_2. Performance comparison for separations of (**c**) CO_2/CH_4 and (**d**) $(CO_2 + H_2S)/CH_4$ with other potential membranes and the 2008 Robeson upper bound of CO_2/CH_4 (Black line).

4. Conclusions

In summary, molecular simulations are performed to investigate the adsorption and diffusion behaviors of several gases in the PDASA-doped GO membranes. Doping a tiny amount (3.0 wt%) of PDASA into GO channels effectively promotes the adsorption ability of acid gases, with the solubility coefficient of H_2S and CO_2 improving almost 2.5 times, while the adsorption abilities of CH_4 and N_2 are almost invariable with the doped PDASA. Theoretical analysis demonstrates that the isotherms of CH_4 and N_2 show weak adsorption, following the Langmuir model, while acid gases exhibit multilayer adsorption in GO membranes, which is relatively complex and described by the Redlich-Peterson model. Molecular insights reveal that the strong adsorption of acid gases in GO membranes is ascribed to their high isosteric heat, great binding affinity and hydrogen bonds. While their diffusion in GO channels is restrained by doping PDASA due to the narrowing of the passage. Even so, the permeability of acid gases and their ideal selectivities over CH_4 are greatly enhanced over Robeson upper bound by doping a tiny amount of PDASA, which suggests that this removal process of acid gases is primarily dominated by preferential adsorption. From the bottom-up, this molecular understanding provides a strategy to develop high-performance GO membranes toward acid gas treatment. Such fundamental insights show the great potential of 2D membranes in the practical application of natural gas processing.

Supplementary Materials: The following supporting information can be downloaded at: https://www.mdpi.com/article/10.3390/membranes12111155/s1, Figure S1: The lg (MSD)-lg (t) curve for the transport of variable gases through pure GO and GO-7.5wt%PDASA membranes.; Table S1: Fitting parameters of α, γ, β, n and correlation coefficient (R^2) for adsorption isotherms of different gases.; Table S2: The solubility co-efficient, diffusion coefficient and permeability with the typically reported unit.; Table S3: Perfor-mance comparison for separations of CO2/CH4 and (CO2+H2S)/CH4. (References [50–59] are cited in the supplementary materials.)

Author Contributions: Conceptualization, Q.L. and G.L.; methodology, Q.L. and G.L.; software, Q.L. and Z.Y.; validation, G.L., R.X. and J.Z.; formal analysis, L.S.; investigation, Q.L. and L.S.; resources, Z.Y. and J.Z.; data curation, Q.L. and R.X.; writing—original draft preparation, Q.L.; writing—review and editing, G.L.; visualization, G.L.; supervision, G.L.; project administration, G.L.; funding acquisition, Q.L. and G.L. All authors have read and agreed to the published version of the manuscript.

Funding: This research was funded by the Anhui Provincial Natural Science Foundation, grant number 2108085QB50, the University Natural Science Research Project of Anhui Province, grant number KJ2020A0286, and the National Natural Science Foundation of China, grant number 22278210. The numerical calculations in this paper have been done on the supercomputing system in the Supercomputing Center of University of Science and Technology of China.

Institutional Review Board Statement: Not applicable.

Informed Consent Statement: Not applicable.

Data Availability Statement: The data presented in this study are available on request from the corresponding author.

Conflicts of Interest: The authors declare no conflict of interest.

Nomenclature

Symbols

Q_{ex}, Q_{ab}	excess, absolute adsorption capacity (mmol·g^{-1})
ρ_g	gas density (mmol·cm^{-3})
V_f, V_o	free, occupied volumes (cm^3·g^{-1}) per unit mass
δ, β and γ	fitting parameters in the adsorption model
Q_L	maximal adsorption capacity in the Langmuir model
K_L	adsorption equilibrium constant in the Langmuir model
n	empirical constant in the Redlich-Peterson model
R^2	correlation coefficient
$g(r)$	radial distribution function (RDF)
MSD	mean square displacement (nm^2)
AFV	accessible free volume (%)
S_i	solubility coefficient (cm^3(STP)·cm^{-3}·mmHg)
D_i	diffusion coefficient (10^{-7} cm^2·s^{-1})
P_i	permeability coefficient (Barrer)
$\alpha_{i/j}$	gas selectivity of species i over j

References

1. Plant, G.; Kort, E.A.; Brandt, A.R.; Chen, Y.; Fordice, G.; Negron, A.M.G.; Schwietzke, S.; Smith, M.; Zavala-Araiza, D. Inefficient and unlit natural gas flares both emit large quantities of methane. *Science* **2022**, *377*, 1566–1571. [CrossRef] [PubMed]
2. Anyanwu, J.-T.; Wang, Y.; Yang, R.T. CO_2 capture (including direct air capture) and natural gas desulfurization of amine-grafted hierarchical bimodal silica. *Chem. Eng. J.* **2021**, *427*, 131561. [CrossRef]
3. Park, J.; Yoon, S.; Oh, S.-Y.; Kim, Y.; Kim, J.-K. Improving energy efficiency for a low-temperature CO_2 separation process in natural gas processing. *Energy* **2020**, *214*, 118844. [CrossRef]
4. Behmadi, R.; Mokhtarian, M.; Davoodi, A.; Hosseinpour, S. Desulfurization of natural gas condensate using polyethylene glycol and water intercalated activated γ-bauxite. *J. Clean. Prod.* **2022**, *376*, 134230. [CrossRef]
5. He, Z.; Gupta, K.M.; Linga, P.; Jiang, J. Molecular Insights into the Nucleation and Growth of CH_4 and CO_2 Mixed Hydrates from Microsecond Simulations. *J. Phys. Chem. C* **2016**, *120*, 25225–25236. [CrossRef]
6. He, Z.; Mi, F.; Ning, F. Molecular insights into CO_2 hydrate formation in the presence of hydrophilic and hydrophobic solid surfaces. *Energy* **2021**, *234*, 121260. [CrossRef]
7. He, Z.; Linga, P.; Jiang, J. What are the key factors governing the nucleation of CO_2 hydrate? *Phys. Chem. Chem. Phys.* **2017**, *19*, 15657–15661. [CrossRef]
8. Cavaignac, R.S.; Ferreira, N.L.; Guardani, R. Techno-economic and environmental process evaluation of biogas upgrading via amine scrubbing. *Renew. Energy* **2021**, *171*, 868–880. [CrossRef]

9. Cheng, Y.; Wang, Z.; Zhao, D. Mixed Matrix Membranes for Natural Gas Upgrading: Current Status and Opportunities. *Ind. Eng. Chem. Res.* **2018**, *57*, 4139–4169. [CrossRef]
10. Chen, X.; Liu, G.; Jin, W. Natural gas purification by asymmetric membranes: An overview. *Green Energy Environ.* **2020**, *6*, 176–192. [CrossRef]
11. Mohsenpour, S.; Guo, Z.; Almansour, F.; Holmes, S.M.; Budd, P.M.; Gorgojo, P. Porous silica nanosheets in PIM-1 membranes for CO_2 separation. *J. Membr. Sci.* **2022**, *661*, 120889. [CrossRef]
12. Han, Y.; Ho, W.S.W. Polymeric membranes for CO_2 separation and capture. *J. Membr. Sci.* **2021**, *628*, 119244. [CrossRef]
13. Ahmad, M.Z.; Peters, T.A.; Konnertz, N.M.; Visser, T.; Téllez, C.; Coronas, J.; Fila, V.; de Vos, W.M.; Benes, N.E. High-pressure CO_2/CH_4 separation of Zr-MOFs based mixed matrix membranes. *Sep. Purif. Technol.* **2020**, *230*, 115858. [CrossRef]
14. Liu, G.; Cadiau, A.; Liu, Y.; Adil, K.; Chernikova, V.; Carja, I.-D.; Belmabkhout, Y.; Karunakaran, M.; Shekhah, O.; Zhang, C.; et al. Enabling Fluorinated MOF-Based Membranes for Simultaneous Removal of H_2S and CO_2 from Natural Gas. *Angew. Chem.* **2018**, *130*, 15027–15032. [CrossRef]
15. Singh, S.; Varghese, A.M.; Reinalda, D.; Karanikolos, G.N. Graphene-based membranes for carbon dioxide separation. *J. CO2 Util.* **2021**, *49*, 101544. [CrossRef]
16. Liu, Q.; Chen, M.; Mao, Y.; Liu, G. Theoretical study on Janus graphene oxide membrane for water transport. *Front. Chem. Sci. Eng.* **2020**, *15*, 913–921. [CrossRef]
17. Chen, L.; Shi, G.; Shen, J.; Peng, B.; Zhang, B.; Wang, Y.; Bian, F.; Wang, J.; Li, D.; Qian, Z.; et al. Ion sieving in graphene oxide membranes via cationic control of interlayer spacing. *Nature* **2017**, *550*, 380–383. [CrossRef] [PubMed]
18. Zheng, J.; Wang, R.; Ye, Q.; Chen, B.; Zhu, X. Multilayered graphene oxide membrane with precisely controlled interlayer spacing for separation of molecules with very close molecular weights. *J. Membr. Sci.* **2022**, *657*, 120678. [CrossRef]
19. Liang, F.; Liu, Q.; Zhao, J.; Guan, K.; Mao, Y.; Liu, G.; Gu, X.; Jin, W. Ultrafast water-selective permeation through graphene oxide membrane with water transport promoters. *AIChE J.* **2019**, *66*, e16812. [CrossRef]
20. Mohsenpour, S.; Ameen, A.W.; Leaper, S.; Skuse, C.; Almansour, F.; Budd, P.M.; Gorgojo, P. PIM-1 membranes containing—Graphene oxide for CO_2 separation. *Sep. Purif. Technol.* **2022**, *298*, 121447. [CrossRef]
21. Wang, T.; Cheng, C.; Wu, L.-g.; Shen, J.-n.; Van der Bruggen, B.; Chen, Q.; Chen, D.; Dong, C.-y. Fabrication of polyimide membrane incorporated with functional graphene oxide for CO_2 separation: The effects of GO surface modification on membrane performance. *Environ. Sci. Technol.* **2017**, *51*, 6202–6210. [CrossRef] [PubMed]
22. Robeson, L.M. The upper bound revisited. *J. Membr. Sci.* **2008**, *320*, 390–400. [CrossRef]
23. Zhu, X.; Tian, C.; Do-Thanh, C.-L.; Dai, S. Two-Dimensional Materials as Prospective Scaffolds for Mixed-Matrix Membrane-Based CO_2 Separation. *ChemSusChem* **2017**, *10*, 3304–3316. [CrossRef]
24. Yan, F.; Guo, Y.; Wang, Z.; Zhao, L.; Zhang, X. Efficient separation of CO_2/CH_4 by ionic liquids confined in graphene oxide: A molecular dynamics simulation. *Sep. Purif. Technol.* **2022**, *289*, 120736. [CrossRef]
25. Khakpay, A.; Rahmani, F.; Nouranian, S.; Scovazzo, P. Molecular insights on the CH_4/CO_2 separation in nanoporous graphene and graphene oxide separation platforms: Adsorbents versus membranes. *J. Phys. Chem. C* **2017**, *121*, 12308–12320.
26. Zheng, X.; Ban, S.; Liu, B.; Chen, G. Strain-controlled graphdiyne membrane for CO_2/CH_4 separation: First-principle and molecular dynamic simulation. *Chin. J. Chem. Eng.* **2020**, *28*, 1898–1903. [CrossRef]
27. Kallo, M.T.; Lennox, M.J. Understanding $CO_2/CH4$ separation in pristine and defective 2D MOF CuBDC nanosheets via nonequilibrium molecular dynamics. *Langmuir* **2020**, *36*, 13591–13600.
28. Wang, Z.; Yan, F.; Bai, L.; Zhang, X.; Liu, X.; Zhang, X. Insight into $CO_2/CH4$ separation performance in ionic liquids/polymer membrane from molecular dynamics simulation. *J. Mol. Liq.* **2022**, *357*, 119119.
29. Liu, Q.; Zhu, H.; Liu, G.; Jin, W. Efficient separation of (C1–C2) alcohol solutions by graphyne membranes: A molecular simulation study. *J. Membr. Sci.* **2021**, *644*, 120139. [CrossRef]
30. Liu, Q.; Cheng, L.; Liu, G. Enhanced Selective Hydrogen Permeation through Graphdiyne Membrane: A Theoretical Study. *Membranes* **2020**, *10*, 286. [CrossRef]
31. Liu, Q.; Liu, Y.; Liu, G. Simulation of cations separation through charged porous graphene membrane. *Chem. Phys. Lett.* **2020**, *753*, 137606. [CrossRef]
32. Liu, Q.; Wu, Y.; Wang, X.; Liu, G.; Zhu, Y.; Tu, Y.; Lu, X.; Jin, W. Molecular dynamics simulation of water-ethanol separation through monolayer graphene oxide membranes: Significant role of O/C ratio and pore size. *Sep. Purif. Technol.* **2019**, *224*, 219–226. [CrossRef]
33. Liu, Q.; Gupta, K.M.; Xu, Q.; Liu, G.; Jin, W. Gas permeation through double-layer graphene oxide membranes: The role of interlayer distance and pore offset. *Sep. Purif. Technol.* **2018**, *209*, 419–425. [CrossRef]
34. Siepmann, J.I.; Frenkel, D. Configurational bias Monte Carlo: A new sampling scheme for flexible chains. *Mol. Phys.* **1992**, *75*, 59–70. [CrossRef]
35. Braga, C.; Travis, K.P. A configurational temperature Nosé-Hoover thermostat. *J. Chem. Phys.* **2005**, *123*, 134101. [CrossRef]
36. Sun, H. COMPASS: An ab Initio Force-Field Optimized for Condensed-Phase ApplicationsOverview with Details on Alkane and Benzene Compounds. *J. Phys. Chem. B* **1998**, *102*, 7338–7364. [CrossRef]
37. Essmann, U.; Perera, L.; Berkowitz, M.L.; Darden, T.; Lee, H.; Pedersen, L.G. A smooth particle mesh Ewald method. *J. Chem. Phys.* **1995**, *103*, 8577–8593. [CrossRef]

38. Lin, Y.; Pan, D.; Li, J.; Zhang, L.; Shao, X. Application of Berendsen barostat in dissipative particle dynamics for nonequilibrium dynamic simulation. *J. Chem. Phys.* **2017**, *146*, 124108. [CrossRef]
39. Lopez-Echeverry, J.S.; Reif-Acherman, S.; Araujo-Lopez, E. Peng-Robinson equation of state: 40 years through cubics. *Fluid Phase Equilib.* **2017**, *447*, 39–71. [CrossRef]
40. Chen, L.; Liu, K.; Jiang, S.; Huang, H.; Tan, J.; Zuo, L. Effect of adsorbed phase density on the correction of methane excess adsorption to absolute adsorption in shale. *Chem. Eng. J.* **2021**, *420*, 127678. [CrossRef]
41. De Lorenzo, L.; Tocci, E.; Gugliuzza, A.; Drioli, E. Pure and modified co-poly(amide-12-b-ethylene oxide) membranes for gas separation studied by molecular investigations. *Membranes* **2012**, *2*, 346–366. [CrossRef]
42. Zhou, J.; Zhu, X.; Hu, J.; Liu, H.; Hu, Y.; Jiang, J. Mechanistic insight into highly efficient gas permeation and separation in a shape-persistent ladder polymer membrane. *Phys. Chem. Chem. Phys.* **2014**, *16*, 6075–6083. [CrossRef]
43. Tocci, E.; Gugliuzza, A.; De Lorenzo, L.; Macchione, M.; De Luca, G.; Drioli, E. Transport properties of a co-poly(amide-12-b-ethylene oxide) membrane: A comparative study between experimental and molecular modelling results. *J. Membr. Sci.* **2008**, *323*, 316–327. [CrossRef]
44. Alafnan, S.; Awotunde, A.; Glatz, G.; Adjei, S.; Alrumaih, I.; Gowida, A. Langmuir adsorption isotherm in unconventional resources: Applicability and limitations. *J. Pet. Sci. Eng.* **2021**, *207*, 109172. [CrossRef]
45. Liu, Q.; Guo, H.; Shan, Y. Adsorption of fluoride on synthetic siderite from aqueous solution. *J. Fluor. Chem.* **2010**, *131*, 635–641. [CrossRef]
46. Tocci, E.; De Lorenzo, L.; Bernardo, P.; Clarizia, G.; Bazzarelli, F.; Mckeown, N.B.; Carta, M.; Malpass-Evans, R.; Friess, K.; Pilnáček, K.; et al. Molecular Modeling and Gas Permeation Properties of a Polymer of Intrinsic Microporosity Composed of Ethanoanthracene and Tröger's Base Units. *Macromolecules* **2014**, *47*, 7900–7916. [CrossRef]
47. Kalam, S.; Abu-Khamsin, S.A.; Kamal, M.S.; Patil, S. Surfactant adsorption isotherms: A review. *ACS Omega* **2021**, *6*, 32342–32348. [CrossRef]
48. Javanainen, M.; Hammaren, H.; Monticelli, L.; Jeon, J.-H.; Miettinen, M.S.; Martinez-Seara, H.; Metzler, R.; Vattulainen, I. Anomalous and normal diffusion of proteins and lipids in crowded lipid membranes. *Faraday Discuss.* **2012**, *161*, 397–417. [CrossRef]
49. Mazo, M.; Khudobin, R.; Balabaev, N.; Belov, N.; Ryzhikh, V.; Nikiforov, R.; Chatterjee, R.; Banerjee, S. Structure and free volume of fluorine-containing polyetherimides with Pendant di-tert-butyl groups investigated by molecular dynamics simulation. *Polymer* **2022**, *14*, 125318. [CrossRef]
50. Qian, Q.; Wright, A.M.; Lee, H.; Dincă, M.; Smith, Z.P. Low-temperature $H_2S/CO_2/CH_4$ separation in mixed-matrix membranes containing MFU-4. *Chem. Mater.* **2021**, *33*, 6825–6831. [CrossRef]
51. Liu, G.; Cadiau, A.; Liu, Y.; Adil, K.; Chernikova, V.; Carja, I.-D.; Belmabkhout, Y.; Karunakaran, M.; Shekhah, O.; Zhang, C.; et al. Enabling fluorinated MOF-based membranes for simultaneous removal of H_2S and CO_2 from natural gas. *Angew. Chem. Int. Ed.* **2018**, *57*, 14811–14816. [CrossRef]
52. Ahmad, M.Z.; Peters, T.A.; Konnertz, N.M.; Visser, T.; Téllez, C.; Coronas, J.; Fila, V.; de Vos, W.M.; Benes, N.E. High-pressure CO_2/CH_4 separation of Zr-MOFs based mixed matrix membranes. *Sep. Purif. Technol.* **2020**, *230*, 115858. [CrossRef]
53. Yahaya, G.O.; Hayek, A.; Alsamah, A.; Shalabi, Y.A.; Ben Sultan, M.M.; Alhajry, R.H. Copolyimide membranes with improved H_2S/CH_4 selectivity for high-pressure sour mixed-gas separation. *Sep. Purif. Technol.* **2021**, *272*, 118897. [CrossRef]
54. Hayek, A.; Alsamah, A.; Alaslai, N.; Maab, H.; Qasem, E.A.; Alhajry, R.H.; Alyami, N.M. Unprecedented Sour Mixed-Gas Permeation Properties of Fluorinated Polyazole-Based Membranes. *ACS Appl. Polym. Mater.* **2020**, *2*, 2199–2210. [CrossRef]
55. Hayek, A.; Yahaya, G.O.; Alsamah, A.; Alghannam, A.A.; Jutaily, S.A.; Mokhtari, I. Pure- and sour mixed-gas transport properties of 4,4'-methylenebis(2,6-diethylaniline)-based copolyimide membranes. *Polymer* **2019**, *166*, 184–195. [CrossRef]
56. Hayek, A.; Alsamah, A.; Yahaya, G.O.; Qasem, E.A.; Alhajry, R.H. Post-synthetic modification of CARDO-based materials: Application in sour natural gas separation. *J. Mater. Chem. A* **2020**, *8*, 23354–23367. [CrossRef]
57. Alghannam, A.A.; Yahaya, G.O.; Hayek, A.; Mokhtari, I.; Saleem, Q.; Sewdan, D.A.; Bahamdan, A.A. High pressure pure- and mixed sour gas transport properties of Cardo-type block co-polyimide membranes. *J. Membr. Sci.* **2018**, *553*, 32–42. [CrossRef]
58. Liu, Y.; Liu, Z.; Liu, G.; Qiu, Y.; Bhuwania, N.; Chinn, D.; Koros, W.J. Surprising plasticization benefits in natural gas upgrading using polyimide membranes. *J. Membr. Sci.* **2020**, *593*, 117430. [CrossRef]
59. Liu, G.; Chernikova, V.; Liu, Y.; Zhang, K.; Belmabkhout, Y.; Shekhah, O.; Zhang, C.; Yi, S.; Eddaoudi, M.; Koros, W.J. Mixed matrix formulations with MOF molecular sieving for key energy-intensive separations. *Nat. Mater.* **2018**, *17*, 283–289. [CrossRef]

Article

Analysis of the Influence of Process Parameters on the Properties of Homogeneous and Heterogeneous Membranes for Gas Separation

Daniel Polak and Maciej Szwast *

Faculty of Chemical and Process Engineering, Warsaw University of Technology, Warynskiego 1, 00-645 Warsaw, Poland
* Correspondence: maciej.szwast@pw.edu.pl; Tel.: +48-22-234-64-16

Abstract: Heterogeneous membranes, otherwise known as Mixed Matrix Membranes (MMMs), which are used in gas separation processes, are the subject of growing interest. This is due to their potential to improve the process properties of membranes compared to those of homogeneous membranes, i.e., those made of polymer only. Using such membranes in a process involves subjecting them to varying temperatures and pressures. This paper investigates the effects of temperature and feed pressure on the process properties of homogeneous and heterogeneous membranes. Membranes made of Pebax®2533 copolymer and containing additional fillers such as SiO_2, ZIF−8, and POSS-Ph were investigated. Tests were performed over a temperature range of 25–55 °C and a pressure range of 2–8 bar for N_2, CH_4, and CO_2 gases. It was found that temperature positively influences the increase in permeability, while pressure influences permeability depending on the gas used, which is related to the effect of pressure on the solubility of the gas in the membrane.

Keywords: mixed matrix membranes; temperature dependence; pressure dependence; permeability; diffusivity; solubility

Citation: Polak, D.; Szwast, M. Analysis of the Influence of Process Parameters on the Properties of Homogeneous and Heterogeneous Membranes for Gas Separation. *Membranes* **2022**, *12*, 1016. https://doi.org/10.3390/membranes12101016

Academic Editors: Annarosa Gugliuzza and Cristiana Boi

Received: 28 September 2022
Accepted: 14 October 2022
Published: 19 October 2022

Publisher's Note: MDPI stays neutral with regard to jurisdictional claims in published maps and institutional affiliations.

Copyright: © 2022 by the authors. Licensee MDPI, Basel, Switzerland. This article is an open access article distributed under the terms and conditions of the Creative Commons Attribution (CC BY) license (https://creativecommons.org/licenses/by/4.0/).

1. Introduction

Membrane gas separation is a process increasingly used in industrial processes [1]. In many cases, it is displacing other, more classical gas separation processes, such as adsorption, absorption, or cryogenic treatment. Using membrane techniques, good results are obtained in the separation of air components [2–4], biogas components [5–7], the separation of helium from natural gas [8,9], hydrogen recovery [10–12], natural gas sweetening [13–15], or air dehydration [16] and natural gas dehydration [17,18]. In order to improve the efficiency of the membrane process and to optimize the energy consumption of such a process, single or multi-stage plants are used, along with recirculation of selected streams [19–21]. Process efficiency and performance are also affected by process conditions, in particular, operating pressure and temperature [22–24]. The very important aspect is the selection of the correct membrane for the specific application. It is the selectivity of the membrane towards selected components of the gas mixture and the permeability of the membrane that determines its suitability for a particular application. However, in this paper, we focus on the influence of operational conditions on membrane performance. We can imagine such processes where high pressure or high temperature are required. For example, the natural gas is obtained under a pressure of several dozen bar. Its purification with membranes could, therefore, take place at high pressures, without the need to reduce it first. In turn, synthesis gas is also obtained at a pressure of several dozen bar and, additionally, at a temperature of several hundred degrees Celsius. Its treatment by membranes could, again, take place with only partial cooling. Other such examples of elevated temperature gases are any flue gases. We should note that the membranes discussed in this paper are made of polymers, which naturally limits the pressures and temperatures at which they can be used.

In membrane gas separation processes, dense polymeric membranes perform best. However, it has been known for years that such membranes have their limitations, in that it is difficult to simultaneously achieve high membrane selectivity and high product flux [25–27]. A method for overcoming these limitations is the manufacturing of polymer-based heterogeneous membranes, also known in the literature as Mixed Matrix Membranes [28–30]. In such membranes, solid particles such as silica, organometallic structures, nanotubes, or nanowires are dispersed in a polymer matrix. The presence of these fillers in the polymer matrix causes changes in the physico-chemical properties of the material and, thus, affects the process properties (selectivity and permeability) of the membrane [31,32]. As a result of such changes, membranes made from polymers and fillers improve their process properties and are no longer subject to the same limitations as membranes made solely from polymer. This is the reason why there is a growing interest in the manufacture, research and use of heterogeneous membranes. The properties of heterogeneous membranes are influenced both by the type of filler used and by the concentration of this filler in the matrix. A number of works have been devoted to this issue. Membrane properties, as mentioned earlier, are also affected by process parameters such as temperature or feed pressure. This issue has received much less attention in the literature, which is the motivation for the research and analysis undertaken in this paper. Indeed, it can be found in the literature that operating conditions (pressure, temperature) affect the process [33–38]. However, these mentions in the literature are made as general remarks, or as remarks referring to the process as such or in the context of mathematical modelling. There is no mention in the literature regarding the influence of process conditions on the various parameters describing membrane properties. Understanding the relationship between the effect of process conditions on membrane properties will also allow for better planning of the membrane gas separation process.

This paper presents a study of the process properties of homo- and heterogeneous flat membranes made of block copolymer and three types of fillers, namely SiO_2, ZIF−8, and POSS-Ph. The effects of temperature and feed pressure on membrane permeability and on diffusion and solubility coefficients were investigated.

2. Materials and Methods

2.1. Flat Membranes

The flat membranes were manufactured from a block copolymer with the trade name Pebax®2533 (Arkema, France). For heterogeneous membranes, the fillers were SiO_2 nanoparticles (nanopowder 10–20 nm, Sigma-Aldrich, Poznan, Poland), ZIF−8 (Basolite Z1200 by BASF, Sigma-Aldrich, Poznan, Poland) or POSS-Ph (PSS-Octaphenyl-substituted, Sigma-Aldrich, Poznan, Poland). The rationale for undertaking research with these particular compounds and the structural formulae of these materials can be found in another paper of ours [31]. In that paper, material studies of such heterogeneous membranes are presented.

Preparation of the flat membranes started with dissolving the polymer granules in a solvent, which, in this case, was 2−butanol (Sigma-Aldrich, Poznan, Poland). In preparing homogeneous membranes, a 7 wt% solution of polymer in solvent was prepared each time. The weighed solvent and polymer were placed in an oil bath and stirred vigorously at 80 °C until the solution was completely dissolved and homogenized, which took at least 24 h. For the manufacture of heterogeneous membranes, weighed amounts of fillers, i.e., SiO_2, ZIF−8, or POSS-Ph, respectively, were gradually added to the polymer solution at this stage. The amount of additives is specified as a mass percentage relative to the mass of the polymer in solution. When adding the fillers, the temperature of the solution was kept constant by constantly stirring it. The membrane-forming solution (mixture) was then left for a further 24 h at constant temperature and stirred continuously, this time at a lower intensity. In the case of homogeneous membranes, the step of adding fillers was omitted. Just before the membrane was formed, the solution was transferred to the ultrasonic bath for a few minutes. Finally, the solution was poured onto a heated glass plate and spread over it with a casting knife (Elcometer, Manchester, UK). The membrane thus prepared, while still in the liquid state, was left under controlled conditions until the solvent evaporated and the membrane solidified. The membrane was removed from the glass plate during the ultrapure water bath. After drying, the membrane was tested.

The thickness of the fabricated membrane was measured by scanning electron microscopy (SEM) using a PhenomPro instrument (PhenomWorld, Eindhoven, The Netherlands).

2.2. Time Lag Method

The time lag method was used to measure membrane properties such as diffusion coefficient, solubility coefficient, or permeability [39,40]. A stand of our own design was used, which included a diaphragm module for a flat diaphragm, a vacuum pump, a connection to a gas pressure cylinder, pressure transmitters, a thermostatic device, and a computer with software.

The basic time lag method allows the diffusion coefficient (D) of the gas in the membrane to be determined. However, analysis of the rate of permeate pressure increase over time also allows the permeability of the membrane (P) to be determined, and consequently, by dividing the permeability by the diffusion coefficient, it allows the solubility coefficient (S) to be determined indirectly. This is consistent with Equation (1):

$$P = S \cdot D \tag{1}$$

The pure gases in vessels were supplied by Air Products (Warsaw, Poland).

3. Results and Discussion

The following chapter presents an analysis of the effects of feed pressure and process temperature on the separation properties of fabricated membranes with flat geometries. For this stage of the study, a homogeneous membrane made of Pebax®2533 was used. Two membranes with different concentrations of each filler used were selected, namely SiO_2, ZIF−8, and POSS-Ph. Time-lag tests were performed for three different temperatures (25 °C, 45 °C, 55 °C) and three different feed pressures (2 bar, 4 bar, 8 bar). Pure gases of N_2, CH_4, and CO_2 were used.

The resulting permeability values are shown in Figures 1–3 and in Tables A1–A4. In presenting the results of the gas permeation measurements of the developed membranes, a barrer unit was used, which is a non-SI unit, but is generally accepted in membrane-related literature. The conversion into SI units can be done as follows:

$$1 \text{ barrer} = 3.35 \cdot 10^{-16} \frac{\text{mol} \cdot \text{m}}{\text{m}^2 \cdot \text{s} \cdot \text{Pa}} \tag{2}$$

An analysis of Figures 1–3 and the corresponding Tables A1–A4 allows the following preliminary observations to be noted. The effect of feed temperature on membrane permeability is clearly discernible. Higher permeability values are observed for higher temperatures. This is consistent with the van't Hoff–Arrhenius equation [41,42]. Furthermore, the mobility of the polymer chains increases with increasing temperature [43,44]. As a result, the transport of gas molecules across the membrane is facilitated. However, it should be noted that the magnitude of the variation in membrane permeability over the tested temperature range of 25 °C–55 °C depends on the gas and the filler used. The largest percentage increase in permeation with increasing temperature was obtained for N_2, followed by CH_4 and CO_2. This relationship was obtained for all membrane types tested. In turn, for different fillers, the largest percentage increase in permeation as a function of temperature was obtained for ZIF−8 (ca. 40%), followed by SiO_2 (ca. 27%) and POSS-Ph (ca. 22%) in comparison to the homogenous membrane.

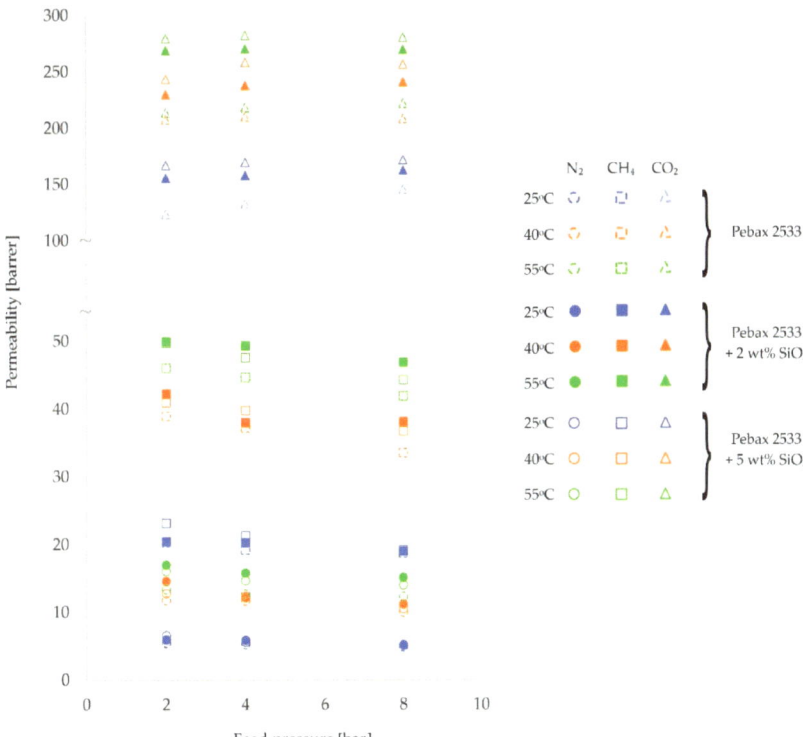

Figure 1. Change in permeability values of homogeneous membrane and heterogeneous membranes containing SiO_2 for different feed pressure and temperature values.

In contrast, changes in permeability values as a function of pressure are not so clear-cut. For N_2 and CH_4 gases, permeability decreases slightly with increasing pressure, while for CO_2 gas the situation is the opposite or, at the very least, no effect of pressure on the change in permeability is observed. Such observations can be made for both homogeneous and heterogeneous membranes. Explaining this effect requires further analysis, in particular, investigating the effect of pressure on diffusion and solubility coefficients. However, it is known that CO_2 molecules, due to the fact that they possess a quadra-pole moment, can interact differently with polymer chains than molecules of other gases [45–47]. The effect observed for N_2 and CH_4 gases, i.e., a decrease in permeability with increasing pressure, can be explained by the compression of the polymer chains of which the membrane is made [24,48,49]. The trend lines for the individual pressure-dependent permeability variations, not drawn on the graphs in Figures 1–3, are virtually parallel for a given gas and constant temperature. This means that the presence of nanoparticles filling the polymer matrix in a heterogeneous membrane has no additional effect on the compression (or prevention thereof) of the polymer chains in the membrane.

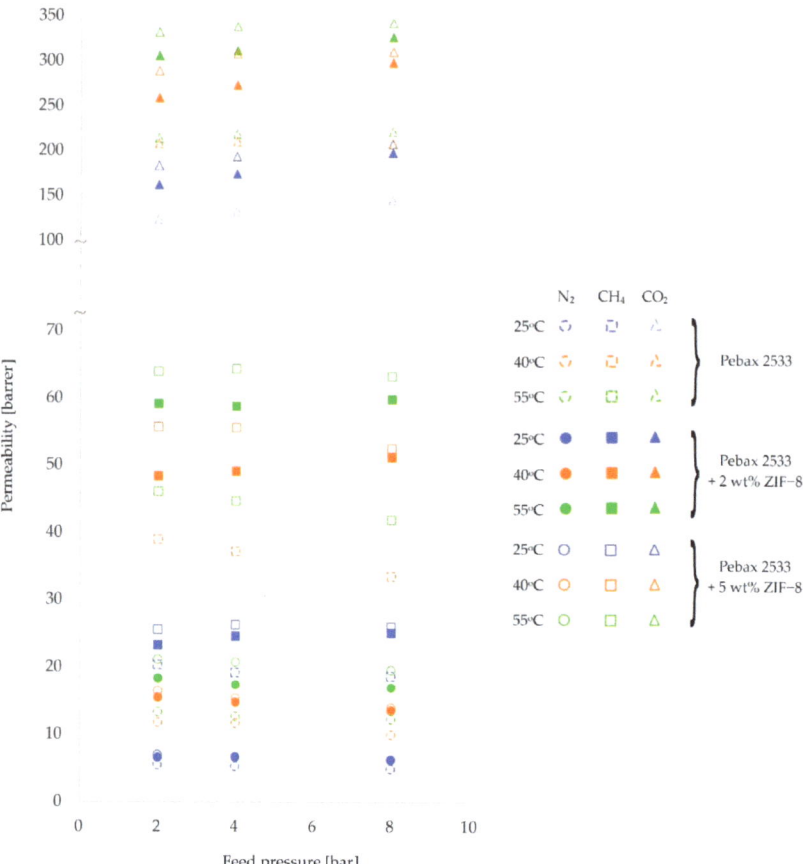

Figure 2. Change in permeability values of homogeneous membrane and heterogeneous membranes containing ZIF−8 for different feed pressure and temperature values.

By analyzing Figures 1–3, it is also possible to see the effect of the presence and concentration of fillers in the polymer matrix of the membrane on permeability. This issue is well known and described in the literature and will not be considered in this paper.

Further analysis of the effects of temperature and feed pressure on membrane properties will look at changes in the gas diffusion coefficient through homogeneous membrane and heterogeneous membranes. For the purposes of analysis, Figures 4–6 and the corresponding Tables A5–A8 have been drawn up, which contain the values of the diffusion coefficients measured for different membranes, different gases, and at different process parameter values.

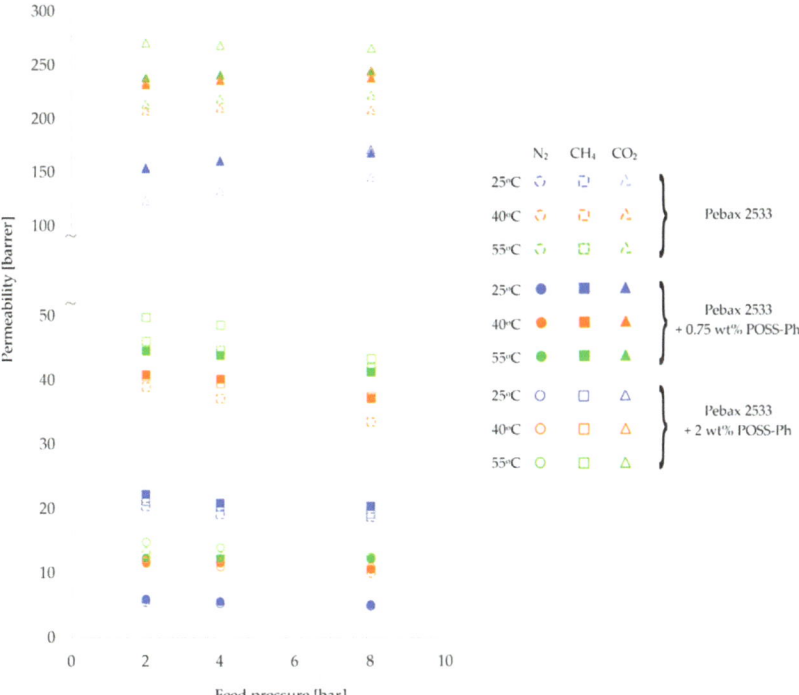

Figure 3. Change in permeability values of homogeneous membrane and heterogeneous membranes containing POSS-Ph for different feed pressure and temperature values.

On the basis of these diffusion coefficient measurements, it can be concluded that for all of the gases tested, there is an improvement in their diffusivity with increasing temperature for all of the membranes tested. In addition, it can be observed that for heterogeneous membranes there is a lower percentage increase in diffusivity than is the case for homogeneous structures. This is related to the interactions between the filler particles and the polymer chains, which reduce the potential for their mobility to increase with increasing process temperature. This is particularly evident for CO_2, but may be due to another adverse phenomenon. As the temperature increases, there is an increase in the free volumes in the polymer structure [50,51], which can result in an intensification of the contact between the filler particles and the polymer. The transport of this gas through the membrane is then impeded. It should also be noted that if the interactions described above were not present, the increase in gas diffusivity with temperature should be greatest for CO_2. This is because the molecules of this gas have the smallest kinetic diameter of the gases studied [24,52].

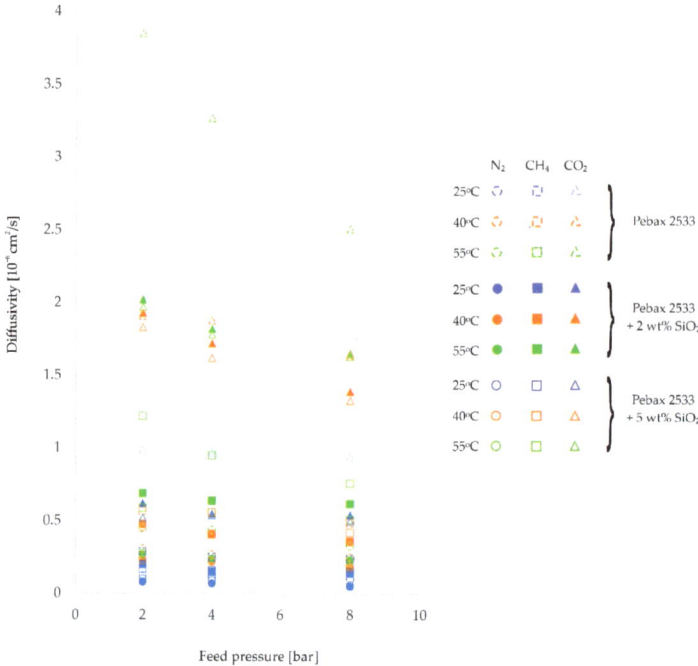

Figure 4. Change in diffusion coefficient values of homogeneous membrane and heterogeneous membranes containing SiO$_2$ for different feed pressure and temperature values.

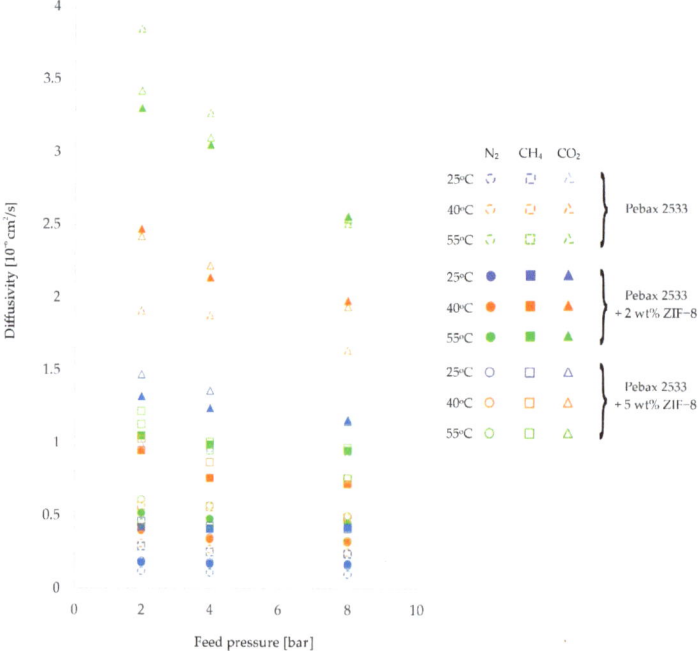

Figure 5. Change in diffusion coefficient values of homogeneous membrane and heterogeneous membranes containing ZIF−8 for different feed pressure and temperature values.

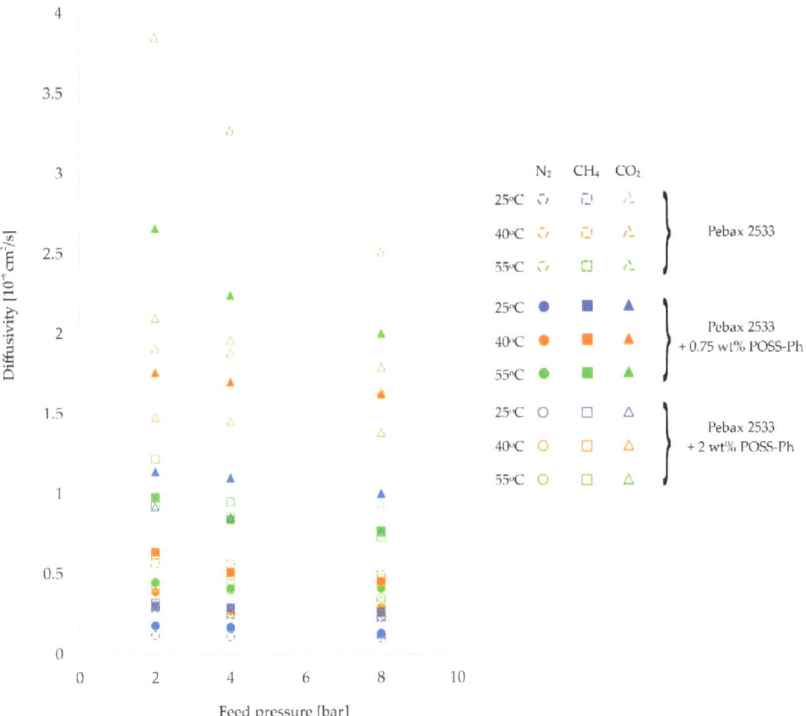

Figure 6. Change in diffusion coefficient values of homogeneous membrane and heterogeneous membranes containing POSS-Ph for different feed pressure and temperature values.

In addition, from the values obtained for the diffusion coefficients, it can be seen that they decrease with increasing feed pressure for all membranes and gases tested. This effect is related to the compression of the polymer as a result of the applied pressure, as already mentioned. Thus, the distances between the polymer chains and their mobility are reduced, which adversely affects the transport stage of the molecules by diffusion. An analysis of the diffusion component of the ideal selectivity coefficient $\alpha_{Di/j}$, calculated as the quotient of the diffusivity coefficient of one gas and the diffusivity coefficient of the other gas, reveals no clear change in the value of this parameter and no unambiguous trend describing the influence of the feed pressure on this parameter. This means that the compression of the polymer due to the applied pressure limits the transport of the tested gases to a similar extent. Similarly, the presence of an inorganic additive does not affect the magnitude and trend of gas diffusion changes with feed pressure.

The second parameter on which the rate of gas permeation through the membranes produced depends is the solubility coefficient, see Equation (1). The resulting values of this magnitude for different process conditions are shown in Figures 7–9 and in Tables A9–A12, for the homogeneous membrane and heterogeneous membranes, respectively. It should be recalled at this point that the solubility values contained in this paper were determined as a quotient of the measured values of permeability and diffusion coefficient, and not determined in separate measurements.

From the results shown in Figures 7–9, it can be seen that the solubility of gases on the surface of the fabricated materials decreases with increasing temperature. In contrast, as the feed pressure increases, the solubility of the gases in the membrane material increases. This is in line with predictions [53,54].

At this point, it is possible to return to the analysis of the effects of temperature and pressure on membrane permeability for CO_2, which, for this gas, was characterized by a smaller effect of temperature and a different trend in the effect of pressure than for N_2 and CH_4 gases.

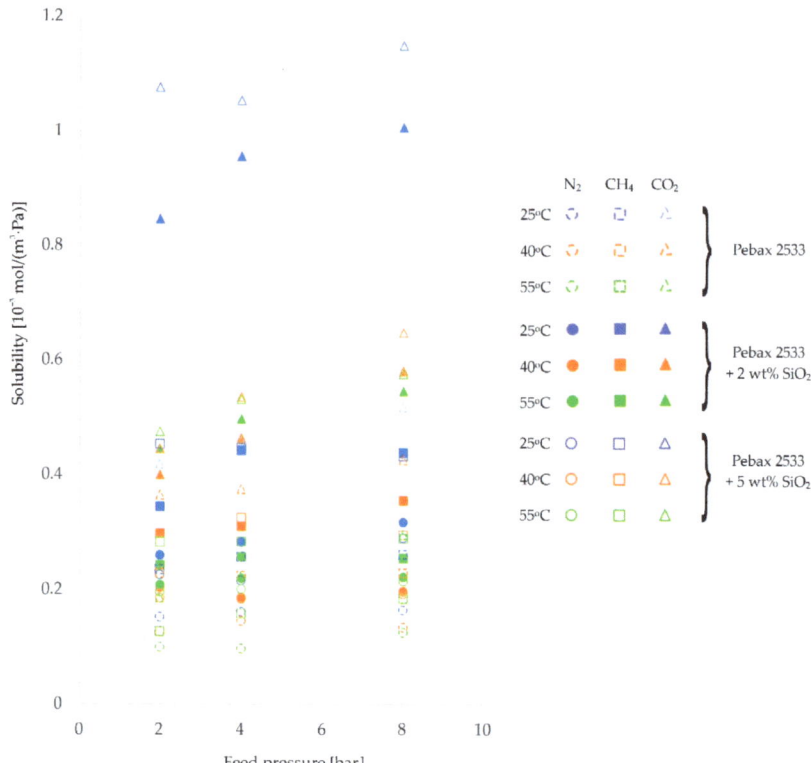

Figure 7. Change in solubility coefficient values of homogeneous membrane and heterogeneous membranes containing SiO_2 for different feed pressure and temperature values.

Comparing the results obtained at 25 °C and 55 °C for the membranes tested, the following average percentage increases in permeability were recorded: for N_2 +62%, for CH_4 +57%, and for CO_2 +40%. The markedly smaller increase in permeability for CO_2 with increasing temperature is associated with a greater decrease in its solubility at the membrane surface (N_2 −20%, CH_4 −36%, CO_2 −64%, respectively) and a smaller improvement in its diffusivity (N_2 +67%, CH_4 +67%, CO_2 +62%, respectively) relative to the other gases with increasing temperature. The effects observed are related to the lower molar heat of condensation of CO_2 relative to the other gases, and therefore also to the lower value of the enthalpy of dissolution, which additionally takes on a negative value for this gas [55]. This manifests itself in two effects. Firstly, a negative value of the enthalpy of dissolution means that the solubility of a given gas on the membrane surface deteriorates with increasing temperature, and secondly, the lower its value, the more difficult it is for the gas to condense [56]. This effect explains the different trend in changes in permeability with pressure to CO_2 gas noted earlier.

By dividing the permeability values for individual gases, the value of the ideal separation factor is obtained. Using the N_2/CO_2 mixture as an example, an analysis of the influence of process conditions on the ideal separation factor will be presented. The data shown in Figure 10 were obtained from the data shown in Figures 1–3 and in Tables A1–A4.

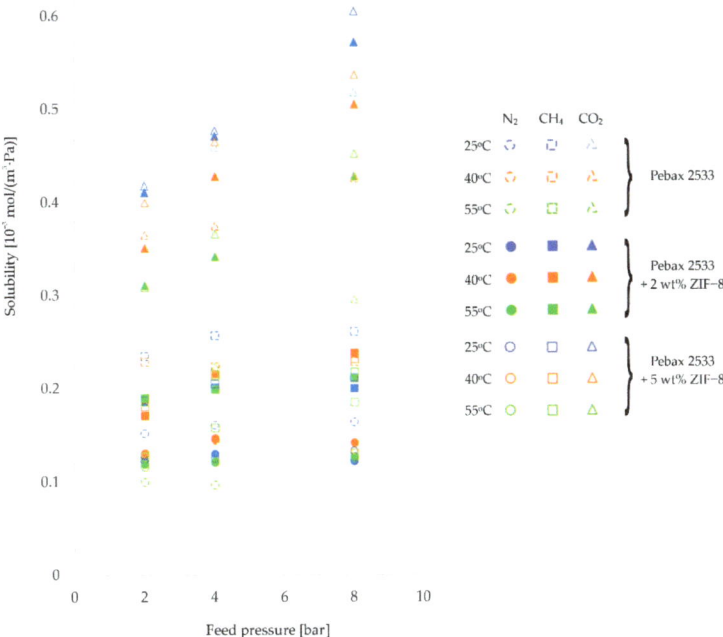

Figure 8. Change in solubility coefficient values of homogeneous membrane and heterogeneous membranes containing ZIF−8 for different feed pressure and temperature values.

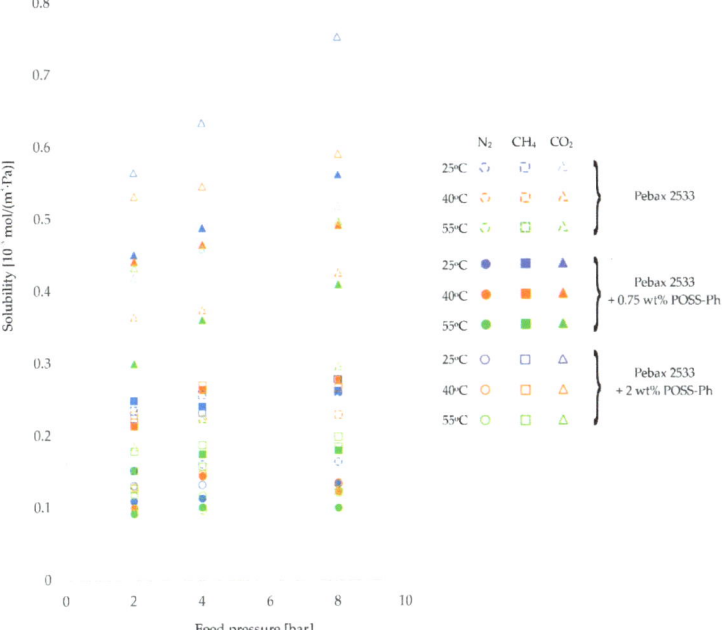

Figure 9. Change in solubility coefficient values of homogeneous membrane and heterogeneous membranes containing POSS-Ph for different feed pressure and temperature values.

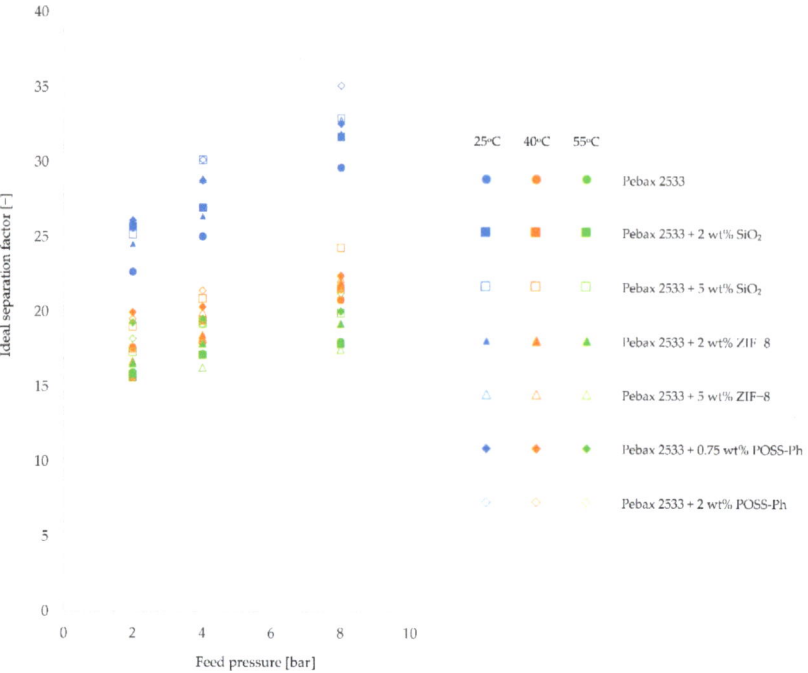

Figure 10. Change in ideal separation factor values for a N_2/CO_2 mixture of homogeneous membrane and heterogeneous membranes containing SiO_2, ZIF−8, and POSS-Ph for different feed pressure and temperature values.

Analyzing Figure 10, a strong correlation can be seen between the ideal separation factor and the feed pressure value. As the pressure increases, the ideal separation factor increases. This applies to all tested membranes, both homogeneous and heterogeneous. The trends for all these membranes are similar and the increase in the value of the ideal separation factor is approx. 40% when changing the feed pressure from 2 bar to 8 bar. A strong correlation is also observed when analyzing the influence of temperature on the value of the ideal separation factor. Here, however, the increasing temperature causes the value of the ideal separation factor to drop. The trends for the different types of membranes are similar. The decrease in the value of the ideal separation factor is ca. 30–45% (depending on type of filler) when changing the temperature from 25 °C to 55 °C. Physicochemical explanations of the observed effects should be sought in consideration of the influence of individual factors on the values of permeability, diffusivity, and solubility.

In the literature, one can find many works devoted to research on gas permeability through membranes made of Pebax®2533 copolymer or its modifications [57–69]. Therefore, it may be interesting for the readers to compare the results in the literature and the results obtained in the research of this work on the Robeson charts [25,27]. Figure 11 shows the comparison of results in the literature obtained for the pure Pebax®2533 polymer and its modifications with the authors' results, with the influence of pressure (Figure 11A) and temperature (Figure 11B) on the process parameters of the membranes. The considerations are limited to one gas mixture, namely the CO_2/N_2 mixture.

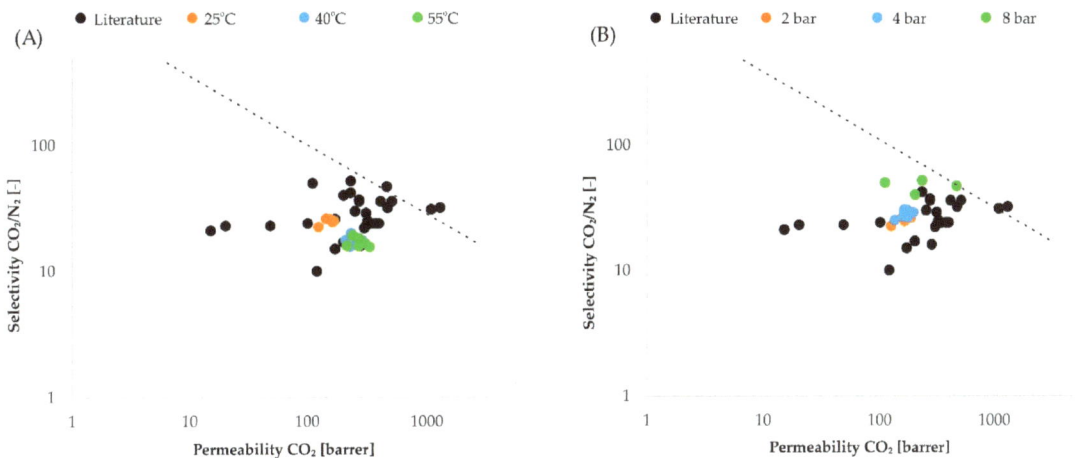

Figure 11. Comparison of results in the literature with the authors' results: (**A**) temperature effect at constant pressure 1 bar; (**B**) pressure effect at constant temperature 25 °C.

4. Conclusions

On the basis of the considerations presented above regarding the influence of process parameters on the diffusivity and solubility of gases, it is possible to determine the reasons for the changes in permeability of the membranes tested to the selected gases.

The value of membrane permeability to gases is influenced by both diffusion coefficient and solubility. An increase in temperature positively increases diffusivity, but negatively affects solubility. An increase in pressure negatively increases diffusivity, but positively increases solubility.

In general, an increase in temperature improves the permeability of membranes, meaning that the effect of temperature on diffusivity in this case outweighs the effect of temperature on solubility. In contrast, an increase in pressure worsens membrane permeability to N_2 and CH_4 gases, while it slightly improves permeability to CO_2. For N_2 and CH_4 gases, the effect of temperature on diffusivity appeared to be greater than the effect of temperature on solubility. For CO_2, on the other hand, the effect of temperature on solubility appeared to be the prevailing effect, which is due to the properties of the gas rather than the membrane.

Author Contributions: Conceptualization, D.P. and M.S.; methodology, D.P. and M.S.; results analysis, D.P. and M.S.; investigation, D.P; resources, M.S.; writing—original draft preparation, M.S. and D.P.; supervision, M.S.; funding acquisition, M.S. All authors have read and agreed to the published version of the manuscript.

Funding: This research received no external funding.

Institutional Review Board Statement: Not applicable.

Data Availability Statement: Not applicable.

Conflicts of Interest: The authors declare no conflict of interest.

Appendix A

Table A1. Permeability of homogeneous membranes to different gases measured for different feed pressure and temperature values.

Membrane Type	Feed Pressure [bar]	Permeability P_i [barrer]								
		N_2 Process Temperature [°C]			CH_4 Process Temperature [°C]			CO_2 Process Temperature [°C]		
		25	40	55	25	40	55	25	40	55
PEBAX 2533	2	5.5 ±0.1	11.8 ±0.3	13.4 ±0.3	20.4 ±0.3	39.0 ±0.6	46.1 ±0.7	123.7 ±1.0	207.9 ±1.7	214.0 ±1.7
	4	5.3 ±0.1	11.7 ±0.3	12.7 ±0.3	19.2 ±0.3	37.2 ±0.6	44.7 ±0.7	132.8 ±1.1	210.4 ±1.7	218.7 ±1.7
	8	4.9 ±0.1	10.0 ±0.3	12.3 ±0.3	18.7 ±0.3	33.5 ±0.5	41.9 ±0.7	145.4 ±1.2	208.4 ±1.7	222.0 ±1.8

Table A2. Permeability of heterogeneous membranes containing SiO_2 to different gases measured for different feed pressure and temperature values.

Membrane Type	Feed Pressure [bar]	Permeability P_i [barrer]								
		N_2 Process Temperature [°C]			CH_4 Process Temperature [°C]			CO_2 Process Temperature [°C]		
		25	40	55	25	40	55	25	40	55
PEBAX 2533 +2 wt% SiO_2	2	6.0 ±0.2	14.6 ±0.4	17.0 ±0.4	20.5 ±0.3	42.3 ±0.7	50.0 ±0.8	155.7 ±1.2	230.2 ±1.8	269.2 ±2.2
	4	5.9 ±0.1	12.2 ±0.3	15.8 ±0.4	20.3 ±0.3	38.0 ±0.6	49.4 ±0.8	158.0 ±1.3	238.1 ±1.9	270.6 ±2.2
	8	5.1 ±0.1	11.2 ±0.3	15.1 ±0.4	19.0 ±0.3	38.1 ±0.6	46.9 ±0.8	162.3 ±1.3	240.9 ±1.9	269.7 ±2.2
PEBAX 2533 +5 wt% SiO_2	2	6.6 ±0.2	12.8 ±0.3	16.1 ±0.4	23.2 ±0.4	41.0 ±0.7	49.8 ±0.8	167.0 ±1.3	243.8 ±2.0	279.8 ±2.2
	4	5.6 ±0.1	12.4 ±0.3	14.7 ±0.4	21.3 ±0.3	39.8 ±0.6	47.6 ±0.8	169.5 ±1.4	258.6 ±2.1	282.5 ±2.3
	8	5.2 ±0.1	10.5 ±0.3	14.0 ±0.4	19.2 ±0.3	36.7 ±0.6	44.3 ±0.7	171.5 ±1.4	256.4 ±2.1	280.4 ±2.2

Table A3. Permeability of heterogeneous membranes containing ZIF−8 to different gases measured for different feed pressure and temperature values.

Membrane Type	Feed Pressure [bar]	Permeability P_i [barrer]								
		N_2 Process Temperature [°C]			CH_4 Process Temperature [°C]			CO_2 Process Temperature [°C]		
		25	40	55	25	40	55	25	40	55
PEBAX 2533 +2 wt% ZIF−8	2	6.6 ±0.2	15.5 ±0.4	18.4 ±0.5	23.3 ±0.4	48.4 ±0.8	59.2 ±0.9	162.0 ±1.3	259.1 ±2.1	305.9 ±2.4
	4	6.6 ±0.2	14.8 ±0.4	17.4 ±0.4	24.6 ±0.4	49.1 ±0.8	58.8 ±0.9	174.2 ±1.4	273.3 ±2.2	311.6 ±2.5
	8	6.2 ±0.2	13.6 ±0.3	17.0 ±0.4	25.1 ±0.4	51.2 ±0.8	59.9 ±1.0	198.1 ±1.6	298.6 ±2.4	327.1 ±2.6

Table A3. Cont.

Membrane Type	Feed Pressure [bar]	Permeability P_i [barrer]								
		N_2 Process Temperature [°C]			CH_4 Process Temperature [°C]			CO_2 Process Temperature [°C]		
		25	40	55	25	40	55	25	40	55
PEBAX 2533 +5 wt% ZIF−8	2	7.0 ±0.2	16.5 ±0.4	21.2 ±0.5	25.6 ±0.4	55.7 ±0.9	64.0 ±1.0	183.3 ±1.5	289.0 ±2.3	332.0 ±2.7
	4	6.7 ±0.2	15.4 ±0.4	20.8 ±0.5	26.3 ±0.4	55.6 ±0.9	64.4 ±1.0	193.5 ±1.5	308.0 ±2.5	338.4 ±2.7
	8	6.3 ±0.2	14.0 ±0.3	19.6 ±0.5	26.0 ±0.4	52.5 ±0.8	63.3 ±1.0	207.7 ±1.7	310.8 ±2.5	342.7 ±2.7

Table A4. Permeability of heterogeneous membranes containing POSS-Ph to different gases measured for different feed pressure and temperature values.

Membrane Type	Feed Pressure [bar]	Permeability P_i [barrer]								
		N_2 Process Temperature [°C]			CH_4 Process Temperature [°C]			CO_2 Process Temperature [°C]		
		25	40	55	25	40	55	25	40	55
PEBAX 2533 +0.75 wt% POSS-Ph	2	5.9 ±0.1	11.6 ±0.3	12.4 ±0.3	22.3 ±0.4	40.9 ±0.7	44.6 ±0.7	153.5 ±1.2	232.0 ±1.9	238.5 ±1.9
	4	5.6 ±0.1	11.6 ±0.3	12.3 ±0.3	20.9 ±0.3	40.2 ±0.6	43.9 ±0.7	160.3 ±1.3	236.1 ±1.9	241.2 ±1.9
	8	5.1 ±0.1	10.6 ±0.3	12.2 ±0.3	20.4 ±0.3	37.2 ±0.6	41.3 ±0.7	167.8 ±1.3	238.0 ±1.9	244.9 ±2.0
PEBAX 2533 +2 wt% POSS-Ph	2	6.0 ±0.2	12.0 ±0.3	14.9 ±0.4	21.2 ±0.3	40.6 ±0.7	49.8 ±0.8	154.5 ±1.2	234.6 ±1.9	271.0 ±2.2
	4	5.3 ±0.1	11.0 ±0.3	14.0 ±0.4	20.3 ±0.3	39.5 ±0.6	48.6 ±0.8	161.0 ±1.3	236.2 ±1.9	268.6 ±2.1
	8	4.9 ±0.1	10.9 ±0.3	12.5 ±0.3	19.2 ±0.3	37.5 ±0.6	43.4 ±0.7	171.1 ±1.4	244.0 ±2.0	266.0 ±2.1

Table A5. Diffusion coefficient of homogeneous membranes to different gases measured for different feed pressure and temperature values.

Membrane Type	Feed Pressure [bar]	Diffusivity D_i [10^{-6} cm^2/s]								
		N_2 Process Temperature [°C]			CH_4 Process Temperature [°C]			CO_2 Process Temperature [°C]		
		25	40	55	25	40	55	25	40	55
PEBAX 2533	2	0.12 ±0.01	0.31 ±0.02	0.45 ±0.03	0.29 ±0.01	0.57 ±0.04	1.22 ±0.09	0.99 ±0.06	1.91 ±0.11	3.85 ±0.22
	4	0.11 ±0.01	0.27 ±0.02	0.44 ±0.03	0.25 ±0.01	0.56 ±0.04	0.95 ±0.07	0.97 ±0.06	1.88 ±0.11	3.27 ±0.19
	8	0.10 ±0.01	0.25 ±0.02	0.33 ±0.02	0.24 ±0.01	0.49 ±0.03	0.76 ±0.05	0.94 ±0.05	1.64 ±0.09	2.51 ±0.14

Table A6. Diffusion coefficient of heterogeneous membranes containing SiO_2 to different gases measured for different feed pressure and temperature values.

Membrane Type	Feed Pressure [bar]	Diffusivity D_i [10^{-6} cm^2/s]								
		N_2 Process temperature [°C]			CH_4 Process temperature [°C]			CO_2 Process temperature [°C]		
		25	40	55	25	40	55	25	40	55
PEBAX 2533 +2 wt% SiO$_2$	2	0.08 ±0.01	0.24 ±0.02	0.27 ±0.02	0.20 ±0.00	0.48 ±0.03	0.69 ±0.05	0.62 ±0.04	1.93 ±0.11	2.02 ±0.12
	4	0.07 ±0.00	0.22 ±0.02	0.24 ±0.02	0.15 ±0.00	0.41 ±0.03	0.64 ±0.04	0.55 ±0.03	1.72 ±0.10	1.82 ±0.1
	8	0.05 ±0.00	0.19 ±0.01	0.23 ±0.02	0.14 ±0.00	0.36 ±0.03	0.62 ±0.04	0.54 ±0.03	1.39 ±0.08	1.65 ±0.09
PEBAX 2533 +5 wt% SiO$_2$	2	0.10 ±0.01	0.23 ±0.02	0.28 ±0.02	0.17 ±0.00	0.46 ±0.03	0.59 ±0.04	0.52 ±0.03	1.83 ±0.1	1.97 ±0.11
	4	0.09 ±0.01	0.22 ±0.02	0.24 ±0.02	0.16 ±0.00	0.41 ±0.03	0.56 ±0.04	0.54 ±0.03	1.62 ±0.09	1.78 ±0.1
	8	0.06 ±0.00	0.18 ±0.01	0.22 ±0.02	0.15 ±0.00	0.42 ±0.03	0.51 ±0.04	0.50 ±0.03	1.33 ±0.08	1.63 ±0.09

Table A7. Diffusion coefficient of heterogeneous membranes containing ZIF−8 to different gases measured for different feed pressure and temperature values.

Membrane Type	Feed Pressure [bar]	Diffusivity D_i [10^{-6} cm^2/s]								
		N_2 Process Temperature [°C]			CH_4 Process Temperature [°C]			CO_2 Process Temperature [°C]		
		25	40	55	25	40	55	25	40	55
PEBAX 2533 +2 wt% ZIF−8	2	0.18 ±0.01	0.40 ±0.03	0.52 ±0.04	0.42 ±0.01	0.95 ±0.07	1.05 ±0.07	1.32 ±0.08	2.47 ±0.14	3.30 ±0.19
	4	0.17 ±0.01	0.34 ±0.02	0.48 ±0.03	0.41 ±0.01	0.76 ±0.05	0.99 ±0.07	1.24 ±0.07	2.14 ±0.12	3.05 ±0.17
	8	0.17 ±0.01	0.32 ±0.02	0.45 ±0.03	0.42 ±0.01	0.72 ±0.05	0.95 ±0.07	1.16 ±0.07	1.98 ±0.11	2.56 ±0.15
PEBAX 2533 +5 wt% ZIF−8	2	0.19 ±0.01	0.42 ±0.03	0.61 ±0.04	0.46 ±0.01	1.03 ±0.07	1.13 ±0.08	1.47 ±0.08	2.42 ±0.14	3.42 ±0.19
	4	0.18 ±0.01	0.35 ±0.02	0.57 ±0.04	0.43 ±0.01	0.87 ±0.06	1.01 ±0.07	1.36 ±0.08	2.22 ±0.13	3.10 ±0.18
	8	0.16 ±0.01	0.33 ±0.02	0.50 ±0.04	0.41 ±0.01	0.76 ±0.05	0.97 ±0.07	1.15 ±0.07	1.94 ±0.11	2.54 ±0.14

Table A8. Diffusion coefficient of heterogeneous membranes containing POSS-Ph to different gases measured for different feed pressure and temperature values.

Membrane Type	Feed Pressure [bar]	Diffusivity D_i [10^{-6} cm^2/s]								
		N_2 Process Temperature [°C]			CH_4 Process Temperature [°C]			CO_2 Process Temperature [°C]		
		25	40	55	25	40	55	25	40	55
PEBAX 2533 +0.75 wt% POSS-Ph	2	0.18 ±0.01	0.39 ±0.03	0.45 ±0.03	0.30 ±0.01	0.64 ±0.04	0.98 ±0.07	1.14 ±0.06	1.76 ±0.10	2.66 ±0.15
	4	0.17 ±0.01	0.27 ±0.02	0.41 ±0.03	0.29 ±0.01	0.51 ±0.04	0.84 ±0.06	1.10 ±0.06	1.70 ±0.10	2.24 ±0.13
	8	0.13 ±0.01	0.29 ±0.02	0.41 ±0.03	0.26 ±0.01	0.45 ±0.03	0.77 ±0.05	1.00 ±0.06	1.62 ±0.09	2.00 ±0.11
PEBAX 2533 +2 wt% POSS-Ph	2	0.18 ±0.01	0.31 ±0.02	0.42 ±0.03	0.32 ±0.01	0.62 ±0.04	0.93 ±0.07	0.92 ±0.05	1.48 ±0.08	2.10 ±0.12
	4	0.16 ±0.01	0.25 ±0.02	0.40 ±0.03	0.29 ±0.01	0.49 ±0.03	0.86 ±0.06	0.85 ±0.05	1.45 ±0.08	1.96 ±0.11
	8	0.12 ±0.01	0.27 ±0.02	0.35 ±0.02	0.23 ±0.01	0.47 ±0.03	0.73 ±0.05	0.76 ±0.04	1.38 ±0.08	1.79 ±0.10

Table A9. Solubility coefficient of homogeneous membranes to different gases measured for different feed pressure and temperature values.

Membrane Type	Feed Pressure [bar]	Solubility S_i [10^{-3} mol/(m^3·Pa)]								
		N_2 Process Temperature [°C]			CH_4 Process Temperature [°C]			CO_2 Process Temperature [°C]		
		25	40	55	25	40	55	25	40	55
PEBAX 2533	2	0.152 ±0.010	0.127 ±0.008	0.100 ±0.006	0.235 ±0.007	0.229 ±0.007	0.127 ±0.004	0.419 ±0.025	0.365 ±0.022	0.186 ±0.011
	4	0.161 ±0.010	0.145 ±0.009	0.097 ±0.006	0.257 ±0.008	0.224 ±0.007	0.158 ±0.005	0.459 ±0.028	0.375 ±0.022	0.224 ±0.013
	8	0.164 ±0.010	0.134 ±0.008	0.125 ±0.008	0.261 ±0.008	0.229 ±0.007	0.185 ±0.006	0.518 ±0.031	0.426 ±0.026	0.296 ±0.018

Table A10. Solubility coefficient of heterogeneous membranes containing SiO$_2$ to different gases measured for different feed pressure and temperature values.

Membrane Type	Feed Pressure [bar]	Solubility S_i [10^{-3} mol/(m^3·Pa)]								
		N_2 Process Temperature [°C]			CH_4 Process Temperature [°C]			CO_2 Process Temperature [°C]		
		25	40	55	25	40	55	25	40	55
PEBAX 2533 +2 wt% SiO$_2$	2	0.260 ±0.016	0.204 ±0.013	0.209 ±0.013	0.345 ±0.011	0.298 ±0.009	0.243 ±0.008	0.846 ±0.051	0.400 ±0.024	0.446 ±0.027
	4	0.283 ±0.018	0.186 ±0.012	0.220 ±0.014	0.443 ±0.014	0.310 ±0.010	0.257 ±0.008	0.956 ±0.057	0.464 ±0.028	0.498 ±0.030
	8	0.317 ±0.020	0.197 ±0.012	0.222 ±0.014	0.439 ±0.014	0.355 ±0.011	0.255 ±0.008	1.006 ±0.06	0.581 ±0.035	0.546 ±0.033

Table A10. Cont.

Membrane Type	Feed Pressure [bar]	Solubility S_i [10^{-3} mol/(m^3·Pa)]								
		N_2 Process Temperature [°C]			CH_4 Process Temperature [°C]			CO_2 Process Temperature [°C]		
		25	40	55	25	40	55	25	40	55
PEBAX 2533 +5 wt% SiO_2	2	0.226 ±0.014	0.186 ±0.012	0.196 ±0.012	0.454 ±0.014	0.298 ±0.009	0.283 ±0.009	1.076 ±0.065	0.446 ±0.027	0.476 ±0.029
	4	0.216 ±0.014	0.184 ±0.012	0.201 ±0.013	0.447 ±0.014	0.325 ±0.010	0.284 ±0.009	1.053 ±0.063	0.536 ±0.032	0.532 ±0.032
	8	0.290 ±0.018	0.193 ±0.012	0.215 ±0.014	0.433 ±0.013	0.295 ±0.009	0.289 ±0.009	1.149 ±0.069	0.648 ±0.039	0.576 ±0.035

Table A11. Solubility coefficient of heterogeneous membranes containing ZIF−8 to different gases measured for different feed pressure and temperature values.

Membrane Type	Feed Pressure [bar]	Solubility S_i [10^{-3} mol/(m^3·Pa)]								
		N_2 Process Temperature [°C]			CH_4 Process Temperature [°C]			CO_2 Process Temperature [°C]		
		25	40	55	25	40	55	25	40	55
PEBAX 2533 +2 wt% ZIF−8	2	0.123 ±0.008	0.130 ±0.008	0.119 ±0.007	0.186 ±0.006	0.171 ±0.005	0.189 ±0.006	0.411 ±0.025	0.351 ±0.021	0.311 ±0.019
	4	0.130 ±0.008	0.146 ±0.009	0.121 ±0.008	0.201 ±0.006	0.216 ±0.007	0.199 ±0.006	0.471 ±0.028	0.428 ±0.026	0.342 ±0.021
	8	0.122 ±0.008	0.142 ±0.009	0.127 ±0.008	0.200 ±0.006	0.238 ±0.007	0.211 ±0.007	0.572 ±0.034	0.505 ±0.03	0.428 ±0.026
PEBAX 2533 +5 wt% ZIF−8	2	0.124 ±0.008	0.131 ±0.008	0.116 ±0.007	0.187 ±0.006	0.181 ±0.006	0.190 ±0.006	0.418 ±0.025	0.400 ±0.024	0.309 ±0.019
	4	0.125 ±0.008	0.147 ±0.009	0.122 ±0.008	0.205 ±0.006	0.214 ±0.007	0.214 ±0.007	0.477 ±0.029	0.465 ±0.028	0.366 ±0.022
	8	0.133 ±0.008	0.142 ±0.009	0.131 ±0.008	0.212 ±0.007	0.232 ±0.007	0.218 ±0.007	0.605 ±0.036	0.537 ±0.032	0.452 ±0.027

Table A12. Solubility coefficient of heterogeneous membranes containing POSS-Ph to different gases measured for different feed pressure and temperature values.

Membrane Type	Feed Pressure [bar]	Solubility S_i [10^{-3} mol/(m^3·Pa)]								
		N_2 Process Temperature [°C]			CH_4 Process Temperature [°C]			CO_2 Process Temperature [°C]		
		25	40	55	25	40	55	25	40	55
PEBAX 2533 +0.75 wt% POSS-Ph	2	0.109 ±0.007	0.100 ±0.006	0.092 ±0.006	0.249 ±0.013	0.214 ±0.009	0.152 ±0.009	0.451 ±0.027	0.442 ±0.026	0.300 ±0.018
	4	0.113 ±0.007	0.144 ±0.009	0.101 ±0.006	0.241 ±0.008	0.264 ±0.007	0.175 ±0.005	0.488 ±0.029	0.465 ±0.028	0.361 ±0.022
	8	0.133 ±0.008	0.123 ±0.008	0.100 ±0.006	0.262 ±0.008	0.277 ±0.009	0.180 ±0.006	0.562 ±0.034	0.492 ±0.03	0.410 ±0.025

Table A12. Cont.

Membrane Type	Feed Pressure [bar]	Solubility S_i [10^{-3} mol/(m^3·Pa)]								
		N$_2$ Process Temperature [°C]			CH$_4$ Process Temperature [°C]			CO$_2$ Process Temperature [°C]		
		25	40	55	25	40	55	25	40	55
PEBAX 2533 +2 wt% POSS-Ph	2	0.131 ±0.008	0.129 ±0.008	0.118 ±0.007	0.224 ±0.007	0.219 ±0.007	0.179 ±0.006	0.565 ±0.034	0.532 ±0.032	0.433 ±0.026
	4	0.132 ±0.008	0.148 ±0.009	0.118 ±0.007	0.232 ±0.007	0.270 ±0.008	0.188 ±0.006	0.634 ±0.038	0.546 ±0.033	0.459 ±0.028
	8	0.135 ±0.008	0.135 ±0.008	0.121 ±0.008	0.278 ±0.009	0.267 ±0.008	0.199 ±0.006	0.753 ±0.045	0.591 ±0.035	0.497 ±0.03

References

1. Liang, C.Z.; Chung, T.S.; Lai, J.Y. A review of polymeric composite membranes for gas separation and energy production. *Prog. Polym. Sci.* **2019**, *97*, 101141. [CrossRef]
2. Bai, W.; Feng, J.; Luo, C.; Zhang, P.; Wang, H.; Yang, Y.; Zhao, Y.; Fan, H. A comprehensive review on oxygen transport membranes: Development history, current status, and future directions. *Int. J. Hydrog. Energy* **2021**, *46*, 36257–36290. [CrossRef]
3. Hashim, S.M.; Mohamed, A.R.; Bhatia, S. Current status of ceramic-based membranes for oxygen separation from air. *Adv. Colloid Interface Sci.* **2010**, *160*, 88–100. [CrossRef] [PubMed]
4. Han, J.; Bai, L.; Yang, B.; Bai, Y.; Luo, S.; Zeng, S.; Gao, H.; Nie, Y.; Ji, X.; Zhang, S.; et al. Highly selective oxygen/nitrogen separation membrane engineered using a porphyrin-based oxygen carrier. *Membranes* **2019**, *9*, 115. [CrossRef] [PubMed]
5. Baena-Moreno, F.M.; le Saché, E.; Pastor-Perez, L.; Reina, T.R. Membrane-based technologies for biogas upgrading: A review. *Environ. Chem. Lett.* **2020**, *18*, 1649–1658. [CrossRef]
6. Shin, M.S.; Jung, K.H.; Kwag, J.H.; Jeon, Y.W. Biogas separation using a membrane gas separator: Focus on CO$_2$ upgrading without CH$_4$ loss. *Process Saf. Environ. Prot.* **2019**, *129*, 348–358. [CrossRef]
7. Yang, L.; Zhang, S.; Wu, H.; Ye, C.; Liang, X.; Wang, S.; Wu, X.; Wu, Y.; Ren, Y.; Liu, Y.; et al. Porous organosilicon nanotubes in pebax-based mixed-matrix membranes for biogas purification. *J. Membr. Sci.* **2019**, *573*, 301–308. [CrossRef]
8. Dai, Z.; Deng, J.; He, X.; Scholes, C.A.; Jiang, X.; Wang, B.; Guo, H.; Ma, Y.; Deng, L. Helium separation using membrane technology: Recent advances and perspectives. *Sep. Purif.* **2021**, *274*, 119044. [CrossRef]
9. Wang, X.; Shan, M.; Liu, X.; Wang, M.; Doherty, C.M.; Osadchii, D.; Kapteijn, F. High-performance polybenzimidazole membranes for helium extraction from natural gas. *ACS Appl. Mater. Interfaces* **2019**, *11*, 20098–20103. [CrossRef]
10. Shahbaz, M.; Al-Ansari, T.; Aslam, M.; Khan, Z.; Inayat, A.; Athar, M.; Naqvi, S.R.; Ahmed, M.A.; McKay, G. A state of the art review on biomass processing and conversion technologies to produce hydrogen and its recovery via membrane separation. *Int. J. Hydrog. Energy* **2020**, *45*, 15166–15195. [CrossRef]
11. Nayebossadri, S.; Speight, J.D.; Book, D. Hydrogen separation from blended natural gas and hydrogen by Pd-based membranes. *Int. J. Hydrog. Energy* **2019**, *44*, 29092–29099. [CrossRef]
12. Cardoso, S.P.; Azenha, I.S.; Lin, Z.; Portugal, I.; Rodrigues, A.E.; Silva, C.M. Inorganic membranes for hydrogen separation. *Sep. Purif. Rev.* **2018**, *47*, 229–266. [CrossRef]
13. Lei, L.; Lindbråthen, A.; Zhang, X.; Favvas, E.P.; Sandru, M.; Hillestad, M.; He, X. Preparation of carbon molecular sieve membranes with remarkable CO2/CH4 selectivity for high-pressure natural gas sweetening. *J. Membr. Sci.* **2020**, *614*, 118529. [CrossRef]
14. Momeni, M.; Kojabad, M.E.; Khanmohammadi, S.; Farhadi, Z.; Ghalandarzadeh, R.; Babaluo, A.; Zare, M. Impact of support on the fabrication of poly (ether-b-amide) composite membrane and economic evaluation for natural gas sweetening. *J. Nat. Gas Sci. Eng.* **2019**, *62*, 236–246. [CrossRef]
15. Quek, V.C.; Shah, N.; Chachuat, B. Modeling for design and operation of high-pressure membrane contactors in natural gas sweetening. *Chem. Eng. Res. Des.* **2018**, *132*, 1005–1019. [CrossRef]
16. Bui, D.T.; Vivekh, P.; Islam, M.R.; Chua, K.J. Studying the characteristics and energy performance of a composite hollow membrane for air dehumidification. *Appl. Energy* **2022**, *306*, 118161. [CrossRef]
17. Chen, G.Q.; Kanehashi, S.; Doherty, C.M.; Hill, A.J.; Kentish, S.E. Water vapor permeation through cellulose acetate membranes and its impact upon membrane separation performance for natural gas purification. *J. Membr. Sci.* **2015**, *487*, 249–255. [CrossRef]
18. Bolto, B.; Hoang, M.; Xie, Z. A review of water recovery by vapour permeation through membranes. *Water Res.* **2012**, *46*, 259–266. [CrossRef]
19. Ramírez-Santos, Á.A.; Bozorg, M.; Addis, B.; Piccialli, V.; Castel, C.; Favre, E. Optimization of multistage membrane gas separation processes. Example of application to CO$_2$ capture from blast furnace gas. *J. Membr. Sci.* **2018**, *566*, 346–366. [CrossRef]
20. Samei, M.; Raisi, A. Separation of nitrogen from methane by multi-stage membrane processes: Modeling, simulation, and cost estimation. *J. Nat. Gas Sci. Eng.* **2022**, *98*, 104380. [CrossRef]

21. Tao, J.; Wang, J.; Zhu, L.; Chen, X. Integrated design of multi-stage membrane separation for landfill gas with uncertain feed. *J. Membr. Sci.* **2019**, *590*, 117260. [CrossRef]
22. Brunetti, A.; Drioli, E.; Lee, Y.M.; Barbieri, G. Engineering evaluation of CO_2 separation by membrane gas separation systems. *J. Membr. Sci.* **2014**, *454*, 305–315. [CrossRef]
23. He, Y.; Li, G.; Wang, H.; Zhao, J.; Su, H.; Huang, Q. Effect of operating conditions on separation performance of reactive dye solution with membrane process. *J. Membr. Sci.* **2008**, *321*, 183–189. [CrossRef]
24. Szwast, M.; Polak, D.; Marcjaniak, B. The influence of temperature and pressure of the feed on physical and chemical parameters of the membrane made of PEBA copolymer. *Desalin. Water Treat.* **2018**, *128*, 193–198. [CrossRef]
25. Robeson, L.M. Correlation of separation factor versus permeability for polymeric membranes. *J. Membr. Sci.* **1991**, *62*, 165–185. [CrossRef]
26. Freeman, B.D. Basis of permeability/selectivity tradeoff relations in polymeric gas separation membranes. *Macromolecules* **1999**, *32*, 375–380. [CrossRef]
27. Robeson, L.M. The upper bound revisited. *J. Membr. Sci.* **2008**, *320*, 390–400. [CrossRef]
28. Goh, P.S.; Ismail, A.F.; Sanip, S.M.; Ng, B.C.; Aziz, M. Recent advances of inorganic fillers in mixed matrix membrane for gas separation. *Sep. Purif.* **2011**, *81*, 243–264. [CrossRef]
29. Jeazet, H.B.T.; Staudt, C.; Janiak, C. Metal–organic frameworks in mixed-matrix membranes for gas separation. *Dalton Trans.* **2012**, *41*, 14003–14027. [CrossRef]
30. Li, W.; Peng, L.; Li, Y.; Chen, Z.; Duan, C.; Yan, S.; Yuan, B. Hyper cross-linked polymers containing amino group functionalized polyimide mixed matrix membranes for gas separation. *J. Appl. Polym. Sci.* **2022**, *139*, 52171. [CrossRef]
31. Polak, D.; Szwast, M. Material and process tests of heterogeneous membranes containing ZIF-8, SiO_2 and POSS-Ph. *Materials* **2022**, *15*, 6455. [CrossRef] [PubMed]
32. Yazid, A.F.; Mukhtar, H.; Nasir, R.; Mohshim, D.F. Incorporating Carbon Nanotubes in Nanocomposite Mixed-Matrix Membranes for Gas Separation: A Review. *Membranes* **2022**, *12*, 589. [CrossRef] [PubMed]
33. Bhide, B.D.; Stern, S.A. Membrane processes for the removal of acid gases from natural gas. I. Process configurations and optimization of operating conditions. *J. Membr. Sci.* **1993**, *81*, 209–237. [CrossRef]
34. Clarizia, G. Strong and weak points of membrane systems applied to gas separation. *Chem. Eng. Trans.* **2009**, *17*, 1675–1680.
35. Ji, G.; Wang, G.; Hooman, K.; Bhatia, S.; da Costa, J.C.D. Simulation of binary gas separation through multi-tube molecular sieving membranes at high temperatures. *J. Chem. Eng.* **2013**, *218*, 394–404. [CrossRef]
36. Zhao, L.; Riensche, E.; Menzer, R.; Blum, L.; Stolten, D. A parametric study of CO_2/N_2 gas separation membrane processes for post-combustion capture. *J. Membr. Sci.* **2008**, *325*, 284–294. [CrossRef]
37. Sanders, D.F.; Smith, Z.P.; Guo, R.; Robeson, L.M.; McGrath, J.E.; Paul, D.R.; Freeman, B.D. Energy-efficient polymeric gas separation membranes for a sustainable future: A review. *Polymer* **2013**, *54*, 4729–4761. [CrossRef]
38. Castel, C.; Wang, L.; Corriou, J.P.; Favre, E. Steady vs unsteady membrane gas separation processes. *Chem. Eng. Sci.* **2018**, *183*, 136–147. [CrossRef]
39. Frisch, H.L. The time lag in diffusion. *J. Phys. Chem.* **1957**, *61*, 93–95. [CrossRef]
40. Rutherford, S.W.; Do, D.D. Review of time lag permeation technique as a method for characterisation of porous media and membranes. *Adsorption* **1997**, *3*, 283–312. [CrossRef]
41. Laidler, K.J. The development of the Arrhenius equation. *J. Chem. Educ.* **1984**, *61*, 494. [CrossRef]
42. Gajdoš, J.; Galić, K.; Kurtanjek, Ž.; Ciković, N. Gas permeability and DSC characteristics of polymers used in food packaging. *Polym. Test.* **2000**, *20*, 49–57. [CrossRef]
43. Roth, C.B.; Dutcher, J.R. Glass transition and chain mobility in thin polymer films. *J. Electroanal. Chem.* **2005**, *584*, 13–22. [CrossRef]
44. Adam, G.; Gibbs, J.H. On the temperature dependence of cooperative relaxation properties in glass-forming liquids. *J. Chem. Phys.* **1965**, *43*, 139–146. [CrossRef]
45. Lee, J.Y.; Park, C.Y.; Moon, S.Y.; Choi, J.H.; Chang, B.J.; Kim, J.H. Surface-attached brush-type CO_2-philic poly (PEGMA)/PSf composite membranes by UV/ozone-induced graft polymerization: Fabrication, characterization, and gas separation properties. *J. Membr. Sci.* **2019**, *589*, 117214. [CrossRef]
46. Sadeghi, M.; Talakesh, M.M.; Arabi Shamsabadi, A.; Soroush, M. Novel application of a polyurethane membrane for efficient separation of hydrogen sulfide from binary and ternary gas mixtures. *ChemistrySelect* **2018**, *3*, 3302–3308. [CrossRef]
47. Fried, J.R.; Hu, N. The molecular basis of CO2 interaction with polymers containing fluorinated groups: Computational chemistry of model compounds and molecular simulation of poly [bis (2, 2, 2-trifluoroethoxy) phosphazene]. *Polymer* **2003**, *44*, 4363–4372. [CrossRef]
48. Zhao, Y.; Jung, B.T.; Ansaloni, L.; Ho, W.W. Multiwalled carbon nanotube mixed matrix membranes containing amines for high pressure CO2/H2 separation. *J. Membr. Sci.* **2014**, *459*, 233–243. [CrossRef]
49. Frey, S.L.; Zhang, D.; Carignano, M.A.; Szleifer, I.; Lee, K.Y.C. Effects of block copolymer's architecture on its association with lipid membranes: Experiments and simulations. *Chem. Phys.* **2007**, *127*, 114904. [CrossRef]
50. Deng, Q.; Zandiehnadem, F.; Jean, Y.C. Free-volume distributions of an epoxy polymer probed by positron annihilation: Temperature dependance. *Macromolecules* **1992**, *25*, 1090–1095. [CrossRef]
51. Dlubek, G.; Saarinen, K.; Fretwell, H.M. The temperature dependence of the local free volume in polyethylene and polytetrafluoroethylene: A positron lifetime study. *J. Polym. Sci. B: Polym. Phys.* **1998**, *36*, 1513–1528. [CrossRef]

52. Mehio, N.; Dai, S.; Jiang, D.E. Quantum mechanical basis for kinetic diameters of small gaseous molecules. *J. Phys. Chem. A* **2014**, *118*, 1150–1154. [CrossRef]
53. Costello, L.M.; Koros, W.J. Temperature dependence of gas sorption and transport properties in polymers: Measurement and applications. *Ind. Eng. Chem. Res.* **1992**, *31*, 2708–2714. [CrossRef]
54. Davis, P.K.; Lundy, G.D.; Palamara, J.E.; Duda, J.L.; Danner, R.P. New pressure-decay techniques to study gas sorption and diffusion in polymers at elevated pressures. *Ind. Eng. Chem. Res.* **2004**, *43*, 1537–1542. [CrossRef]
55. Finotello, A.; Bara, J.E.; Camper, D.; Noble, R.D. Room-temperature ionic liquids: Temperature dependence of gas solubility selectivity. *Ind. Eng. Chem. Res.* **2008**, *47*, 3453–3459. [CrossRef]
56. Klopffer, M.H.; Flaconneche, B. Transport properties of gases in polymers: Bibliographic review. *Oil Gas Sci. Technol.* **2001**, *56*, 223–244. [CrossRef]
57. Gugliuzza, A.; Drioli, E. Role of additives in the water vapor transport through block co-poly (amide/ether) membranes: Effects on surface and bulk polymer properties. *Eur. Polym. J.* **2004**, *40*, 2381–2389. [CrossRef]
58. Gao, J.; Mao, H.; Jin, H.; Chen, C.; Feldhoff, A.; Li, Y. Functionalized ZIF-7/Pebax®2533 mixed matrix membranes for CO_2/N_2 separation. *Micropor. Mesopor. Mater.* **2020**, *297*, 110030. [CrossRef]
59. Casadei, R.; Giacinti Baschetti, M.; Yoo, M.J.; Park, H.B.; Giorgini, L. Pebax®2533/graphene oxide nanocomposite membranes for carbon capture. *Membranes* **2020**, *10*, 188. [CrossRef]
60. Gugliuzza, A.; Fabiano, R.; Garavaglia, M.G.; Spisso, A.; Drioli, E. Study of the surface character as responsible for controlling interfacial forces at membrane–feed interface. *J. Colloid Interface Sci.* **2006**, *303*, 388–403. [CrossRef]
61. Li, G.; Kujawski, W.; Knozowska, K.; Kujawa, J. Pebax®2533/PVDF thin film mixed matrix membranes containing MIL-101 (Fe)/GO composite for CO_2 capture. *RSC Adv.* **2022**, *12*, 29124. [CrossRef]
62. Li, G.; Kujawski, W.; Knozowska, K.; Kujawa, J. Thin film mixed matrix hollow fiber membrane fabricated by incorporation of amine functionalized metal-organic framework for CO_2/N_2 separation. *Materials* **2021**, *14*, 3366. [CrossRef] [PubMed]
63. Nafisi, V.; Hägg, M.B. Development of dual layer of ZIF-8/PEBAX-2533 mixed matrix membrane for CO_2 capture. *J. Membr. Sci.* **2014**, *459*, 244–255. [CrossRef]
64. Gugliuzza, A.; Drioli, E. Evaluation of CO_2 permeation through functional assembled mono-layers: Relationships between structure and transport. *Polymer* **2005**, *46*, 9994–10003. [CrossRef]
65. Ansari, A.; Navarchian, A.H.; Rajati, H. Permselectivity improvement of PEBAX®2533 membrane by addition of glassy polymers (Matrimid® and polystyrene) for CO_2/N_2 separation. *J. Appl. Polym. Sci.* **2022**, *139*, e51556. [CrossRef]
66. Hassanzadeh, H.; Abedini, R.; Ghorbani, M. CO_2 Separation over N_2 and CH_4 Light Gases in Sorbitol-Modified Poly(ether-block-amide)(Pebax 2533) Membrane. *Ind. Eng. Chem. Res.* **2022**, *61*, 13669–13682. [CrossRef]
67. De Luca, G.; Gugliuzza, A.; Drioli, E. Competitive hydrogen-bonding interactions in modified polymer membranes: A density functional theory investigation. *J. Phys. Chem. B* **2009**, *113*, 5473–5477. [CrossRef]
68. Rahman, M.M.; Filiz, V.; Shishatskiy, S.; Abetz, C.; Neumann, S.; Bolmer, S.; Khan, M.M.; Abetz, V. PEBAX® with PEG functionalized POSS as nanocomposite membranes for CO_2 separation. *J. Membr. Sci.* **2013**, *437*, 286–297. [CrossRef]
69. Lee, S.; Park, S.C.; Kim, T.Y.; Kang, S.W.; Kang, Y.S. Direct molecular interaction of CO_2 with KTFSI dissolved in Pebax 2533 and their use in facilitated CO_2 transport membranes. *J. Membr. Sci.* **2018**, *548*, 358–362. [CrossRef]

Article

Pathway for Water Transport through Breathable Nanocomposite Membranes of PEBAX with Ionic Liquid [$C_{12}C_1$im]Cl

Ziqi Cheng [1,2], Shen Li [1,2], Elena Tocci [3], Giacomo Saielli [4,5], Annarosa Gugliuzza [3,*] and Yanting Wang [1,2,6,*]

[1] CAS Key Laboratory of Theoretical Physics, Institute of Theoretical Physics, Chinese Academy of Sciences, Beijing 100190, China
[2] School of Physical Sciences, University of Chinese Academy of Sciences, Beijing 100049, China
[3] National Research Council—Institute on Membrane Technology (CNR-ITM), Via Pietro Bucci 17C, 87036 Rende, Italy; e.tocci@itm.cnr.it
[4] National Research Council—Institute on Membrane Technology (CNR-ITM), Unit of Padova, Via Marzolo, 1, 35131 Padova, Italy; giacomo.saielli@unipd.it
[5] Department of Chemical Sciences, University of Padova, Via Marzolo, 1, 35131 Padova, Italy
[6] Center for Theoretical Interdisciplinary Sciences, Wenzhou Institute, University of Chinese Academy of Sciences, Wenzhou 325001, China
* Correspondence: a.gugliuzza@itm.cnr.it (A.G.); wangyt@itp.ac.cn (Y.W.)

Citation: Cheng, Z.; Li, S.; Tocci, E.; Saielli, G.; Gugliuzza, A.; Wang, Y. Pathway for Water Transport through Breathable Nanocomposite Membranes of PEBAX with Ionic Liquid [$C_{12}C_1$im]Cl. *Membranes* 2023, 13, 749. https://doi.org/10.3390/membranes13090749

Academic Editors: Isabel Coelhoso and Sébastien Déon

Received: 26 July 2023
Revised: 18 August 2023
Accepted: 20 August 2023
Published: 22 August 2023

Copyright: © 2023 by the authors. Licensee MDPI, Basel, Switzerland. This article is an open access article distributed under the terms and conditions of the Creative Commons Attribution (CC BY) license (https://creativecommons.org/licenses/by/4.0/).

Abstract: Water transport through membranes is an attractive topic among the research dedicated to dehydration processes, microenvironment regulation, or more simply, recovery of freshwater. Herein, an atomistic computer simulation is proposed to provide new insights about a water vapor transport mechanism through PEBAX membranes filled with ionic liquid (IL) [$C_{12}C_1$im]Cl. Starting from experimental evidence that indicates an effective increase in water permeation as the IL is added to the polymer matrix (e.g., up to $85 \cdot 10^{-3}$ (g·m)/(m²·day) at 318.15 K for PEBAX@2533 membranes loaded with 70% of IL), molecular dynamics simulations are proposed to explore the key role of IL in water transport inside membranes. The polar region composed of anions and cationic head groups of the IL is demonstrated to serve as the pathway for water transport through the membrane. Water molecules always stay near the pathway, which becomes wider and thus has a larger water-accessible area with increasing IL concentration. Hence, the diffusion coefficients of water molecules and ions increase as the IL concentration increases. The simulation provides useful indications about a microscopic mechanism that regulates the transport of water vapor through a kind of PEBAX/IL membrane, resulting in full agreement with the experimental evidence.

Keywords: ionic liquid crystal; PEBAX membrane; breathability; molecular dynamics simulation

1. Introduction

PEBAX@2533 [80PTMO/PA12] is an elastomeric block copolyamide with high processability [1,2], mechanical strength [3], and high-performing permeability to vapors and quadrupolar and condensable gases [4–7]. These features make it particularly attractive and competitive for applications in wearable textiles [8], industrial equipment [4,9,10], microelectronics [11], and environmental protection [12]. This copolymer is widely used in the preparation of breathable films for regulation of microclimate [13,14], which is regarded as a micro-space air stream between two different neighboring gaps. The regulation of microclimate is of great importance in the preservation of works of art [15] since rapid changes in relative humidity can destroy or irreversibly compromise the integrity of cultural heritage. Similarly, a lack of balance between temperature and humidity can affect microclimate conditions in textiles [8,16,17], causing discomfort and chances of skin rashes, itching, and allergies. Last but not least, dehydration is also applied to preventing corrosion from condensed water in source gas streams [18,19]. The membrane technology can provide suitable and intelligent solutions and benefits in the fields wherein removal or regulation of water vapor concentration is desired [20]. Despite the fact that a large number of materials have been proposed to intensify the transport of water vapor through the PEBAX membranes [14,21–24], most of them do not satisfy today's large demand of adapting themselves according to external changes in a fast and reversible way. This aspect is relevant to the design of intelligent systems that could restore and maintain constant desired microenvironments in a reversible and highly reproducible way.

Ionic liquids (ILs) are room-temperature molten salts with unique properties such as negligible vapor pressure, good thermal stability and non-flammability, as well as high ionic conductivity and a wide window of electrochemical stability [25–27]. They also have the ability to dissolve most organic and some inorganic materials as well as biopolymers [28]. Ionic liquid crystals (ILCs), namely ILs in their liquid crystal state, are a preferred family of materials for their reversible assembling ability [29,30]. Long-chain imidazolium-based ILs are well-known compounds capable of forming ILCs and continuously receive a great deal of attention as materials that combine the unique solvent properties of ILs with the long-range partial order of LCs [31]. Effective use of ILCs as electrolytes in dye-sensitized solar cells [32,33], electro-fluorescence switches [34], electrolytes for Li-ion batteries [35], and electrochemical sensors [36], to mention but a few, has been clearly demonstrated. When they are embedded in polymeric membranes, ILCs can play a key role since in the ionic fluid phase they have special conductive properties of mass and charge, while in the ordered fluid phase they have preferential directions that enable fine tuning of the transport properties of the geometry and structure of the polymer and of external parameters [37].

A lot of studies have been dedicated to the analysis of structure and dynamics in bulk ionic liquids [38–43], whilst the number of papers dedicated to the confinement of these materials in organic and inorganic matrixes and to changes in IL structure and interfaces is somewhat limited [44]. When filling polymeric matrixes with ILCs, changes in the final properties of the nanocomposite systems can be experimentally detected by suitable tools [45]; however, the relationship between the low dimensionality of interface systems and fluidity of the interfaces becomes quite difficult to understand. Indeed, the shape of the ILC interfaces and the orientation and structure of fluid molecules constantly change, producing nanoscale effects whose amplification is perceived on a larger scale. As known, IL compounds have a unique chemical structure with a large amount of possible organic cation and anion species from which to select. These ions can interact with organic polymers and inorganic nanofillers, and their intermolecular interactions can modify the polymer nanocomposites and enhance the interfacial connection between the polymer matrix and the nanofillers [46]. The modification of nanofillers by IL is a promising method to prepare multifunctional polymer nanocomposites, which can be regarded as a practical route for the development of adaptable and monitorable systems [44]. Much experimental evidence has shown that the addition of IL to PEBAX can lead to the formation of more amorphous and less crystalline membranes [47,48] due to the fact that ILs increase the tendency of polymers to form new bonds and form a homogeneous mixture with the IL. Ample experimental work has shown that membranes of PEBAX or other polymers with ILs have excellent gas permeability and selectivity [49–52]. In various types of ILs, imidazolium-based ILs are often used to modify nanofillers because they work in a wide variety of chemical structures [53].

In this work, we perform atomistic molecular dynamics (MD) simulations that provide new insights into the reorganization of 1-dodecyl-3-methylimidazolium chloride ([$C_{12}C_1$im]Cl, abbreviated as C_{12}) confined in PEBAX@2533 elastomeric membranes and into the establishment of local interfacial forces and tunneling events that assist water mass transfer through nanocomposite matrixes. As a case study, we analyze PEBAX@2533 membranes embedding a large amount of [$C_{12}C_1$im]Cl, which exhibit an increasing capability to transfer a large quantity of water vapor with concentration and temperature at high reproducibility and in a reversible way. The choice of PEBAX as a host matrix is due to the fact that large amounts of nanofillers can be embedded in a nanofilm without compromising manageability thanks to its elastomeric properties.

Starting from experimental evidence of the changes in water permeability, our intent is to study, at the molecular scale, the relationships established between IL concentration and diffusing water molecules, in order to assess the behavior of ILs when confined in polymeric networks and approaching polar penetrant such as water. This study provides new useful insights about the action of ionic groups in water transport while also suggesting the ability of IL to self-assemble into somewhat ordered regions when constrained in an elastomeric host polymer network. Through our simulation results, we have found that the polar parts of the ILs form a polar network in the membrane, serving as a pathway for water to penetrate. The pathway becomes wider as the concentration of IL in the membrane increases; thus, a larger water-accessible area is provided and the diffusion of water molecules becomes faster. This effect increases further with temperature due to thermal motion of water molecules, which facilitates the diffusion through polar pathways. Undoubtedly, the choice of confining an IL in an elastomeric polymer network such as PEBAX provides more freedom of movement and rearrangement, yielding high reproducibility and reversibility, especially at higher temperatures. Moreover, with molecular mechanisms clarified, thermosensitive materials such as ILs are expected to provide a great chance of synchronizing transport properties with the external environment.

This study is therefore a critical and due step to identify driving forces for the design of responsive breathable membranes necessary for microclimate regulation and controlled dehydration. More specifically, this study can be regarded as a preliminary step towards controlling, at the molecular scale, water permeation through membranes for the development of intelligent devices that enable the regulation of humidity in microenvironments.

2. Experimental and Simulation Methods

2.1. Materials

The thermoplastic elastomer poly(ether-block-amide) (PEBAX@2533, Arkema, Milan, Italy) was used as a precursor of dense membranes (10 wt.%), and 1-dodecyl-3-methylimidazolium cation of $[C_{12}C_1im]Cl$ was purchased from IoLiTec (Heilbronn, Germany) and used as the nanofiller at different concentrations of 30, 50, and 70% (w/w), and 1-Proponal (PrOH) and n-butanol (BuOH) (99.5%, Carlo Erba, Milan, Italy) were used in a mixture (1:3 v/v %) to dissolve polymer and nanofiller. All materials were used as received.

2.2. Membrane Preparation and Breathability Tests

Nanocomposite membranes were prepared through the blend of an elastomeric poly(ether-block-amide) (PEBAX@2533) with -dodecyl-3-methylimidazolium cation of $[C_{12}C_1im]Cl$ according to dry phase inversion [22]. The membranes were prepared by dissolving the IL at different contents in a mixture of PrOH/BuOH, and the polymer was then added to homogeneous solutions under vigorous stirring. Clear solutions were poured in Petri glasses, and the solvents were evaporated under controlled environmental conditions (T = 293.15 K and RH < 50%) to obtain membranes with a thickness of 50 μm. The membranes were air-dried for 4–5 days at room temperature and then stored in an oven at 40 °C for 3 days and under vacuum at room temperature for a further 7 days to remove solvent traces. The membrane breathability, herein expressed as water permeation $(g·m)/(m^2·day)$, was tested within a range of temperatures (298.15–318.15 K) according to the right cup method (ASTM E96B) [8,54]. Water breathability was expressed as the amount of water transmitted over time through a specific area of the membrane. The breathability was normalized per the thickness of each membrane for comparison.

2.3. Simulation Methods

All-atom MD simulations were performed to study the structural and dynamical characteristics of water molecules in PEBAX membranes filled with $[C_{12}C_1im]Cl$. The molecular structures are shown in Figure 1.

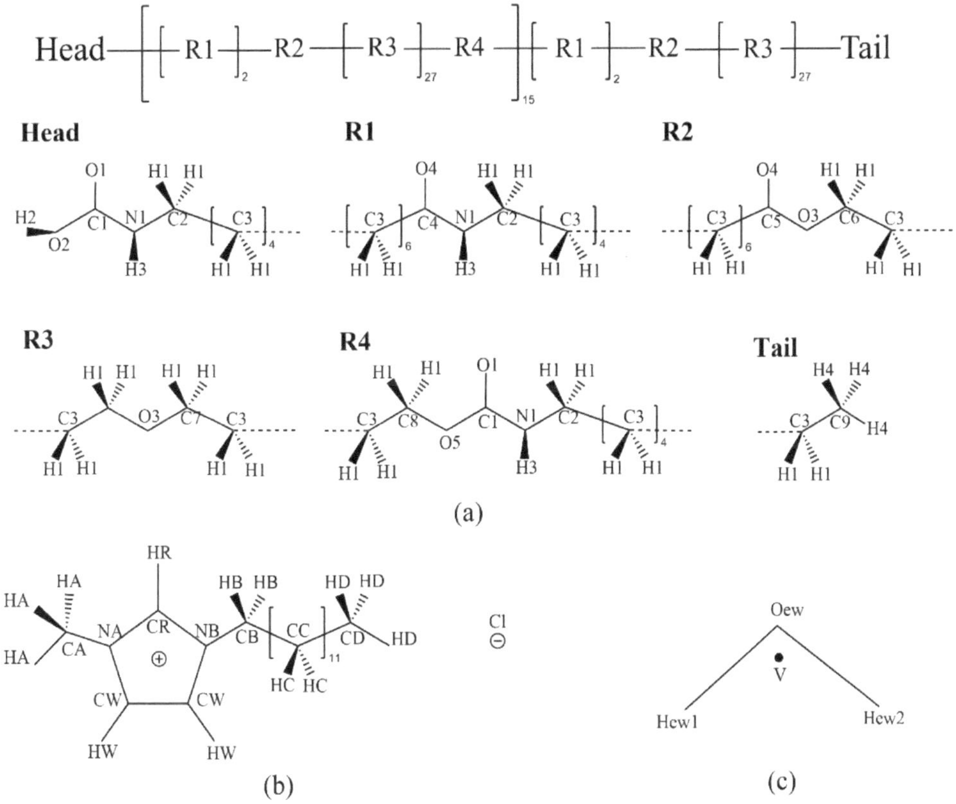

Figure 1. Molecular structures of PEBAX (**a**), [C$_{12}$C$_1$im]Cl (**b**), and TIP4P/EW water model (**c**).

The force field parameters for PEBAX and [C$_{12}$C$_1$im]Cl were all taken from the OPLS-AA force field [55,56]. The partial charges of all the atoms in PEBAX and [C$_{12}$C$_1$im]Cl were calculated with the RESP method [57]. As suggested by Chaban et al. [58], the net charges of [C$_{12}$C$_1$im]Cl were reduced by a factor of 0.8 to conveniently mimic ion polarizability. All partial charges we have calculated are listed in Table 1, and the associated atom names are marked in Figure 1. The TIP4P/EW force field was adopted to model water molecules (Figure 1c), in which the position of the virtual site 'V' $r_V = r_{Oew} + a(r_{Hew1} - r_{Oew}) + b(r_{Hew2} - r_{Oew})$ with $a = b = 0.106676721$, where r_{Oew}, r_{Hew1}, and r_{Hew2} are the positions of the oxygen and hydrogen atoms. All force field parameters of TIP4P/EW are listed in Table 2, from which the effective diameter of a water molecule can be estimated as about 2.8 angstroms [59]. To simulate membranes with different concentrations of [C$_{12}$C$_1$im]Cl, three different systems were constructed, whose data are summarized in Table 3. The number of water molecules remains 1/3 of [C$_{12}$C$_1$im]Cl ion-pair numbers, and five additional systems with a molecular ratio of water to C$_{12}$ of 1/3, 2/3, 1, 2, and 10 were also simulated, keeping the 30% w/w as a typical C$_{12}$ concentration (Table 3). Six BEBAX chains are enough for investigating their local structural features with respect to water and IL molecules.

Table 1. The partial charges of PEBAX and [$C_{12}C_1$im]Cl.

Atom Name	Partial Charge/e	Atom Name	Partial Charge/e
C1	0.885	O1	−0.585
C2	0.104	O2	−0.612
C3	−0.106	O3	−0.396
C4	0.642	O4	−0.531
C5	0.702	O5	−0.266
C6	0.119	O6	−0.414
C7	0.027	N1	−0.699
C8	−0.159	NA	0.176
CA	−0.28	NB	0.176
CB	−0.136	H1	0.053
CC	−0.096	H2	0.423
CD	−0.192	H3	0.378
CR	−0.072	H4	0.064
CW	−0.192	HA	0.144
Cl	−0.8	HB	0.144
HW	0.216	HC	0.048
HR	0.168	HD	0.064

Table 2. Force field parameters of the TIP4P/EW water model.

Atom Name	Partial Charge/e	
Oew	0	
V	−1.04844	
Hew1	0.52422	
Hew2	0.52422	
Valence bond	**Bond length/nm**	k_{bond}/kJ mol^{-1} nm^{-2}
Oew-Hew1	0.09572	502,416.0
Oew-Hew2	0.09572	502,416.0
Valence angle	**Angle/°**	k_{angle}/kJ mol^{-1} rad^{-2}
Hew1-Oew-Hew2	104.52	628.02

Table 3. Number of different molecules for the simulated systems.

[$C_{12}C_1$im]Cl Concentrations in Weight	PEBAX Chains	[$C_{12}C_1$im]Cl Ion Pairs	Water Molecules	Water/C_{12} Ratio
30%	6	375	125	1/3
50%	6	870	290	1/3
70%	6	2040	680	1/3
30%	6	375	125	1/3
30%	6	375	250	2/3
30%	6	375	375	1
30%	6	375	750	2
30%	6	375	3750	10

All the systems simulated in this work are in a cubic box with periodic boundary conditions applied to all three dimensions. A cutoff distance of 12 Å was used to treat the van der Waals (VDW) and the real part of the electrostatic interactions, and the particle-mesh Ewald method [60] was applied to calculating the electrostatic interactions. For three different membranes without water, initial configurations were prepared by a simulated annealing procedure in an *NVT* ensemble for 6 ns from 1000 K down to 600 K, continuously. Another simulated annealing procedure in an *NPT* ensemble was then followed with the following sequential steps: 4 ns at 600 K and 1 bar, 6 ns at 500 K and 100 bar, 6 ns at 450 K and 50 bar, 6 ns at 400 K and 1 bar, 10 ns at 350 K and 1 bar, and 10 ns at 298 K and 1 bar.

To prevent water molecules from erroneously affecting the structure of the nanocomposite films at high temperatures, water molecules are randomly inserted into the membrane configuration at 300 K and 0.5 ns in an *NPT* ensemble to obtain the appropriate system size. Water molecules are then fully mixed with the membrane in an *NVT* ensemble after lowering the temperature from 500 K down to 450, 400, 350, 328, and 298 K, with a simulation duration of 1 ns at each temperature. In order to obtain accurate kinetic data, a 60 ns simulation in an *NPT* ensemble was performed at 298 K and 318 K, respectively, and data were sampled in the last 20 ns simulation.

In these runs, the system temperature and pressure were kept constant by using the Nosé–Hoover thermostat [61] with a time constant of 0.5 ps and the Parrinello–Rahman barostat [62] with a time constant of 2 ps, respectively. All the simulations were performed with GROMACS 2020 software [63] with a time step of 1 fs.

3. Experimental Results

Dense nanocomposite PEBAX membranes filled with $[C_{12}C_1im]Cl$ were prepared via dry phase inversion with the intent to investigate how ILCs can affect water vapor transport through them. Changes in permeation, namely breathability, were investigated according to the right cup method procedure [8,54], and membranes filled with three different amounts of filler (30, 50, 70 w/w) were tested by running the temperature from 298.15 to 318.15 K and then down to 298.15 K again (Figure 2). Interestingly, the increase in the $[C_{12}C_1im]Cl$ content produces amplified water transport, which is further raised with temperature. This effect appears to be more marked at the highest concentration of filler, indicating a wider assistance to water vapor permeation (Figure 2a). Moreover, the reversibility of the process can be detected as the temperature decreases from 318.15 to 298.15 K (Figure 2b), suggesting the IL as a powerful choice for making the membranes responsive to changes in temperature.

In all cases, a small amount of hysteresis can be observed between running uphill and downhill temperatures due to a probable reorganization of the materials assembled inside an elastomeric membrane. As proof, the third run uphill at 298.15 K indicates a substantial overlapping of permeation properties for the membranes containing different amounts of $[C_{12}C_1im]Cl$ (Figure 2c). This implies a certain fluidity of ILC in this kind of constrained polymeric system and a great freedom to rearrange itself quickly and reversibly. On the other hand, the elasticity of the host polymer is expected to facilitate the self-assembly of the IL during confinement as well as its rearrangement with temperature (Figure 2b).

It is well known that PEBAX@2533 has a great ability to embed large amounts of nanofiller [13,64] and dissolve polar penetrants through solution–diffusion mechanisms [65,66]. Considering the elastomeric features of this polymer [67], this behavior is not surprising and suggests a major accessibility of water molecules to hydrophilic sites, wherein they can be allocated temporarily. An increase in thermal motion is further expected to allow water to be accessible to a larger number of sorption regions and diffuse itself through broader fluctuating free gaps generated by higher mobility of the polymer segment chains. In a previous work [46], the DFT calculation demonstrated that a low concentration of a non-ionic organic nanofiller in PEBAX causes a competition of interaction energies determining higher availability of polymer polar moieties as a domino effect, while at a higher concentration the polar moieties are saturated and disallow water sorption [46].

In the present study, a significant increase in water permeation is instead observed with raising IL concentration. However, it is crucial to understand how and if the IL self-assembles in a constrained environment and which chemical moieties are involved in water transfer. To understand which forces and events address the behavior of water molecules during penetration in mixed matrices, a computational investigation at the micro- and nano-scale is hence necessary.

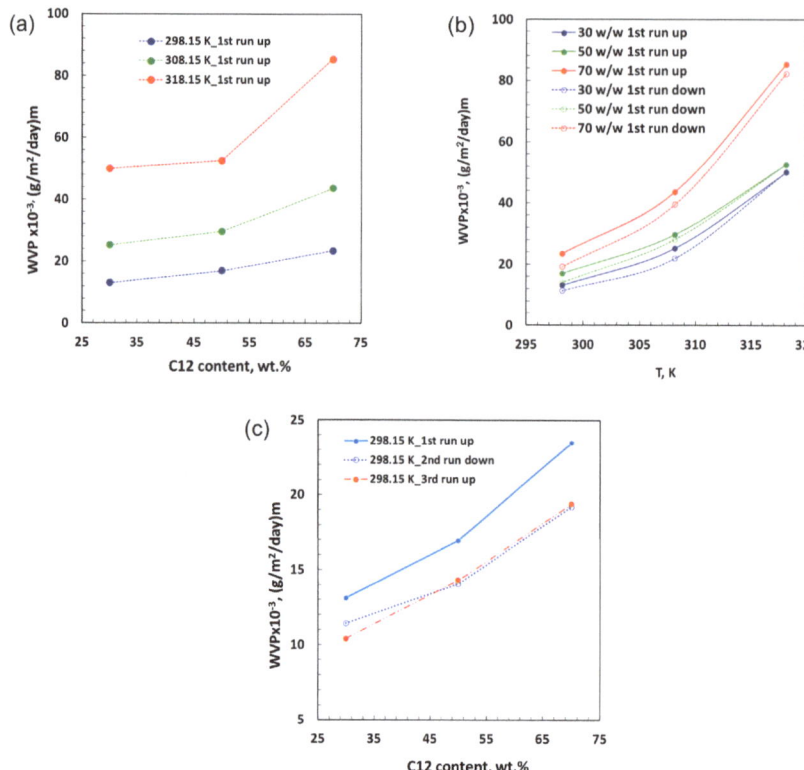

Figure 2. (a) Water permeation through PEBAX membranes containing different amounts of [$C_{12}C_1$im]Cl. (b) Estimation of water permeation with running uphill and downhill temperatures. (c) Reproducibility of water permeation properties estimated at 298 K for all nanocomposite membranes.

With this purpose, MD simulations were performed to corroborate this experimental evidence, serving as a case study and shedding light on the role that polar intermolecular interactions and the charged pathway play in the transport of water molecules through polymeric matrixes containing a typical IL such as [$C_{12}C_1$im]Cl.

4. Simulation Results

4.1. Water Pathway Formed by IL in Nanocomposite Membrane

After the initial build-up structures of PEBAX filled with [$C_{12}C_1$im]Cl and water were equilibrated, the rearrangements of these molecules inside simulation boxes were examined. It was found that the IL cations were reorganized in designated regions compared to the structures of pure [$C_{12}C_1$im]Cl, orienting the polar head towards the core of the agglomerates composed of anions and cationic head groups, whereas the hydrophobic tails are oriented outwards (Figure 3a,b) due to the large amount of hydrocarbon chains in both [$C_{12}C_1$im]Cl and PEBAX that tend to mix with each other. Water molecules are more concentrated around Cl^- and the cationic head groups that form the water pathway, and only a few are present in the polymer region (Figures 3b and 4a). On the other hand, amides tend to aggregate with each other, which may promote the cross-linking of polymer chains and thus improve the strength of the membrane (Figures 3b and 4b). Moreover, the correlation between amides and the head of [$C_{12}C_1$im] (Figure 4b) improves the interfacial interaction and filler–polymer interface compatibility in the blend membrane. As ILs can form hydrogen bonds with the PA hard segments of PEBAX [68], the enhanced hydrogen bonding leads to better dispersion of the filler in the polymer.

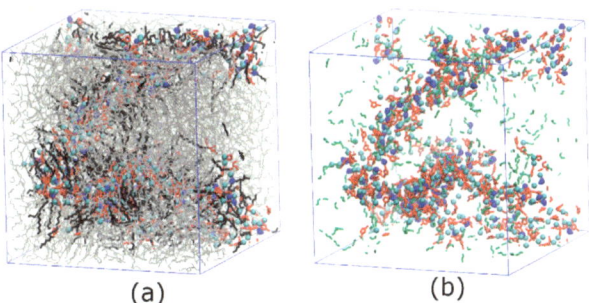

Figure 3. The snapshots of membranes with PEBAX and 30% w/w [C$_{12}$C$_1$im]Cl. The mole ratio of water to [C$_{12}$C$_1$im]Cl is 1:3. (**a**) All components. (**b**) Polar groups (head of cation, Cl$^-$, water, and amide) only. The gray lines are PEBAX chains, the black lines are [C$_{12}$C$_1$im] side chains, the red rings are cationic head groups, the cyan balls are chloride ions, the blue beans are water molecules, and the green sticks are amide bonds in PEBAX.

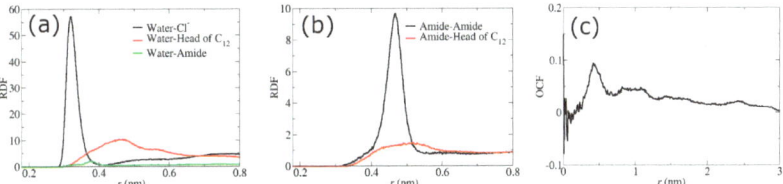

Figure 4. RDFs of (**a**) water with Cl$^-$, head of [C$_{12}$C$_1$im], and amides; (**b**) amides with amides and head of [C$_{12}$C$_1$im]; (**c**) orientational correlation function of [C$_{12}$C$_1$im] side chains, respectively, for the membrane with PEBAX and 30% w/w [C$_{12}$C$_1$im]Cl at 298 K.

It is worth mentioning that although the pure [C$_{12}$C$_1$im]Cl system exhibits an ILC state [69] in the temperature range of this study, due to the similarity between the alkyl cationic side chains of [C$_{12}$C$_1$im]Cl and PEBAX, the side chains themselves do not have long-range orientational correlations (Figure 4c), indicating that the ILs do not form the ILC structure inside PEBAX.

From the above simulation results, we conclude that the polar parts of the IL, composed of anions and cationic head groups, form a continuous polar water pathway inside PEBAX, which facilitates the transport of water molecules through the nanocomposite membrane of PEBAX with [C$_{12}$C$_1$im]Cl. The PEBAX polymers do not participate directly in forming the water pathway.

4.2. Influence of [C$_{12}$C$_1$im]Cl Concentration and Temperature

The network formed by the polar region of [C$_{12}$C$_1$im]Cl becomes larger with increasing [C$_{12}$C$_1$im]Cl concentration (Figure 5). Figure 6 shows the mean square displacements (MSDs), defined as $MSD(t) = \langle |r(t_0 + t) - r(t_0)|^2 \rangle$, where t is the time interval, $r(t_0)$ is the position of the atom at time t_0, and $\langle \cdots \rangle$ denotes the ensemble average, for water molecules with different IL concentrations at 298 K. It can be seen that water molecules diffuse faster in the membrane with a larger IL concentration due to the fact that the polar region of the IL is larger. The smoothness of the MSD curves indicates that the simulation time is long enough to obtain reliable dynamics. The diffusion coefficient for certain types of atoms can be calculated by fitting the slope of the corresponding MSD as $D = MSD/6t$.

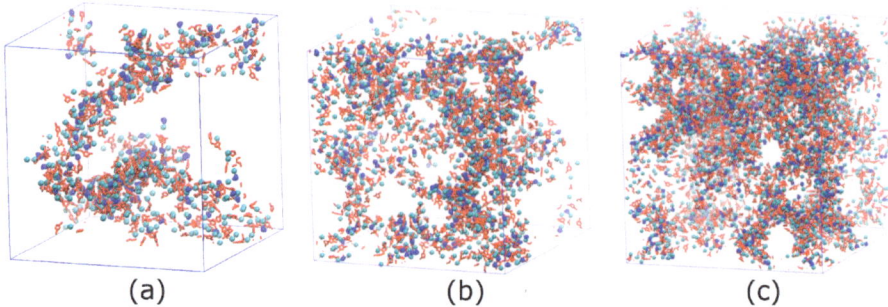

Figure 5. The snapshots of PEBAX membranes with 30% *w/w* (**a**), 50% *w/w* (**b**), and 70% *w/w* (**c**) [$C_{12}C_1$im]Cl at 298 K. Red rings represent cationic head groups, cyan balls represent chloride ions, and blue beans represent water molecules.

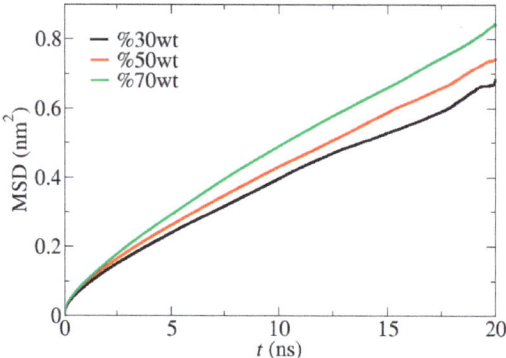

Figure 6. Mean square displacements for water molecules in the PEBAX membranes with 30% *w/w*, 50% *w/w*, and 70% *w/w* [$C_{12}C_1$im]Cl, respectively, at 298 K.

With the diffusion coefficients listed, Table 4 indicates that the diffusion of all molecules, i.e., water, Cl^- and N1, increases with the IL concentration, that the diffusion of water is much greater than for Cl^- and N1, and that the diffusion coefficients of chloride ion and water molecules increase with temperature significantly and are close to those in an aqueous solution at 318 K.

Table 4. Diffusion coefficients of water molecules, Cl^-, and the N1 atoms in [$C_{12}C_1$im].

Diffusion Coefficient 10^{-5} cm^2/s		30 *w/w*%	50 *w/w*%	70 *w/w*%
Water molecules	298 K	0.00520	0.00563	0.00658
	318 K	0.01026	0.01449	0.01718
Cl^-	298 K	0.00032	0.00048	0.00062
	318 K	0.00083	0.00139	0.00182
N1	298 K	0.00038	0.00051	0.00060
	318 K	0.00072	0.00118	0.00163

The above results can be understood as follows. Since water molecules always tend to stay in the polar region of [$C_{12}C_1$im]Cl, a higher concentration of [$C_{12}C_1$im]Cl in the nanocomposite membrane leads to a larger accessible area of polar water pathway and thus to faster diffusion of water through the hydrophilic sites. On the other hand, the polar water pathway assists the diffusion of water to provide an amplification of water transfer through the nanocomposite membranes. Additionally, because the thermal motion of water molecules increases with temperature, it is easier to move through the polar region at a higher temperature, which was also confirmed by our experimental evidence.

4.3. Influence of Water Concentration

To understand how water concentration influences the polar water pathway described above, we simulated the nanocomposite membrane with PEBAX and 30% w/w [$C_{12}C_1$im]Cl along with a various number of water molecules at T = 298 K. As can be seen in Figure 7a–e, most water molecules were distributed in the polar region of [$C_{12}C_1$im]Cl. Moreover, as shown in Figure 7e, when water molecules are excessive, they will gather together to form droplet-like local structures rather than being distributed almost evenly and individually along the polar water pathway.

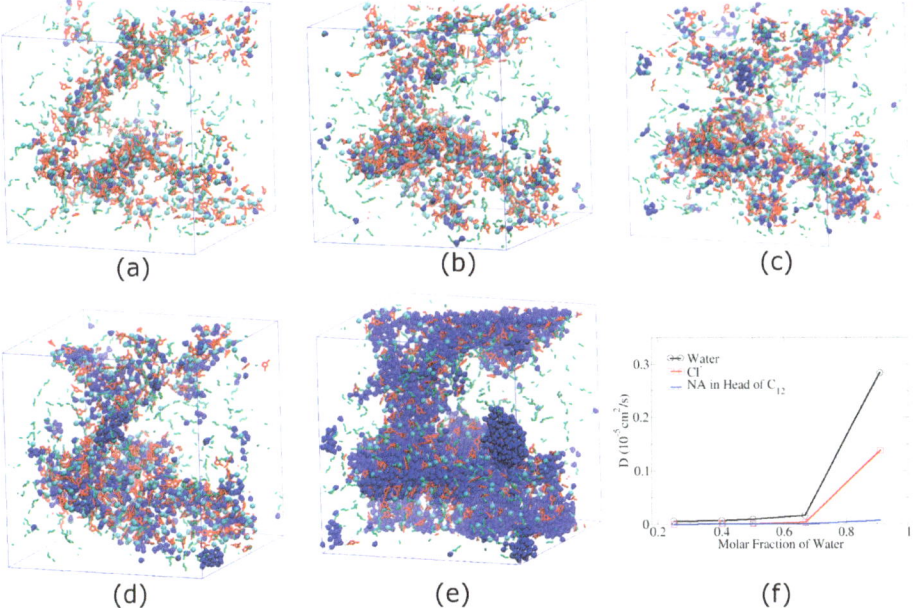

Figure 7. Snapshots of miscible membranes with PEBAX and 30% w/w [$C_{12}C_1$im]Cl at 298 K, whose mole ratios of water to [$C_{12}C_1$im]Cl are 1:3 (**a**), 2:3 (**b**), 1:1 (**c**), 2:1 (**d**), and 10:1 (**e**). The gray lines represent [$C_{12}C_1$im]Cl side chains, red rings represent cationic head groups, purple balls represent chloride ions, blue beans represent water molecules, and green beans represent amide bonds in PEBAX. The other parts of PEBAX are not shown. (**f**) Diffusion coefficients of water, Cl$^-$, and the IL cationic head group at 298 K.

The diffusion coefficients of water molecules and chloride ions change slowly when the number of water molecules is small (Figure 7f). However, when the number of water molecules is large enough to form droplets, the diffusion coefficients of chloride ions and water molecules increase significantly and are close to those in an aqueous solution.

The tendency of the diffusivity of components in the system changing with the concentration of water molecules (Figure 7f) is basically consistent with the research results of Jiang et al.'s work [70] on water molecules in pure ILs, which reveals that the diffusion of water and ions is very slow but increases significantly when the water content is greater than 50% mole fraction.

These results indicate that when the water concentration is relatively low, water molecules distribute evenly and individually along the water pathway, the structural topology of the membrane is not altered, and the dynamics increase almost linearly with water concentration; when the water concentration is so high that water molecules are excessive, water molecules tend to form local water droplets, and correspondingly, the diffusivities of water molecules and anions increase drastically.

5. Conclusions

In $[C_{12}C_1im]Cl$ and PEBAX nanocomposite membranes, at the nanometer scale, charged anions and cationic head groups of $[C_{12}C_1im]Cl$ form a continuous polar network. Most water molecules disperse in the polar region of $[C_{12}C_1im]Cl$, and only very few are distributed near amide regions of the polymer. Therefore, we conclude that the polar region of the IL composed of anions and cationic head groups forms a water pathway for water to be transported through the IL-based polymeric membranes, and the PEBAX polymers do not participate in the water pathway directly. When the IL concentration increases, the polar network becomes larger, providing more hydrophilic sites where water molecules can be allocated. Consequently, the diffusion of water molecules increases with IL concentration and also increases with temperature due to faster thermal motion. The increase in water concentration changes the structural and dynamical properties of the system almost linearly when water molecules are not excessive, and drastically thereafter.

The above results suggest that the presence of hydrophilic $[C_{12}C_1im]Cl$ regions inside polymer networks improves water adsorption and diffusion, yielding amplification of the mass transfer through the membrane, in full accordance with the experimental evidence. This study is a preliminary investigation on behavior of an IL in constrained host polymer matrices wherein water molecules are diffused. It provides useful indications about the behavior of materials confined in predefined volumetric space, suggesting that these kinds of materials are promising for realizing thermo-regulated permeable membranes.

Author Contributions: Conceptualization, A.G. and G.S.; investigation, Z.C., S.L. and A.G.; data analysis, Y.W., A.G., E.T. and G.S.; writing—draft, Y.W. and A.G.; writing—review and editing, Y.W., A.G., G.S. and E.T.; funding acquisition, G.S. and Y.W. All authors have read and agreed to the published version of the manuscript.

Funding: This research was funded by the bilateral agreement CNR-NSFC 2021-2022 (NSFC grant No. 22011530390), the National Natural Science Foundation of China (Nos. 11774357 and 11947302), and Wenzhou Institute, University of Chinese Academy of Sciences (No. WIUCASQD2023009).

Institutional Review Board Statement: Not applicable.

Data Availability Statement: Not applicable.

Acknowledgments: The allocations of computer time on a Tianhe-2 supercomputer and on the HPC cluster of ITP-CAS are appreciated.

Conflicts of Interest: The authors declare no conflict of interest.

References

1. McKeen, L.W. Thermoplastic Elastomers. In *Fatigue and Tribological Properties of Plastics and Elastomers*; Elsevier: Amsterdam, The Netherlands, 2016; pp. 279–289.
2. Schar, M.; Zweifel, L.; Arslan, D.; Grieder, S.; Maurer, C.; Brauner, C. Fused Filament Fabrication of Bio-Based Polyether-Block-Amide Polymers (PEBAX) and Their Related Properties. *Polymers* **2022**, *14*, 5092. [CrossRef] [PubMed]
3. Flesher, J.R. Pebax® polyether block amide—A new family of engineering thermoplastic elastomers. In Proceedings of the High Performance Polymers: Their Origin and Development: Proceedings of the Symposium on the History of High Performance Polymers at the American Chemical Society Meeting, New York, NY, USA, 15–18 April 1986; pp. 401–408.
4. Jonquieres, A.; Clément, R.; Lochon, P. Permeability of block copolymers to vapors and liquids. *Prog. Polym. Sci.* **2002**, *27*, 1803–1877. [CrossRef]
5. Hasan, M.R.; Zhao, H.; Steunou, N.; Serre, C.; Malankowska, M.; Téllez, C.; Coronas, J. Optimization of MIL-178(Fe) and Pebax® 3533 loading in mixed matrix membranes for CO_2 capture. *Int. J. Greenh. Gas Control* **2022**, *121*, 103791. [CrossRef]

6. Potreck, J.; Nijmeijer, K.; Kosinski, T.; Wessling, M. Mixed water vapor/gas transport through the rubbery polymer PEBAX® 1074. *J. Membr. Sci.* **2009**, *338*, 11–16. [CrossRef]
7. Liu, L.; Chakma, A.; Feng, X. Propylene separation from nitrogen by poly(ether block amide) composite membranes. *J. Membr. Sci.* **2006**, *279*, 645–654. [CrossRef]
8. Gugliuzza, A.; Drioli, E. A review on membrane engineering for innovation in wearable fabrics and protective textiles. *J. Membr. Sci.* **2013**, *446*, 350–375. [CrossRef]
9. Niaounakis, M. Sports/Toys/Board Games. In *Biopolymers: Applications and Trends*; Elsevier: Amsterdam, The Netherlands, 2015; pp. 427–443.
10. Barbucci, A.; Delucchi, M.; Cerisola, G. Organic coatings for concrete protection: Liquid water and water vapour permeabilities. *Prog. Org. Coat.* **1997**, *30*, 293–297. [CrossRef]
11. An, L.; Zhang, N.; Zeng, X.; Zhong, B.; Yu, Y. Quasi-isotropically thermoconductive, antiwear and insulating hierarchically assembled hexagonal boron nitride nanosheet/epoxy composites for efficient microelectronic cooling. *J. Colloid Interface Sci.* **2022**, *608*, 1907–1918. [CrossRef]
12. Yoon, K.; Hsiao, B.S.; Chu, B. Functional nanofibers for environmental applications. *J. Mater. Chem.* **2008**, *18*, 5326–5334. [CrossRef]
13. Gugliuzza, A.; Drioli, E. New performance of a modified poly(amide-12-b-ethyleneoxide). *Polymer* **2003**, *44*, 2149–2157. [CrossRef]
14. Gugliuzza, A.; Drioli, E. Role of additives in the water vapor transport through block co-poly(amide/ether) membranes: Effects on surface and bulk polymer properties. *Eur. Polym. J.* **2004**, *40*, 2381–2389. [CrossRef]
15. Camuffo, D. *Microclimate for Cultural Heritage: Measurement, Risk Assessment, Conservation, Restoration, and Maintenance of Indoor and Outdoor Monuments*; Elsevier: Amsterdam, The Netherlands, 2019.
16. Zhong, W.; Xing, M.M.; Pan, N.; Maibach, H.I. Textiles and human skin, microclimate, cutaneous reactions: An overview. *Cutan. Ocul. Toxicol.* **2006**, *25*, 23–39. [CrossRef] [PubMed]
17. Roh, J.-S.; Kim, S. All-fabric intelligent temperature regulation system for smart clothing applications. *J. Intell. Mater. Syst. Struct.* **2015**, *27*, 1165–1175. [CrossRef]
18. Díaz Rincón, M.; Jiménez-Junca, C.; Roa Duarte, C. A novel absorption process for small-scale natural gas dew point control and dehydration. *J. Nat. Gas Sci. Eng.* **2016**, *29*, 264–274. [CrossRef]
19. Mokhatab, S.; Poe, W.A.; Speight, J.G. *Handbook of Natural Gas Transmission and Processing*; Gulf Professional Publishing: Houston, TX, USA, 2006; pp. 323–364.
20. Gugliuzza, A.; Pingitore, V.; Miriello, D.; Drioli, E. Functional carbon nanotubes for high-quality charge transfer and moisture regulation through membranes: Structural and functional insights. *Phys. Chem. Chem. Phys.* **2015**, *17*, 12919–12926. [CrossRef]
21. Prefol, T.; Gain, O.; Sudre, G.; Gouanve, F.; Espuche, E. Development of Breathable Pebax((R))/PEG Films for Optimization of the Shelf-Life of Fresh Agri-Food Products. *Membranes* **2021**, *11*, 692. [CrossRef]
22. Gugliuzza, A.; Fabiano, R.; Garavaglia, M.G.; Spisso, A.; Drioli, E. Study of the surface character as responsible for controlling interfacial forces at membrane-feed interface. *J. Colloid Interface Sci.* **2006**, *303*, 388–403. [CrossRef]
23. Akhtar, F.H.; Kumar, M.; Peinemann, K.-V. Pebax®1657/Graphene oxide composite membranes for improved water vapor separation. *J. Membr. Sci.* **2017**, *525*, 187–194. [CrossRef]
24. Li, W.; Chang, Z.; Lin, L.; Xu, X. Effect of montmorillonite on PEBAX® 1074-based mixed matrix membranes to be used in humidifiers in proton exchange membrane fuel cells. *e-Polymers* **2020**, *20*, 171–184. [CrossRef]
25. Tang, C.; Wang, Y. Phase behaviors of ionic liquids attributed to the dual ionic and organic nature. *Commun. Theor. Phys.* **2022**, *74*, 097601. [CrossRef]
26. Wilkes, J.S.; Zaworotko, M.J. Air and water stable 1-ethyl-3-methylimidazolium based ionic liquids. *J. Chem. Soc. Chem. Commun.* **1992**, *13*, 965–967. [CrossRef]
27. Wilkes, J.S. A short history of ionic liquids—From molten salts to neoteric solvents. *Green Chem.* **2002**, *4*, 73–80. [CrossRef]
28. Rogers, R.D.; Seddon, K.R. Ionic Liquids—Solvents of the Future? *Science* **2003**, *302*, 792–793. [CrossRef] [PubMed]
29. Liu, B.; Yang, T.; Mu, X.; Mai, Z.; Li, H.; Wang, Y.; Zhou, G. Smart Supramolecular Self-Assembled Nanosystem: Stimulus-Responsive Hydrogen-Bonded Liquid Crystals. *Nanomaterials* **2021**, *11*, 448. [CrossRef]
30. Wang, R.; Fang, C.; Yang, L.; Li, K.; Zhu, K.; Liu, G.; Chen, J. The Novel Ionic Liquid and Its Related Self-Assembly in the Areas of Energy Storage and Conversion. *Small Sci.* **2022**, *2*, 2200048. [CrossRef]
31. Ji, Y.; Shi, R.; Wang, Y.; Saielli, G. Effect of the chain length on the structure of ionic liquids: From spatial heterogeneity to ionic liquid crystals. *J. Phys. Chem. B* **2013**, *117*, 1104–1109. [CrossRef]
32. Fabregat-Santiago, F.; Bisquert, J.; Palomares, E.; Otero, L.; Kuang, D.; Zakeeruddin, S.M.; Grätzel, M. Correlation between photovoltaic performance and impedance spectroscopy of dye-sensitized solar cells based on ionic liquids. *J. Phys. Chem. C* **2007**, *111*, 6550–6560. [CrossRef]
33. Gorlov, M.; Kloo, L. Ionic liquid electrolytes for dye-sensitized solar cells. *Dalton Trans.* **2008**, *20*, 2655–2666. [CrossRef]
34. Lee, W.R.; Kim, H.J.; Kim, J.Y.; Kim, T.H.; Ahn, K.D.; Kim, E. Electro-fluorescence switching of bis-imidazolium ionic liquids. *J. Nanosci. Nanotechnol.* **2008**, *8*, 4630–4634. [CrossRef]
35. Lewandowski, A.; Świderska-Mocek, A. Ionic liquids as electrolytes for Li-ion batteries—An overview of electrochemical studies. *J. Power Sources* **2009**, *194*, 601–609. [CrossRef]
36. Wei, D.; Ivaska, A. Applications of ionic liquids in electrochemical sensors. *Anal. Chim. Acta* **2008**, *607*, 126–135. [CrossRef]
37. Salikolimi, K.; Sudhakar, A.A.; Ishida, Y. Functional Ionic Liquid Crystals. *Langmuir* **2020**, *36*, 11702–11731. [CrossRef] [PubMed]

38. Cao, W.; Wang, Y.; Saielli, G. Metastable State during Melting and Solid-Solid Phase Transition of [C(n)Mim][NO(3)] (n = 4–12) Ionic Liquids by Molecular Dynamics Simulation. *J. Phys. Chem. B* **2018**, *122*, 229–239. [CrossRef] [PubMed]
39. Cao, W.; Wang, Y. Phase Behaviors of Ionic Liquids Heating from Different Crystal Polymorphs toward the Same Smectic-A Ionic Liquid Crystal by Molecular Dynamics Simulation. *Crystals* **2019**, *9*, 26. [CrossRef]
40. Saielli, G. MD simulation of the mesomorphic behaviour of 1-hexadecyl-3-methylimidazolium nitrate: Assessment of the performance of a coarse-grained force field. *Soft Matter* **2012**, *8*, 10279–10287. [CrossRef]
41. Quevillon, M.J.; Whitmer, J.K. Charge Transport and Phase Behavior of Imidazolium-Based Ionic Liquid Crystals from Fully Atomistic Simulations. *Materials* **2018**, *11*, 64. [CrossRef]
42. Peng, H.; Kubo, M.; Shiba, H. Molecular dynamics study of mesophase transitions upon annealing of imidazolium-based ionic liquids with long-alkyl chains. *Phys. Chem. Chem. Phys.* **2018**, *20*, 9796–9805. [CrossRef]
43. Saielli, G.; Bagno, A.; Wang, Y. Insights on the isotropic-to-smectic A transition in ionic liquid crystals from coarse-grained molecular dynamics simulations: The role of microphase segregation. *J. Phys. Chem. B* **2015**, *119*, 3829–3836. [CrossRef]
44. Dai, Z.; Noble, R.D.; Gin, D.L.; Zhang, X.; Deng, L. Combination of ionic liquids with membrane technology: A new approach for CO_2 separation. *J. Membr. Sci.* **2016**, *497*, 1–20. [CrossRef]
45. Gugliuzza, A.; Drioli, E. PVDF and HYFLON AD membranes: Ideal interfaces for contactor applications. *J. Membr. Sci.* **2007**, *300*, 51–62. [CrossRef]
46. De Luca, G.; Gugliuzza, A.; Drioli, E. Competitive hydrogen-bonding interactions in modified polymer membranes: A density functional theory investigation. *J. Phys. Chem. B* **2009**, *113*, 5473–5477. [CrossRef] [PubMed]
47. Yang, B.; Bai, L.; Wang, Z.; Jiang, H.; Zeng, S.; Zhang, X.; Zhang, X. Exploring NH3 Transport Properties by Tailoring Ionic Liquids in Pebax-Based Hybrid Membranes. *Ind. Eng. Chem. Res.* **2021**, *60*, 9570–9577. [CrossRef]
48. Ghasemi Estahbanati, E.; Omidkhah, M.; Ebadi Amooghin, A. Preparation and characterization of novel Ionic liquid/Pebax membranes for efficient CO_2/light gases separation. *J. Ind. Eng. Chem.* **2017**, *51*, 77–89. [CrossRef]
49. Fam, W.; Mansouri, J.; Li, H.; Chen, V. Improving CO_2 separation performance of thin film composite hollow fiber with Pebax®1657/ionic liquid gel membranes. *J. Membr. Sci.* **2017**, *537*, 54–68. [CrossRef]
50. Martínez-Izquierdo, L.; Téllez, C.; Coronas, J. Highly stable Pebax® Renew® thin-film nanocomposite membranes with metal organic framework ZIF-94 and ionic liquid [Bmim][BF4] for CO_2 capture. *J. Mater. Chem. A* **2022**, *10*, 18822–18833. [CrossRef]
51. Shahrezaei, K.; Abedini, R.; Lashkarbolooki, M.; Rahimpour, A. A preferential CO_2 separation using binary phases membrane consisting of Pebax®1657 and [Omim][PF6] ionic liquid. *Korean J. Chem. Eng.* **2019**, *36*, 2085–2094. [CrossRef]
52. Pardo, F.; Zarca, G.; Urtiaga, A. Effect of feed pressure and long-term separation performance of Pebax-ionic liquid membranes for the recovery of difluoromethane (R32) from refrigerant mixture R410A. *J. Membr. Sci.* **2021**, *618*, 118744. [CrossRef]
53. Noorhisham, N.A.; Amri, D.; Mohamed, A.H.; Yahaya, N.; Ahmad, N.M.; Mohamad, S.; Kamaruzaman, S.; Osman, H. Characterisation techniques for analysis of imidazolium-based ionic liquids and application in polymer preparation: A review. *J. Mol. Liq.* **2021**, *326*, 115340. [CrossRef]
54. Akhtar, F.H.; Vovushua, H.; Villalobos, L.F.; Shevate, R.; Kumar, M.; Nunes, S.P.; Schwingenschlögl, U.; Peinemann, K.-V. Highways for water molecules: Interplay between nanostructure and water vapor transport in block copolymer membranes. *J. Membr. Sci.* **2019**, *572*, 641–649. [CrossRef]
55. Lopes, J.N.C.; Deschamps, J.; Pádua, A.l.A.H. Modeling Ionic Liquids Using a Systematic All-Atom Force Field. *J. Phys. Chem. B* **2004**, *108*, 2038–2047. [CrossRef]
56. Lopes, J.N.C.; Pádua, A.l.A.H. Molecular Force Field for Ionic Liquids Composed of Triflate or Bistriflylimide Anions. *J. Phys. Chem. B* **2004**, *108*, 16893–16898. [CrossRef]
57. Woods, R.; Chappelle, R. Restrained electrostatic potential atomic partial charges for condensed-phase simulations of carbohydrates. *J. Mol. Struct. THEOCHEM* **2000**, *527*, 149–156. [CrossRef]
58. Chaban, V.V.; Voroshylova, I.V.; Kalugin, O.N. A new force field model for the simulation of transport properties of imidazolium-based ionic liquids. *Phys. Chem. Chem. Phys.* **2011**, *13*, 7910–7920. [CrossRef] [PubMed]
59. Horn, H.W.; Swope, W.C.; Pitera, J.W.; Madura, J.D.; Dick, T.J.; Hura, G.L.; Head-Gordon, T. Development of an improved four-site water model for biomolecular simulations: TIP4P-Ew. *J. Chem. Phys.* **2004**, *120*, 9665–9678. [CrossRef] [PubMed]
60. Darden, T.; York, D.; Pedersen, L. Particle mesh Ewald: An $N \cdot \log(N)$ method for Ewald sums in large systems. *J. Chem. Phys.* **1993**, *98*, 10089–10092. [CrossRef]
61. Nosé, S. A molecular dynamics method for simulations in the canonical ensemble. *Mol. Phys.* **1984**, *52*, 255–268. [CrossRef]
62. Parrinello, M.; Rahman, A. Polymorphic transitions in single crystals: A new molecular dynamics method. *J. Appl. Phys.* **1981**, *52*, 7182–7190. [CrossRef]
63. Abraham, M.J.; Murtola, T.; Schulz, R.; Páll, S.; Smith, J.C.; Hess, B.; Lindahl, E. GROMACS: High performance molecular simulations through multi-level parallelism from laptops to supercomputers. *SoftwareX* **2015**, *1*, 19–25. [CrossRef]
64. Tocci, E.; Gugliuzza, A.; De Lorenzo, L.; Macchione, M.; De Luca, G.; Drioli, E. Transport properties of a co-poly(amide-12-b-ethylene oxide) membrane: A comparative study between experimental and molecular modelling results. *J. Membr. Sci.* **2008**, *323*, 316–327. [CrossRef]
65. Gugliuzza, A.; De Luca, G.; Tocci, E.; De Lorenzo, L.; Drioli, E. Intermolecular interactions as controlling factor for water sorption into polymer membranes. *J. Phys. Chem. B* **2007**, *111*, 8868–8878. [CrossRef]

66. Gugliuzza, A.; Drioli, E. Evaluation of CO_2 permeation through functional assembled mono-layers: Relationships between structure and transport. *Polymer* **2005**, *46*, 9994–10003. [CrossRef]
67. Sheth, J.P.; Xu, J.; Wilkes, G.L. Solid state structure–property behavior of semicrystalline poly (ether-block-amide) PEBAX® thermoplastic elastomers. *Polymer* **2003**, *44*, 743–756. [CrossRef]
68. Huang, G.; Isfahani, A.P.; Muchtar, A.; Sakurai, K.; Shrestha, B.B.; Qin, D.; Yamaguchi, D.; Sivaniah, E.; Ghalei, B. Pebax/ionic liquid modified graphene oxide mixed matrix membranes for enhanced CO_2 capture. *J. Membr. Sci.* **2018**, *565*, 370–379. [CrossRef]
69. Bradley, A.; Hardacre, C.; Holbrey, J.; Johnston, S.; McMath, S.; Nieuwenhuyzen, M. Small-angle X-ray scattering studies of liquid crystalline 1-alkyl-3-methylimidazolium salts. *Chem. Mater.* **2002**, *14*, 629–635. [CrossRef]
70. Jiang, W.; Wang, Y.T.; Voth, G.A. Molecular dynamics simulation of nanostructural organization in ionic liquid/water mixtures. *J. Phys. Chem. B* **2007**, *111*, 4812–4818. [CrossRef]

Disclaimer/Publisher's Note: The statements, opinions and data contained in all publications are solely those of the individual author(s) and contributor(s) and not of MDPI and/or the editor(s). MDPI and/or the editor(s) disclaim responsibility for any injury to people or property resulting from any ideas, methods, instructions or products referred to in the content.

Review

Thorium Removal, Recovery and Recycling: A Membrane Challenge for Urban Mining

Geani Teodor Man [1,2], Paul Constantin Albu [3], Aurelia Cristina Nechifor [1], Alexandra Raluca Grosu [1], Szidonia-Katalin Tanczos [4], Vlad-Alexandru Grosu [5,*], Mihail-Răzvan Ioan [3] and Gheorghe Nechifor [1,*]

1. Analytical Chemistry and Environmental Engineering Department, University Politehnica of Bucharest, 011061 Bucharest, Romania; man_geani@yahoo.com (G.T.M.); aureliacristinanechifor@gmail.com (A.C.N.); andra.grosu@upb.ro (A.R.G.)
2. National Research and Development Institute for Cryogenics and Isotopic Technologies—ICSI, 240050 Râmnicu Valcea, Romania
3. Radioisotopes and Radiation Metrology Department (DRMR), IFIN Horia Hulubei, 023465 Măgurele, Romania; paulalbu@gmail.com (P.C.A.); razvan.ioan@nipne.ro (M.-R.I.)
4. Department of Bioengineering, University Sapientia of Miercurea-Ciuc, 500104 Miercurea Ciuc, Romania; tczszidonia@yahoo.com
5. Department of Electronic Technology and Reliability, Faculty of Electronics, Telecommunications and Information Technology, University Politehnica of Bucharest, 061071 Bucharest, Romania
* Correspondence: vlad.grosu@upb.ro (V.-A.G.); ghnechifor@gmail.com (G.N.)

Citation: Man, G.T.; Albu, P.C.; Nechifor, A.C.; Grosu, A.R.; Tanczos, S.-K.; Grosu, V.-A.; Ioan, M.-R.; Nechifor, G. Thorium Removal, Recovery and Recycling: A Membrane Challenge for Urban Mining. *Membranes* **2023**, *13*, 765. https://doi.org/10.3390/membranes13090765

Academic Editors: Clàudia Fontàs, Annarosa Gugliuzza and Cristiana Boi

Received: 12 July 2023
Revised: 16 August 2023
Accepted: 24 August 2023
Published: 29 August 2023

Copyright: © 2023 by the authors. Licensee MDPI, Basel, Switzerland. This article is an open access article distributed under the terms and conditions of the Creative Commons Attribution (CC BY) license (https://creativecommons.org/licenses/by/4.0/).

Abstract: Although only a slightly radioactive element, thorium is considered extremely toxic because its various species, which reach the environment, can constitute an important problem for the health of the population. The present paper aims to expand the possibilities of using membrane processes in the removal, recovery and recycling of thorium from industrial residues reaching municipal waste-processing platforms. The paper includes a short introduction on the interest shown in this element, a weak radioactive metal, followed by highlighting some common (domestic) uses. In a distinct but concise section, the bio-medical impact of thorium is presented. The classic technologies for obtaining thorium are concentrated in a single schema, and the speciation of thorium is presented with an emphasis on the formation of hydroxo-complexes and complexes with common organic reagents. The determination of thorium is highlighted on the basis of its radioactivity, but especially through methods that call for extraction followed by an established electrochemical, spectral or chromatographic method. Membrane processes are presented based on the electrochemical potential difference, including barro-membrane processes, electrodialysis, liquid membranes and hybrid processes. A separate sub-chapter is devoted to proposals and recommendations for the use of membranes in order to achieve some progress in urban mining for the valorization of thorium.

Keywords: thorium removal; thorium recovery; thorium recycling; thorium separation; thorium transport; thorium separation processes; thorium membrane separation; thorium membrane concentration; thorium determination

1. Introduction

Thorium is a relatively exotic element, although it is known to have a significant natural abundance compared to lead [1]. With a component of the actinide series at position #90 and a weight of the gram atom equaling 232.03, it is unstable (radioactive) in all its isotopes except isotope ^{232}Th [2]. The half-time of ^{232}Th is so long that it is considered stable when joining uranium, which also occurs naturally [3]. The interest in thorium as a nuclear material should have resulted in an increased interest both on the part of researchers and on the part of energy producers [4]. However, the particularity of thorium is that it cannot sustain a chain reaction by itself, as is the case with uranium and plutonium [5], but fission can be produced under the influence of neutrons from an external source [6]. If the thorium atoms absorb a neutron, they turn into a heavier isotope, which then rapidly disintegrates into an isotope of the element protactinium and further into a fissioned uranium isotope under the incidence of bombardment with another neutron [7]. Because its disintegration line does not end with a material usable in the military industry, the interest in this nuclear fuel remains low [8].

Thus, the number of existing publications highlighted in Google Scholar [9] or SCOPUS [10], selected according to a specific algorithm [11] and using various keywords of scientific interest, is relatively moderate or even low (Table 1).

Table 1. The number of publications highlighted in Google Scholar on various keywords related to thorium.

Keywords *	Scholar Google Publication Number in Different Periods			SCOPUS Publication Number
	Any Time	2014–2023	2021–2023	1995–2023
Thorium separation	162,000	82,000	12,900	2186
Thorium concentration	199,000	12,200	6200	7888
Thorium recovery	79,000	17,900	9200	896
Thorium removal	62,000	17,500	13,800	132
Membrane thorium separation	21,900	10,600	3730	458
Membrane thorium concentration	27,600	19,000	4610	141
Membrane thorium recovery	18,000	8600	3850	27
Membrane thorium removal	21,600	12,000	4500	5
"Thorium separation"	883	244	79	25
"Thorium recovery"	611	204	87	34
"Thorium recycling"	50	18	4	2
"Thorium membrane"	7	2	–	2

* accessed on 24 June 24 2023.

Recent publications are consistent and draw attention to the need to reconsider thorium as a nuclear material with a clear perspective [1–5] but also as an environmental polluter [6–8]. On the other hand, new materials usable in various analytical or technological separative techniques are also studied [12–20].

However, after analyzing Table 1, many would be discouraged to start research on aspects of recovery, recycling, or removal of thorium from various sources, although its common applications have determined its presence in urban waste in surprisingly high concentrations.

This last observation led to the initiation of this paper, whose aim is to warn both researchers and environmental officers regarding the danger of the uncontrolled spread of thorium as well as propose simpler solutions for removal, recovery and recycling, based on processes very close to "urban mining".

The specific objectives of this work are to emphasize the existence of thorium in various materials used over time for common applications, the toxicity and bio-medical implications of thorium, the established technologies for obtaining thorium, the speciation of thorium in aqueous solutions, and the determination of thorium membrane processes with integration perspectives in thorium recovery or removal technologies and proposals regarding this aspect.

2. Applications of Thorium

Thorium, and especially thorium dioxide, has found relatively numerous applications for a radioactive element, even if this radioactivity is weak [21–23]. As various and unexpected, with many having been abandoned, the applications of thorium are so common (Figure 1) that they have become dangerous [21], especially since after the use of various materials and under the conditions of inattention in recycling or selective collection, thorium ends up in the environment [22].

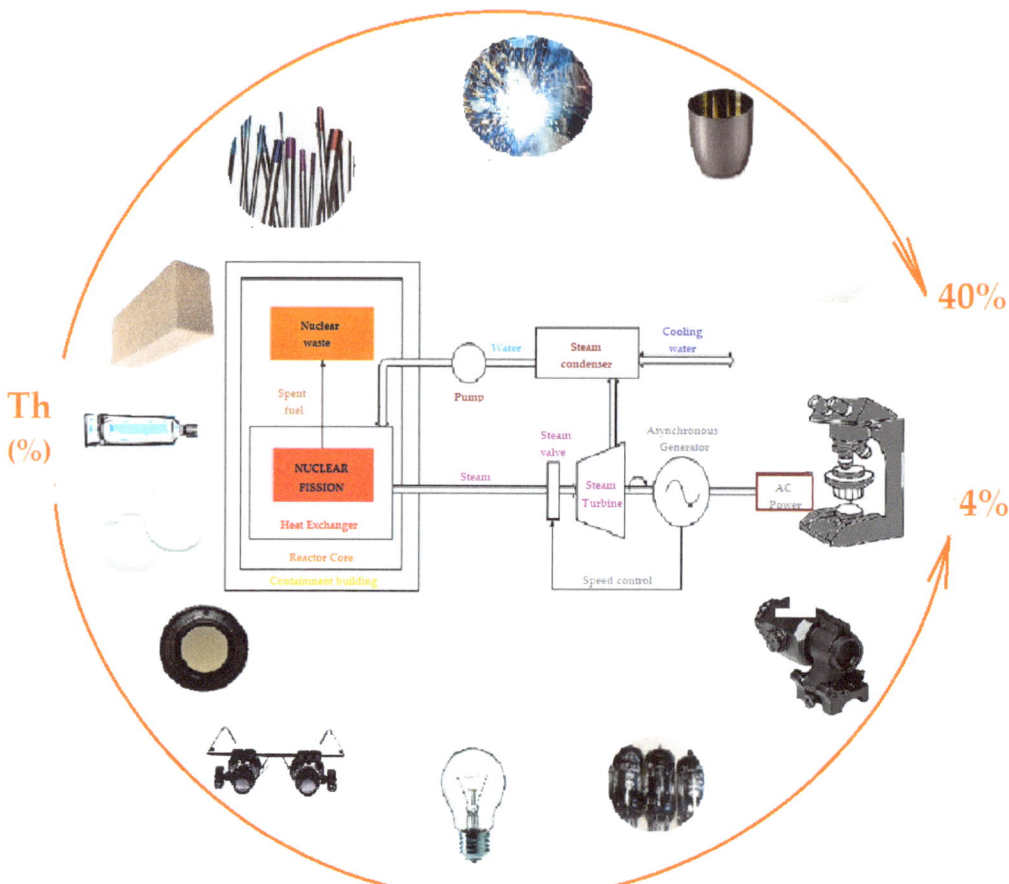

Figure 1. Domestic applications of thorium and thorium dioxide, along with the alleged use in generating energy in nuclear power plants (U–Th cycle).

In addition to its surprising use in toothpaste, in the dating of hominids, as a contrast agent in certain radiological examinations or as a filament in incandescent light bulbs, lamps, lanterns, and thorium mantles [23], it is also used for technical applications in which it is practically irreplaceable: crucibles for high temperatures, welding electrodes and alloys (aluminum, magnesium, steel), lamps for special electronic equipment, mantles in the metallurgical industry, industrial catalysts (ammonia, sulfuric acid, cracking hydrocarbons), the manufacture of thorium-mixed oxide tablets and uranium, oxygen detectors, and lenses for various optical and opto-electronic devices (having excellent wavelength dispersion and high refractive index) [24,25].

We can conclude that thorium, although radioactive, can be found in the aerospace industry, automobile industry, chemical and metallurgical industry, electrotechnical industry, electronics industry, dentistry (cements for dentistry, optical and surgical instruments, manufacturing), and in art objects (alloys, jewelry, sculptures, statues) [26,27], which leads to thorium being an environmental pollutant [28–30].

3. Toxicity and Bio-Medical Implications

Thorium is included on the list of carcinogenic substances [31], even though it decomposes through alfa decay [32], and the emitted alfa radiation cannot penetrate human skin [33].

The dangers associated with its radioactivity, due to the use of thorium in various technologies that capitalize on the high melting of thorium dioxide, lead to the following [34–38]:

- the amounts of thorium in the environment can be accidentally increased during processing;
- humans absorb thorium through food or drinking water (in areas adjacent to mining operations);
- the quantities in the air are very small (insignificant and generally neglected);
- amounts are high near hazardous waste storage or processing sites;
- amounts are high in industrial laboratories or mining laboratories that mill minerals containing thorium.

The medical effects, observed over time, of those who acquire thorium at work are as follows [39–42]:

- greater chance of developing lung disease;
- higher occurrence of lung and pancreatic cancer;
- changes in genetic material;
- higher instance of blood cancer;
- greater chance of developing liver diseases (when injecting thorium for X-rays);
- storage in bones (long-term exposure) can lead to the generation of bone cancer.

Being a heavy metal, the medical effects of thorium as well as the precautions for working with it must be considered [43,44].

At the same time, natural thorium is in secular equilibrium with its descendants, which makes it necessary to consider their radiotoxicity; for this reason, it is classified among the most dangerous radionuclides [45,46].

4. Classical Technology

Thorium is found in monazite (1 to 15%) in concentrations that allow it to be exploited on an industrial scale, through classical technologies [47]. At the same time, thorium appears in mining processes, especially those aimed at obtaining rare earths or uranium [48–50].

The schemes in Figure 2 show the main operations that lead to obtaining thorium from monazite through acid (Figure 2a) or the basic attack (Figure 2b) of impurities; however, in principle, any mineral is considered as a source of thorium, with the series of technological operations being the same [51–58].

In practice, the mineral (source of thorium) is brought to a state of fine grinding in order to be attacked by sulfuric acid or a base (sodium hydroxide), so that the parts of the mineral not containing thorium pass into the solution, while others are removed by filtration. The filtrate containing thorium (colloidal) can be directly processed (when purification is not done in this technology) or precipitated, filtered and finally subjected to purification by extraction in an organic solvent (kerosene) and TBP as a complexant [59], or an organic solvent and an amine or a selective complexant and re-extraction [60].

If the source of thorium is a mineral containing rare earths or the residue obtained during the processing of various minerals in order to obtain rare earth elements (REEs), then the basic procedures used in the separation, concentration and purification of thorium are leaching [61–63], precipitation [64–67], solvent extraction [68,69] and ion exchange [70].

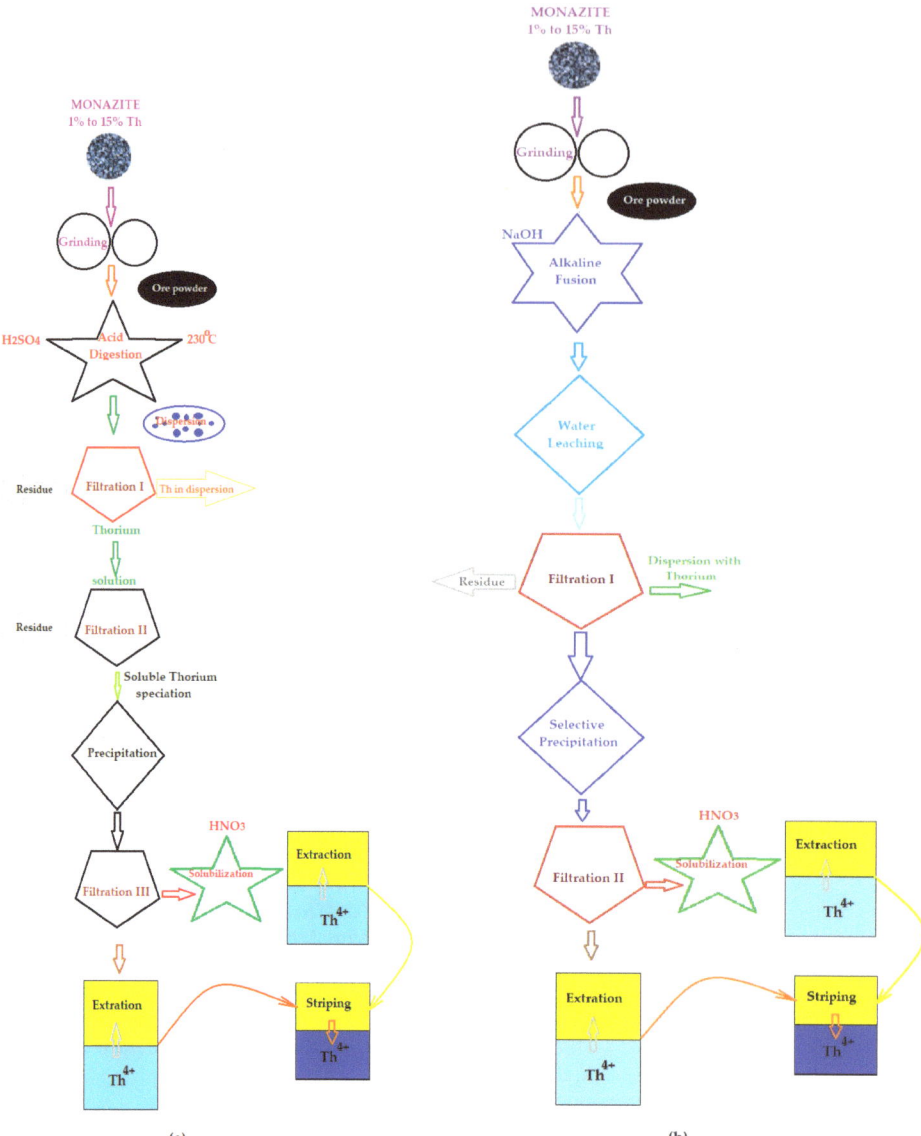

Figure 2. Simplified flowcharts for obtaining thorium from monazite: (**a**) acid digestion; (**b**) alkaline fusion.

Obtaining thorium from pure compounds (halogens, halides) or alloys can be performed using physical (thermal) or chemical (reduction) processes [71,72].

For the current work, which involves obtaining thorium from industrial residues or by-products (waste), the diagrams in Figure 2 present, as narrow technological points, the filtration and extraction operations likely to be replaced to avoid environmental pollution [73,74].

5. Thorium Speciation

Thorium compounds are relatively few compared to other elements, even the less reactive ones [75,76]. Thus, thorium dioxide, halogens or a nitride are encountered, but the speciation of the thorium ion (Th^{4+}) in aqueous solutions is of practical importance, as countless hydroxylated chemical species can be generated: $[ThOH]^{3+}$, $[Th(OH)_2]^{2+}$, $[Th(OH)_3]^{+}$, $[Th(OH)_4]$, $[Th(OH)_2(CO_3)_2]^{2-}$, $[Th_2(OH)_2]^{6+}$ and $[Th(H_2O)_9]^{4+}$ [77–79], hence the importance of the operational parameters (pH, ionic strength, temperature, contact ions in the aqueous solution [80–83]), which would be the object of the study of membrane processes.

When dissolving thorium nitrate (for example) in water, the mentioned hydroxyl species are formed, but also combinations that may include carbon dioxide (present in the environment). Considering the formation of only thorium hydroxides in aqueous solution, a series of chemical species are formed as a result of some equilibria with proton exchange, which is dependent on pH and is shown hypothetically in Figure 3. The degree of formation in solution of various chemical species can be determined exactly if the acidity constants of the chemical species and/or stability constants of the hydroxyl complexes are known [67].

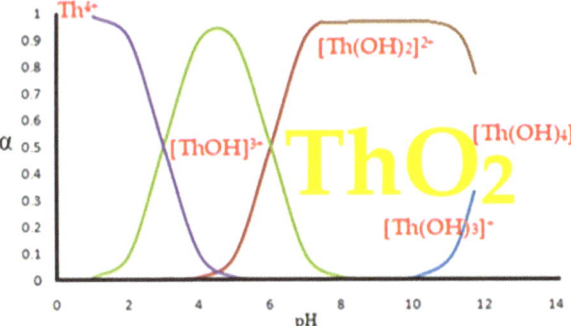

Figure 3. Hypothetical stability diagram of thorium hydroxo-complexes in aqueous medium.

The appearance of thorium dioxide is also related to the pH, ionic strength and temperature of the solution; however, as a solid phase, it depends essentially on the concentration of thorium obtained at a given moment in the phases of a technology, and more importantly, on the distribution of thorium in the environment [80–83].

At the same time, if we consider that thorium is obtained in the source solution as Th^{4+} ion, then a wide series of organic complexants (Figure 4) can contribute to the formation of some speciations involved in the concentration and recovery, especially through extraction, of thorium [84–102].

The speciation of thorium in aqueous solution is important because the various hydroxo-hydroxyl species have different sizes, so membrane processes based on size separation can be chosen accordingly, moving from reverse osmosis to nanofiltration or even to ultrafiltration or colloidal filtration. Certainly, the aspects of chemical speciation of thorium in the presence of some inorganic complexants, but especially organic ones, are much more difficult to exploit, because the new chemical species have various hydrophobic-hydrophilic shells, depending on the considered ligand. These chemical species can be considered for separations with liquid or composite membranes that exhibit selective or even specific interactions with the ligands that incorporate the thorium ion.

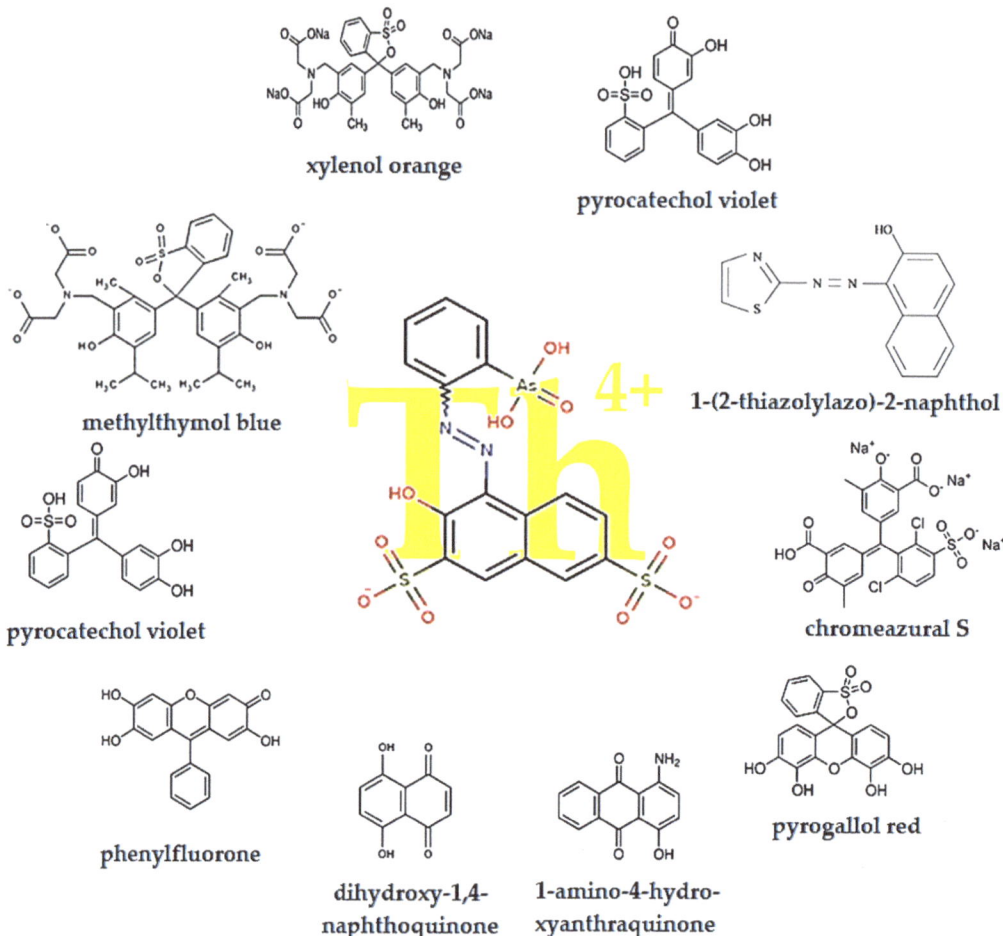

Figure 4. Common organic reagents involved in thorium ion complexation and/or extraction. The colored groups (in red) interact with the thorium ion (in yellow).

6. Thorium Determination

Although a radioactive element, physical–chemical analysis based on specific reactions finds permanent use in various applications [86–99]. The main reagent, studied exhaustively and used with excellent results in various working conditions, is Thorin, 1-(2-Arsonophenylazo)-2-hydroxy-3,6-naphthalene-disulfonic acid sodium salt, 2-(2-Hydroxy-3,6-disulfo-1-naphthylazo)-benzene-arsenic acid sodium salt; Empirical Formula (Hill Notation): $C_{16}H_{11}AsN_2Na_2O_{10}S_2$ [87,88].

Otherwise, the radiation analyses for thorium are few, although they refer to the entire radiation register (α, β or γ) [103–109] (Table 2).

That is why, alongside the highly developed spectrophotometric methods [95], the studied reagents are today widely used for preconcentration [96], with a view to the permanent development of new methods, including electrochemical, optode, electrochemical sensor and coupled spectral methods [110–134].

Table 2. Analysis methods for thorium: characteristics and applications.

Analytical Methods	Samples and/or Applications	Characteristics	Refs.
Radiometric analysis (α, β or γ)	Determination of uranium, thorium, plutonium, americium and curium ultra-traces	Photon–electron rejecting α liquid scintillation	[103]
	Determination for levels of uranium and thorium in water along Oum Er-Rabia River	Alpha track detectors	[104]
	Thorium determination in intercomparison samples and in some Romanian building materials	Gamma ray spectrometry	[105]
	Thorium determination	Miscellaneous techniques	[106,107]
X-ray fluorescence spectrometry	Determination of thorium in natural water	Coupled with preconcentration method	[108]
	Trace element determination in thorium oxide	Total reflection X-ray fluorescence spectrometry	[109]
Inductively Coupled Plasma- (ICP-)	Analysis of rare earth elements, thorium and uranium in geochemical certified reference materials and soils	Mass spectrometry (ICP-MS)	[110]
	Determination of trace element concentrations and stable lead, uranium and thorium isotope ratios in in NORM and NORM-polluted sample leachates	Quadrupole-ICP-MS	[111]
	Chemical separation and determination of seventeen trace metals in thorium oxide matrix using a novel extractant—Cyanex-923	Atomic Emission Spectrometry (AES)	[112]
	Determination of Th and U	AES with MSF	[113]
	Determination of trace thorium in uranium dioxide	AES	[114]
	Determination of REE, U, Th, Ba and Zr in simulated hydrogeological leachates	AES after matrix solvent extraction	[115]
	Determination of thorium and light rare-earth elements in soil water and its high molecular mass organic fractions	MS and on-line-coupled size-exclusion chromatography	[116]
	Determination of trace thorium and uranium impurities in scandium with high matrix	Optical Emission Spectrometry (OES)	[117]
	Determination of thorium (IV), titanium (IV), iron (III), lead (II) and chromium (III) on 2-nitroso-1-naphthol-impregnated MCI GEL CHP20P resin	Preconcentration and MS	[118]
	Trace metal determination in uranium and thorium compounds without prior matrix separation	Electrothermal vaporization and AES	[119]
Atomic Absorption Spectrometry (AAS)	Thorium, zirconium and vanadium as chemical modifiers in the determination of arsenic	Electrothermal atomization	[120]
Cyclic Voltametric (CV)	Application in some nuclear material characterizations	Uranyl ion in sulfuric acid solutions	[121]
Chemically Modified Electrode (CME)	Determination of thorium by adsorptive type	Poly-complex system	[122]
Fluorogenic thorium sensors	Based on 2,6-pyridinedicarboxylic acid-substituted tetraphenylethenes	Induced emission characteristics	[123]
Selective optode	Design and evaluation of thorium (IV)	Membrane was prepared by incorporating 4-(p-nitrophenyl azo)–pyrocatechol	[124]
Micellar electrokinetic chromatographic	Ore and fish samples	Analysis of Th, U, Cu, Ni, Co and FE	[125]
Laser-induced breakdown spectrometry	Determination of trace constituents in thoria	Determination of thorium or uranyl ions	[126,127]
Electrochemical and spectro-electrochemical	Studies of bis(diketonate) thorium (IV) and uranium (IV) porphyrins	Complexes were synthesized using a hexa-aza porphyrin	[128]
Electrochemically modified detector	Elemental analysis of actinides	Graphite electrode with phthalocyanine	[129]
Selective extraction and trace determination of thorium	Synthesis of its application in water samples by spectrophotometry	UiO-66-OH zirconium MOF	[130]
Anodic polarization of thorium	Study of tungsten, cadmium and thorium electrodes	Electrochemical impedance spectroscopy	[131]
High-performance liquid chromatography	Studies on lanthanides, uranium and thorium	Amide-modified reversed phase supports	[132]
Ion exchange	Extraction of thorium on resin	Available extraction chromatographic resin	[133]
	Separation of actinium from proton-irradiated thorium metal	Extraction chromatography	[134]

The presented analytical methods offer, in addition to the information needed for the specific determinations of thorium in various matrices, possible ways of approaching separation through membrane techniques alternative to extraction, ion exchange, adsorption or chromatography.

7. Thorium Separation and/or Pre-Concentration

In analytical or technological research, the concentration of thorium has constituted a special problem, as it accompanies the rare earth elements (REEs), especially uranium [135,136].

The concentration process that is often followed and technologically supported is the pre-concentration of thorium [137–139].

Table 3 presents results that can form the basis of the development of urban thorium mining and that are focused on the concentration, pre-concentration, separation, extraction, ion exchange, sorption and bio-sorption of thorium or thorium–uranium from various samples and aqueous solutions [140–170].

Table 3. Concentration and separation of thorium using various techniques and selective materials.

Processes/Methods/Techniques	Materials	Characteristics	Refs.
Thorium removal	Different adsorbents	Activated carbons and zeolites (natural and synthetic)	[140]
Removal of thorium (IV) from aqueous Solutions	Modification of clinoptilolite as a robust adsorbent	Highly efficient thorium removal material	[141]
Preconcentration of uranium in natural water samples	New polymer with imprinted ions	Determination by digital imaging	[142]
Adsorption of trace thorium (IV) from aqueous solution	Mono-modified β-cyclodextrin polyrotaxane	Using response surface methodology (RSM)	[143]
Preconcentration and separation of actinides	Novel malonamide-grafted polystyrene-divinyl benzene resin	For hexavalent and tetravalent actinides such as U (VI), Th (IV) and Pu (IV)	[144]
Comparative adsorption	Mesoporous Al_2O_3	Selectivity of Th (IV) compared U (VI), La (III), Ce (III), Sm (III) and Gd (III)	[145]
Extraction and precipitation agents	α-aminophosphonates, -phosphinates, and -phosphine oxides	For rare earth metals, thorium, and uranium	[146]
Removal of polyvalent metal ions	Polyurea-crosslinked alginate aerogels	Eu (III) and Th (IV) from aqueous solutions	[147]
Method for separating thorium	Patented Chinese method	Separating cerium-fluoride and thorium	[148]
Extraction and recovery of cerium (IV) and thorium (IV)	α-aminophosphonate extractant	Extraction and recovery of Ce (IV) and Th (IV) from sulphate medium	[149]
Selective extraction and separation	Sulfate medium using Di(2-ethylhexyl)-N-heptylaminomethylphosphonate	Ce (IV) and Th (IV) from RE (III)	[150]
	α-aminophosphonate extractant	Ce (IV) from thorium and trivalent rare earths	[151]
	α-aminophosphonic acid HEHAPP	Heavy rare earths from chloride medium	[152]
	α-aminophosphonic acid extractant HEHAMP	Rare earths from chloride media	[153]
Study of thorium adsorption	PAN/zeolite composite adsorbent	Adsorption model	[154]
	Tulul Al-Shabba Zeolitic Tuff, Jordan	Adsorption of Th (IV) and U (VI)	[155]
	Sodium clinoptilolite	Removal of Th from aqueous solutions	[156]
	Modification of zeolite	Using tandem acid-base treatments	[157]
Selective cloud point extraction of thorium (IV)	Tetraazonium-based ionic liquid	Thorium extraction isotherm	[158]
Removal of thorium (IV) from aqueous solutions	Deoiled karanja seed cake	Optimization using Taguchi method	[159]
Retention of uranyl and thorium ions from radioactive solution	Peat moss	Retention of uranyl and Th ions from radioactive solution	[160]
Photocatalysis and adsorption	Photo-responsive metal-organic frameworks (MOFs)	Design strategies and emerging applications	[161]
Electrochemical and electrolytic separation	Th (IV) and Ce (III) in $ThF_4{}^-CeF_3$-LiCl-KCl quaternary melt	Separation of Th (IV) and Ce (III)	[162]

Table 3. Cont.

Processes/Methods/Techniques	Materials	Characteristics	Refs.
Selective removal	Hybrid mesoporous adsorbent as benzenesulfonamide-derivative@ZrO2	Thorium ions from aqueous solutions	[163]
Extraction	Sodium diethyldithiocarbamate/polyvinyl chloride	Rare earth group separation from lamprophyre dyke leachate	[164]
Fluorescent sensors	Metal-organic framework (MOF)	Hazardous material detection	[165]
Zeolite adsorption	Separation of radionuclides	From a REE-containing solution	[166]
Equilibrium study	Acidic (chelating) and organophosphorus ligands	Equilibrium constants of mixed complexes of REE	[167]
Molecule for solvent extraction of metals	Thenoyltrifluoroacetone	Thorium extraction	[168]
Chemical adsorption	8-Hydroxyquinoline immobilized bentonite	Removal of U and Th from their aqueous solutions	[169]

The remarkable results in the development of organic ligands (Table 2), and especially of selective materials (Table 3), allow a confident approach to the recovery and recycling of thorium from electrical and electronic waste, but more generally (considering the slightly selective separation of waste) of residues that reach the integrated municipal storage and waste platforms (especially from construction).

8. Membrane and Membrane Processes

Membranes and membrane processes can be an attractive alternative for the separation of chemical species containing thorium from various sources, with reduced concentrations in this element. On the one hand, membranes can integrate into the classical technologies for obtaining thorium; on the other hand, the speculations that can be made between thorium and various complexants are compatible with membrane separations.

In order to highlight these aspects, this subchapter presents some characteristics of the main membrane processes.

Membranes and processes have evolved from laboratory-scale installations to industrial ones, having at the same time an increased economic and commercial importance [170]. Membrane processes have not only replaced some of the conventional separation processes but also have produced remarkable results in areas where conventional techniques are exhausted or very expensive [171]. Among the problems that have determined the exponential development of membrane processes are those of environmental protection, since technologies based on membranes and membrane separation techniques are recognized as ecological technologies [172].

8.1. Introduction to Membranes and Membrane Processes

If we focus on membrane processes, it can be stated that the membrane is a window of a multi-component system (Figure 5), with selective permeability for chemical species of the system [173]. This membrane allows the separation of the considered system, consisting of a continuous phase (solvent) in which ionic chemical species, molecules and macromolecules are dissolved. At the same time, molecular aggregates and dispersed particles can be separated into components by classical or membrane processes [174]. In order for the separation process to occur, the system must be subjected to an electrochemical potential difference or driving force ($\Delta\mu$) [175].

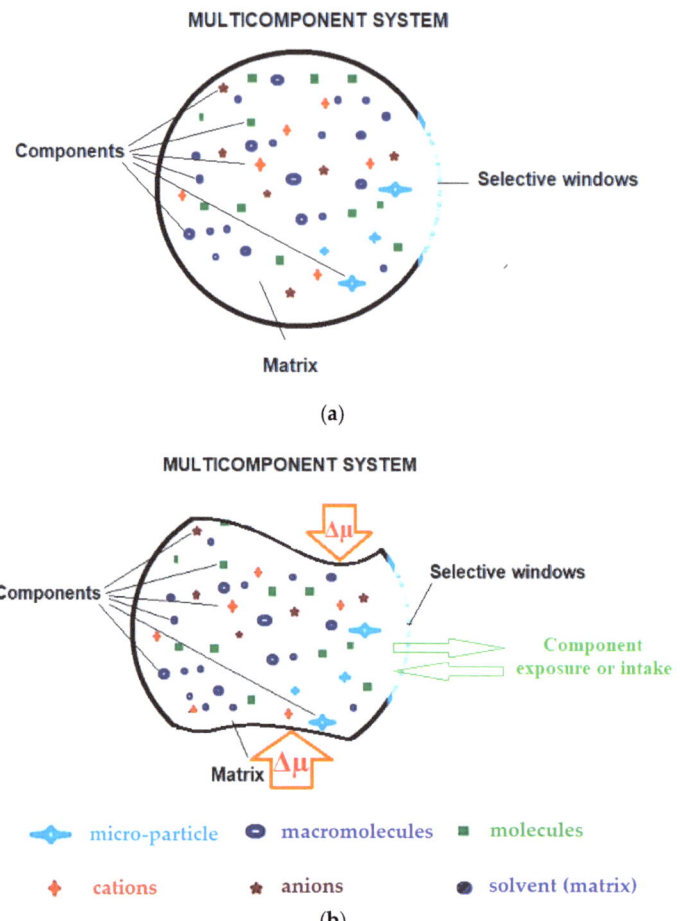

Figure 5. Multicomponent system bordered by a selective window, including ions, small molecules, macromolecules, nanoparticles, microparticles, microorganisms and viruses as suspended particles: (**a**) system in equilibrium; (**b**) system subject to an electrochemical potential difference ($\Delta \mu$). The meaning of shapes and symbols in Figure 5 is as follows.

The most important driving forces on membrane processes are as follows [176]:

- P = transmembrane pressure difference;
- Δc = concentration difference between the two compartments separated by a membrane;
- ΔE = potential difference.

It should be emphasized that in the last decade membrane processes involving potential gradient, thermal, magnetic, and interfacial tension, and volatility have undergone significant development [177].

In this subchapter, we will briefly present the essential aspects of the processes involving pressure or concentration gradient (liquid membranes).

8.2. Barro Membrane Processes

In membrane processes, the pressure difference (Δp) constitutes a technically and economically accessible driving force, leading to many applications, including microfiltration, ultrafiltration, nanofiltration and reverse osmosis (hyperfiltration) [178]. The first and most developed application was the obtaining of drinking water from sea water (Figure 6a), when it was found that, by applying a pressure higher than the osmotic pressure of sea water, most of the solvent passes (96–99%) through a semi-permeable membrane [179]. While these processes have applications on an industrial scale, their introduction in a certain technology presents a flow optimization problem (Figure 6b,c) [178–180], which depends on the load in the chemical species to be removed from the solvent that constitutes the feed [181]. There is the option of operating using dead-end filtration of a cross-flow filtration system [182]. The design of filtration devices may differ; chemical equipment manufacturers compete to create prototypes with increasingly high performance by improving the flow on the membrane (Figure 6d,e) [183,184]. In filtration processes, regardless of preventative efforts, the membrane becomes dirty or clogs, or concentration polarization (solute accumulation) occurs on the layer adjacent to the membrane; thus, process engineering is complemented by the introduction of ultrasonic cleaning devices into the technology, cavitation, magnetic stirring, or pulsatile flow vibrations [185].

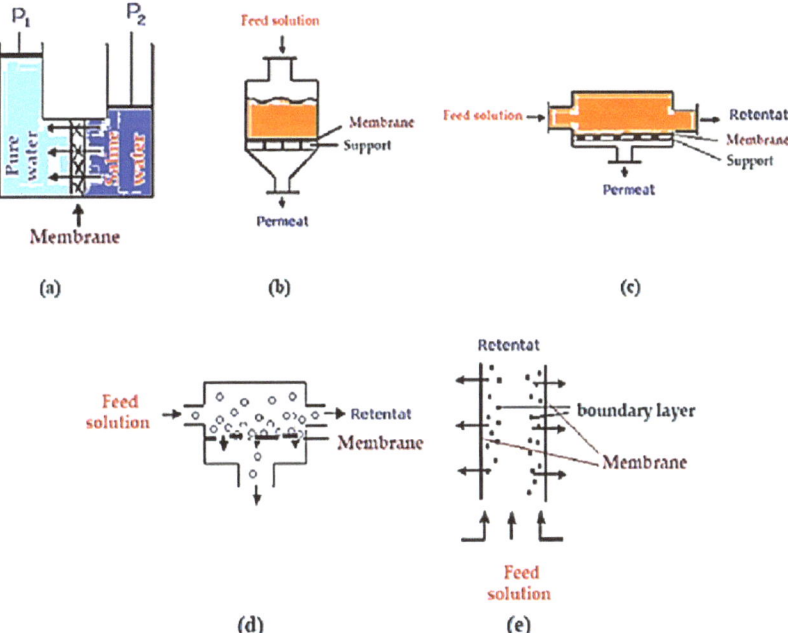

Figure 6. Membrane separation processes under pressure difference: (**a**) obtaining drinking water through reverse osmosis; (**b**) piston type (dead-end filtration); (**c**) tangential flow; (**d**) tangential flow through large sections; (**e**) flow through tubes.

However, in essence, the feeding can be done through large cylindrical, tubular, spiraled or capillary (hollow fiber) spaces, in which, along with the flow through and/or on the membrane and avoiding fouling (contamination, soiling), the aim is to increase the area of the contact surface of the membrane with the dispersed system of feeding (Figure 7) [186]. Of course, the operation can be done by introducing the feeding solution, as in Figure 7, but most often, with the feed solution being dirtier, it is inserted between tubes or fibers for a possible physical cleaning [187–189].

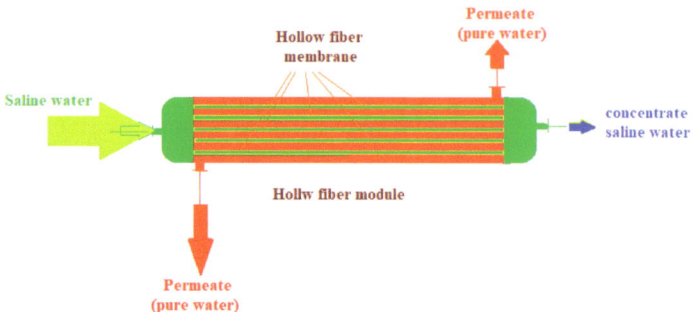

Figure 7. Advanced hollow-fiber filtration module.

A homogenous system can be separated by aggregation (segregation), so that instead of a high-pressure process (Table 4), a lower-pressure one is used [186,187].

Table 4. Characteristics of pressure gradient processes.

Type of Membrane Process	Pore Diameter (nm)	Pressure (Bar)	Obtained Water Content
Reverse osmosis	<0.6	25–60	Pure water (poorly ionized)
Nanofiltration	0.5–10	6–30	Pure water (traces of molecular substances)
Ultrafiltration	7–200	4–15	Pure water, molecular substances and macromolecules
Microfiltration	150–5000	0.1–2.5	Pure water, molecular substances and colloids

The first processes of this kind were promoted by Schamehorn, the ultrafiltration of micellar systems (MUF), which consisted of transforming a solution into an ultra-microdispersed system by adding suitable surfactants, followed by ultrafiltration [190–192].

The condition for using micellar ultrafiltration is that the micelles contain the organic compound, which means an impurity of the concentrate [193].

The variants of ultrafiltration and nanofiltration have undergone significant development due to nano-species and nanomaterials (nanoparticles, nanotubes, nanofibers, proteins, soluble polymers, polyelectrolytes, micelles and vesicles) also being used as carriers (Figure 8) in processes in liquid membranes [194].

The concentration polarization and the diffusion effects related to the sizes of solutes with low molecular masses can influence the working conditions of nano- and ultrafiltration, with the number of additives required being determined experimentally [195].

8.3. Electro-Membrane Processes

Electrodialysis is the most widespread separation process, carried out under an electric potential gradient, which involves ion exchange membranes [196]. In electrodialysis, the extraction, reconcentration and substitution operations are carried out without direct intervention of the electrodes [197]. They are placed at the end of the electrodialysis cells in order to maintain the electric potential difference between the compartments separated by the membranes (Figure 9) [197–199].

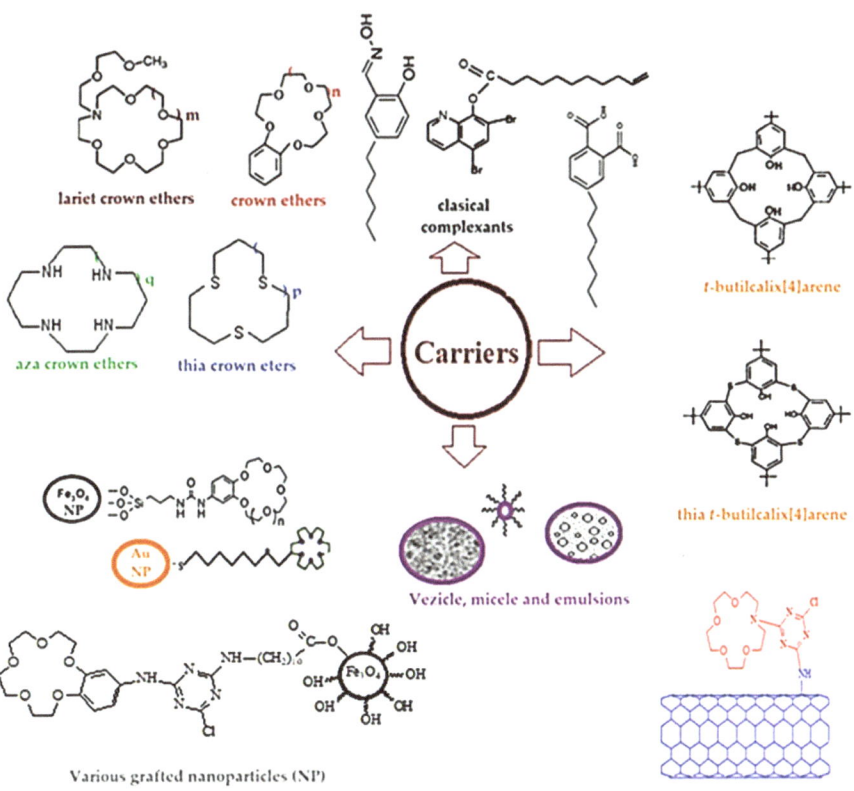

Figure 8. Common types of carriers: macrocyclic compounds, modified classical complexant agents and nano-species [194].

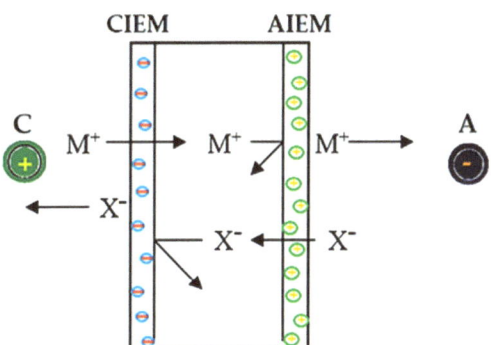

M = cation; X = anion; C = Cathode; A = Anode
AIEM = Anion Ion-Exchange Membrane; CIEM = Cation Ion-Exchange Membrane

Figure 9. Scheme of an electrolysis cell for the concentration of a salt by electrodialysis with two ion exchange membranes.

If we associate an anion exchange membrane with a cathode, it is possible to eliminate an electrolyte, whose cation can be deposited by electrochemical reaction on the cathode [200]. The electrolyte extracted from the diluted circuit by electrodialysis will be recovered in the concentrated circuit according to the principle in Figure 9. This electrolyte will not only be recovered but can be reconcentrated. Recovery and reconcentration are possible because the ions cannot migrate over their compartment, the M cation being retained by the anion exchange membrane, and the X anion by the cation exchange membrane [201,202].

Conducting electrodialysis requires ways to interpose electrodes, aqueous phases to be processed, and membranes, so that the operation can lead at the same time to solute concentrations or to the recovery of deionized water [203–205].

8.4. Membrane Processes Carried out under a Concentration Gradient (Liquid Membrane)

Although the concentration gradient is also found in processes with solid membranes (osmosis, dialysis, forward osmosis), this paper addresses processes with liquid membranes that have a high chance of developing applications in the valorization of thorium [206].

Separation systems with a liquid membrane (LM) or bulk liquid membrane (BLM) are formed by two homogenous liquid phases, immiscible with the membrane, called the source phase (SP) and the receiving phase (RP). The separation of the two liquid membranes is achieved with a third liquid, the membrane (M), which acts as a semi-permeable barrier between the two liquid phases [207–231].

An established graphic but also practical conception of liquid membranes (Figure 10) takes into account the density of the membrane, which is generally an organic solvent or a multicomponent system in which the continuous phase is the organic solvent [208].

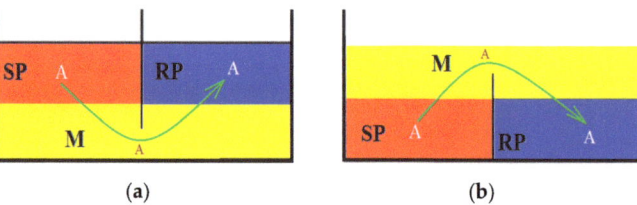

Figure 10. Schematic presentation of membrane systems with an organic solvent: denser (**a**) or less dense (**b**) than aqueous phases. Legend: M = membranes; SP = source phase; RP = receiving phase; A = chemical species of interest for separation [214].

The density of the membrane phase becomes unimportant if the membrane solvent is immobilized in or on a support [209], thus obtaining supported liquid membranes (SLMs). An interesting variant, but not yet sufficiently evaluated in separation processes, is the liquid membrane based on magnetic liquid (ferrofluid) [210], which also has no restrictions on the density of the organic solvent but involves special aspects in terms of stability and the transfer of table [211].

If we focus on BLMs, the technical problems to be solved are as follows: the large volume of solvent used (V), the small mass transfer area (σ), the unit ratio between the volume of the source phase, the volume of the receiving phase (r) and the volume of the membrane organic solvent (OS or M) and, therefore, implicitly, the long operating time (t) [212].

In order to improve the performance, hollow-fiber supported membranes (HFLMs) and emulsion membranes (ELMs) have been greatly developed (Figure 11) [213].

Recently, a BLM system with dispersed phases was studied, in which the aqueous phases of the separation system dispersed in/through the membranes. The membrane is a nanodispersed system of magnetic nanoparticles that have the role of ensuring both convection and transport for ionic chemical species in membranes based on saturated alcohols C_6–C_{12} [214,215]. The most recent design is shown in Figure 12, but other variants using chemical nano-species are also used [216].

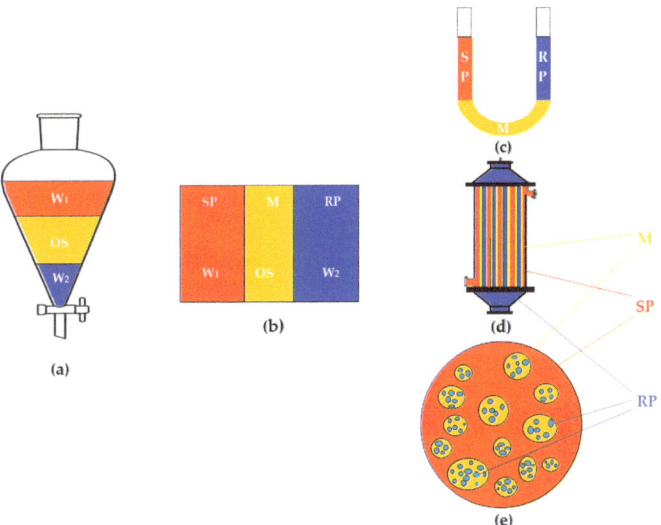

Figure 11. Schematic presentation of extraction and membrane systems with organic solvent: (**a**) water 1 (W1)–organic solvent (OS)–water extraction (W2); (**b**) liquid membranes (LMs); (**c**) bulk liquid membranes (BLMs); (**d**) supported liquid membranes (SLMs); (**e**) emulsion liquid membranes (ELMs). Legend: M = Membrane; SP = Source Phase; RP = Receiving Phase [214,216].

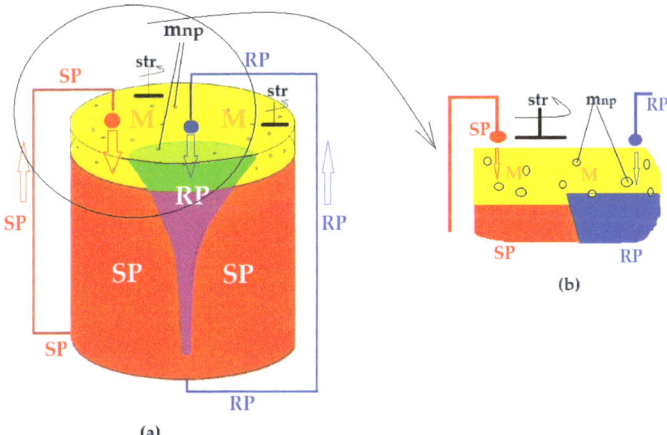

Figure 12. Schematic presentation of the permeation module with dispersed phases: (**a**) front view; (**b**) cross-section detail. Legend: SP—source phase; RP—receiving phase; M—organic solvent membrane; m_{np}—magnetic nanoparticles; str—stirrer with magnetic rods [216,217].

The BLM system with dispersed phases (based on Figure 8 carriers, for example) is close to the performance of liquid membranes on hollow-fiber supports or emulsion-type liquid membranes but has several limitations that restrict its applicability, including the stability of the membrane nanodispersion, control of the size of droplets in recirculating aqueous phases and losses of membrane material (solvent or nanoparticles) [217].

8.5. Transport in Liquid Membranes

The method of achieving the concentration gradient and the nature of the species dissolved in the phases of the membrane system have led to various types of transport through liquid membranes [218–220]. Mainly, however, they can be narrowed down to those specified in Figure 13.

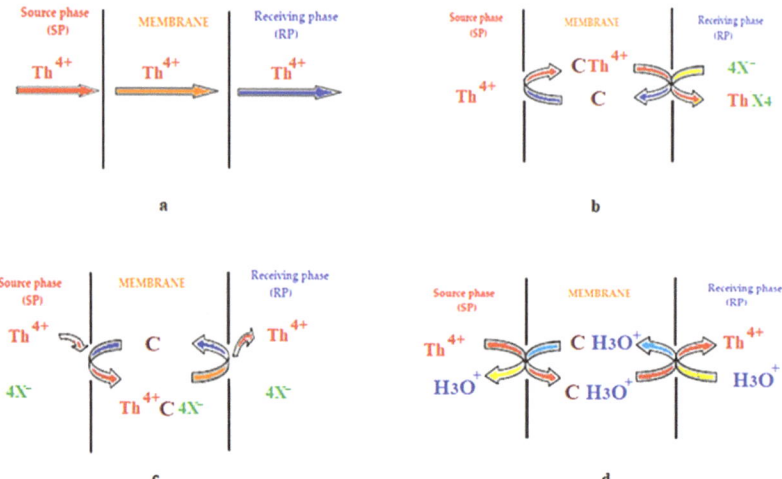

Figure 13. Schematic presentation of the transport mechanism by liquid membranes (C—carrier, X—anion complexant): (**a**) physical "simple" shipping; (**b**) transport with carrier; (**c**) coupled transport; (**d**) counter-transport.

8.5.1. Physical "Simple" Shipping

The simple diffusion type of transport through the solution is usually followed by the permeation of the solute through the liquid membrane due to the concentration gradient (Figure 13a). In this case, the transport of the component from the source phase through the membrane phase occurs with a higher solubility or diffusivity of the solute in the membrane phase. In this type of transport, the mass transfer rate is low and depends on the solubility of the solute in the organic phase, as well as the solubility of the solute in the source and receiver phases [221,222].

8.5.2. Facilitated Transport or Carrier-Mediated Transport

In carrier-mediated transport, a carrier is added to the membrane phase in order to increase the mass transfer rate or separation efficiency of the liquid membrane. It is also known as facilitated transport or transport mediated by a transporter [223]. In this case, the solute dissolved in the source phase, at the source phase–LM interface, reacts chemically with a transporter dissolved in the liquid membrane to form a complex. This complex reacts inversely at the LM–receiving phase interface, releasing the partitioned solute in the receiving phase (Figure 13b). In recent years, this type of transport mediated by a transporter has been intensively developed for the selective transport of cations, anions and neutral species through liquid membranes [224].

8.5.3. Coupled Co- or Counter-Transport

In this type of transport, the transport speed of a certain ion is dependent on the concentration of another ion. In the case of coupled co-transport, the metal ion is transferred together with a counter-anion, with the two species' transport taking place in the same direction. In the coupled counter-transport type, the simultaneous transport of another ion from the receptor phase to the source phase takes place; thus, the transport of the two species takes place in opposite directions [225–227]. Figure 13c,d shows the types of co- and counter-coupled transport of a metal ion.

8.6. Hybrid Membrane Processes

The most common form of treatment of effluents containing heavy metal ions involves the precipitation of metals as a hydroxide, base salt or sulfur. Precipitation is often followed by an additional treatment, such as sedimentation or filtration processes [228–230].

The technique of liquid membranes also presents a huge potential for the application of the removal and valorization of heavy metals, especially for the purpose of environmental protection [221–223].

Currently, the most important commercial application of liquid membrane technologies is the treatment of wastewater and waste [212,224–226].

However, the use of liquid membranes has encountered many obstacles, mainly related to the use of solvents with high toxicity [227,228]; both the reduction of the amount of the membrane solvent required and their replacement with green solvents or nanodispersions have created better opportunities for this process [229,230].

The idea of using nanosystems has led to the development of hybrid processes, which basically follow the mechanism of liquid membranes, but the process design is more advanced [211,231].

9. Problems in Application and Achievement as Well as Development Perspectives of Urban Thorium Mining

The analysis of the processes in which the minerals or waste containing thorium are processed shows that the classic technologies have material losses in the environment, which could be reduced with or through membrane techniques [232–255]. Thus, in the classical thorium recovery technologies, some disadvantages [1,13–20,47–66] of the operations are highlighted (Table 5), which require improvements, especially from the perspective of the loss of thorium in the environment.

Table 5. Possible losses of thorium in the environment and remedial possibilities.

Technological Operation	Losses of Thorium or of Thorium-Contaminated Materials	Means of Remediation or Reduction of Losses
Crushing, grinding	Dust removal Mill shutdown losses Losses when cleaning the machine	Microfilter installation Micro- and ultrafiltration of colloidal washing solutions
Solubilization or leaching	Incomplete solubilization with the chosen reagent Complete solubilization Insufficient concentration of thorium	Solubilization with a complementary reagent Selective reprecipitation and solubilization Concentration by precipitation and microfiltration
Filtration	Thorium retention in the precipitate Reduced concentration of thorium in the filtrate	Washing with solubilizing reagents Reprecipitation and micro- or ultrafiltration
Precipitation	Incomplete precipitation Precipitation of nanometric particles	Nanofiltration or reverse osmosis of the filtrate Colloidal ultrafiltration or nanofiltration
Extraction	Solvent losses Incomplete extraction	Solvent recovery Use of selective extractants
Ion exchange	Blockage of thorium in the ion exchanger (elution inefficiency) Incomplete retention	Change eluent Recovery of ion exchangers for destruction (burning)

The problem of thorium separation, concentration and recycling can be approached by analyzing some of the contributions that offer both priority research directions and viable technical solutions (Table 6) [232–255].

Table 6. Aspects regarding the use of membrane techniques and membrane materials, with possible implications regarding thorium separation.

Membrane Techniques	Materials and Applications	Characteristics	Refs.
Waste Treatment	Liquid radioactive waste treatment		[232]
Liquid Filtration	Membrane surface patterning as a fouling mitigation	Strategy for Processes	[233]
Ionic Liquid	Gas separation membranes		[234]
	Proton exchange membrane in fuel cells		[235]
	Chitosan-based polymers as proton exchange	Roles of Chitosan-Supported Polymers	[236]
	Based electrolytes for energy storage devices		[237]
	Toxicity to living organisms		[238]
Polymer Inclusion Membranes (PIMs)	Sequential determination of Copper (II) and Zinc (II) in natural waters and soil leachates	Chelating Resin	[239]
	Application in the separation of non-ferrous metal ions	Membranes (PIMs) Doped with Alkylimidazole	[240]
	Poly(vinylidene-fluoride-co-hexafluoropropylene) extraction from sulfate solutions	Containing Aliquat® 336 and Dibutyl Phthalate	[241]
Bulk Hybrid Liquid Membranes	Operational limits	Based on Dispersion Systems	[217,242]
	Thorium transport: modeling and experimental validation	Continuous Bulk Liquid Membrane Technique	[243]
Membrane Fabrication	Sustainable membrane development	Polymers and Solvents Used	[244]
Light-Responsive Polymer Membranes	Miscellaneous application	Report Recent Progress In The Research Field	[245]
Adsorptive Membranes and Materials	Modern computer applications	Model for Rare Earth Element Ions	[246]
Nanofiltration	Effect of the adsorption of multicharge cations on the selectivity	NF and Adsorption	[247]
	Extraction of uranium and thorium from aqueous solutions	NF and Extraction	[248]
	Removal of fluoride	By Nature, Diatomite From High-Fluorine Water	[249]
	Removal of radioactive contamination of groundwater, special aspects and advantages	Including RO	[250]
	U from seawater by nanofiltration	Selective Concentration	[251]
Glutathione-Based Magnetic Nanocomposite	Sequestration and recovery of Th ions	Using Recyclable, Low-Cost Materials	[252]
Zeolite Hybrid Adsorbent	Case study of thorium (IV)	Evaluation of Sodium Alginate/Polyvinyl Alcohol/Polyethylene Oxide/ZSM5 Zeolite Hybrid Adsorbent	[253]
Functionalized Maleic-Based Polymer	Thorium (IV) removal from aqueous solutions	Synthesis, Characterization and Evaluation of Thiocarbazide Functionalization	[254]
Electro-deionization (EDI)	Th removal from aqueous solutions by electro-deionization (EDI)	Use of Response Surface Methodology for Optimization of Thorium (IV)	[255]

Recent studies on the separation, concentration, removal or recovery of thorium from aqueous solutions, including by membrane techniques [256–271] (Table 7), have led to promising results, reinforcing the idea that membrane or hybrid processes can contribute to the imaging of the technological recycling of thorium from various residues, especially industrial, on municipal waste processing platforms.

Table 7. Recent materials and processes for thorium recovery.

Processes	Applications	Characteristics	Refs.
Solvent extraction and separation of thorium (IV)	Separation of thorium	From chloride media by a Schiff base	[256]
Leaching and precipitation of thorium ions	Th separation from Cataclastic rocks	Abu Rusheid Area, South Eastern Desert, Egypt	[257]
Ion exchange materials	Process for purification of 225Ac from thorium and radium radioisotopes	Evaluation of inorganic ion exchange materials	[258]
Adsorption	Thorium adsorption	Graphene oxide nanoribbons/manganese dioxide composite material	[259]
Adsorption	Thorium adsorption	Oxidized biochar fibers derived from Luffa cylindrica sponges	[260]
Adsorption	Sorption behavior of thorium (IV)	Activated bentonite	[261]
Adsorption	Adsorption of thorium (IV) response surface modelling and optimization	Amorphous silica	[262]
Adsorption	Th (IV) adsorption	Titanium tetrachloride-modified sodium bentonite	[263]
Adsorption	Evaluation of single and simultaneous thorium and uranium sorption from water systems	Electrospun PVA/SA/PEO/HZSM5 nanofiber	[264]
Synthesis and characterization of poly(TRIM/VPA)-functionalized graphene oxide nanoribbon aerogel	Highly efficient capture of thorium (IV)	Th ions separation from aqueous solutions	[265]
Vinyl-functionalized silica aerogel-like monoliths	Selective separation of radioactive thorium	Thorium separation from monazite	[266]
Recyclable GO@chitosan-based magnetic nanocomposite	Selective removal of uranium	From an aqueous solution of mixed radionuclides of uranium, cesium and strontium	[267]
Study of kinetics, thermodynamics, and isotherms of Sr adsorption	Graphene oxide (GO) and (aminomethyl) phosphonic acid–graphene oxide (AMPA–GO)	Th ion separation	[268]
Bulk liquid membrane containing Alamine 336 as a carrier	Kinetic study of uranium transport	Selectivity of the transport	[269]
Continuous bulk liquid membrane technique	Thorium transport	Modeling and experimental validation	[270]
Kinetic and isotherm analyses using response surface methodology (RSM)	Thorium (IV) adsorptive removal from aqueous solutions	By modified magnetite nanoparticles	[271]

The various compounds [272–277], technologies and processes [272–285] proposed recently, but also some previously used [286–293], can contribute to the construction of a scheme for recuperative separation of thorium on an integrated municipal platform for processing, mainly the waste of electrical devices (lamps, tubes and mantles) and electronics, but also those from the construction industry (welding electrodes, metallic materials and alloys).

In the diagrams in Figure 14, several proposals for technical solutions for urban thorium mining are presented, starting from the raw material: waste that ends up at integrated municipal waste management platforms.

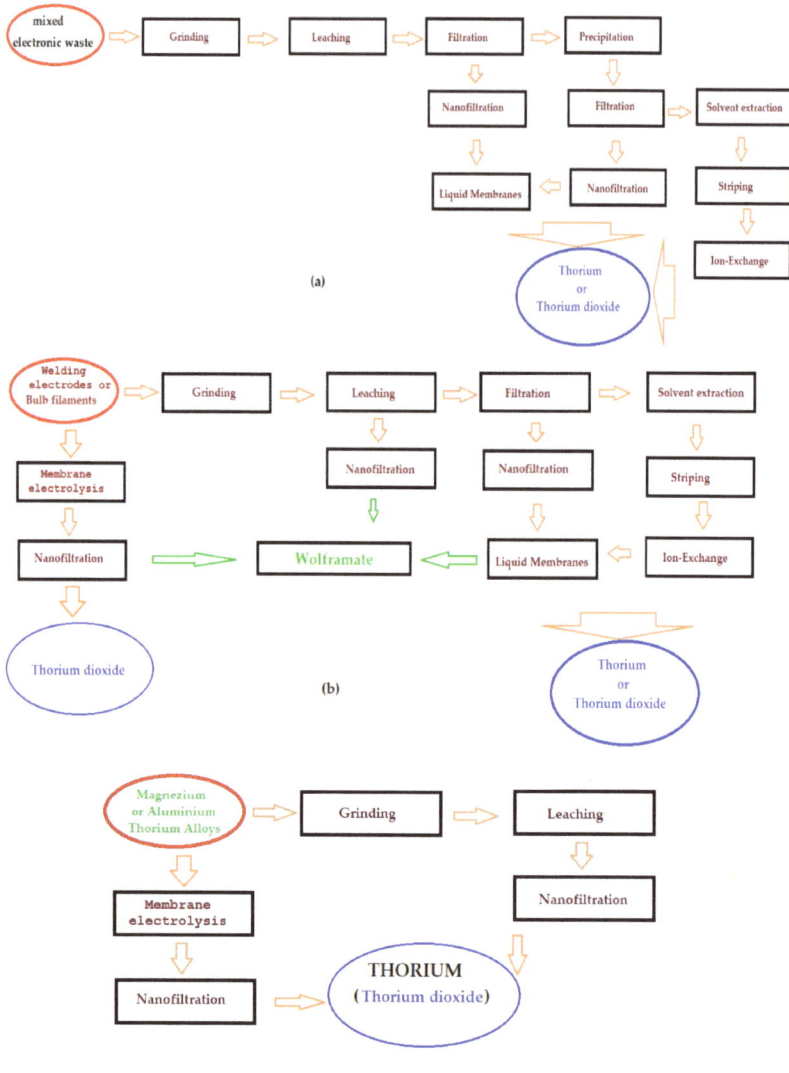

Figure 14. Scheme of proposals for the separation, recovery and recycling of thorium from waste of municipal waste management platforms: (**a**) valorization of thorium from unsorted waste; (**b**) recovery of thorium from electrodes and light bulb filaments; (**c**) valorization of thorium from magnesium or aluminum alloys.

Thus, in a first case (Figure 14a), it is assumed that the waste (assumed to be electrical and electronic waste or metal waste from metal construction materials) contains thorium and is totally unselected. This option would require the use of the classic scheme for the separation of thorium from poor sources, including the following operations: leaching, filtration (sedimentation), precipitation, filtration, solubilization at Th^{4+}, extraction, re-extraction and ion exchange. In this operating scheme, membrane processes that can be integrated to increase the performance of the process are nanofiltration and/or liquid membranes.

A second case may be a raw material containing thorium alloyed with tungsten (filaments of incandescent lamps or other lighting fixtures, welding electrodes or building material alloys). Figure 14b shows the main operations, which consist of leaching, filtration, precipitation, filtration, nanofiltration, solubilization at Th^{4+}, extraction and stripping or membrane electrolysis and nanofiltration.

The third possible case would be a raw material consisting of various wastes from aluminum and magnesium alloys (Figure 14c). Such a waste content can be processed for thorium recovery by membrane electrolysis or acid attack, followed by filtration and nanofiltration.

The proposed operation schemes are highly dependent on the quality of the waste selection that reaches the integrated municipal waste for management and processing platforms.

Certainly, some selection criteria for residues containing thorium can be taken into account in order to approach a treatment scheme as close as possible to the technological flows dedicated to obtaining this element.

Thus, we can consider that the entire deposit where thorium components were found, with a concentration above 1%, can be treated according to the classic acidic or basic digestion schemes.

When the waste deposit receives metal waste (residues of welding electrodes, metal alloys of thorium with aluminum or magnesium), a nitric acid digestion scheme followed by extraction and/or ion exchange will be required.

10. Conclusions

Although a radioactive element and a promising raw material for nuclear power generation, thorium, being a fairly abundant metal (similar to lead), has surprising domestic uses: toothpaste, dental cement, crucibles for high-temperature work, filaments for incandescent bulbs, welding electrodes, aluminum or magnesium alloys, jewelry, sculptures, coats and goggles, devices working at high temperatures and lamps for electronic devices.

Current regulations consider thorium to be a carcinogenic element, and its bio-toxicity and impact on human health (affects internal organs and blood) require the recovery and recycling of thorium, especially in the case of waste from municipal management platforms.

Classical thorium recovery processes require acid or base attack on thorium-containing feedstock, filtration, re-solubilization, extraction and ion exchange.

A variety of complexants and transporters have been used for the separation and preconcentration of thorium (especially for its analysis), which leads to membrane applications (nanofiltration, colloidal ultrafiltration, liquid membranes, emulsion membranes) for thorium utilization.

Membrane processes can intervene throughout the thorium recovery and recycling stream, increasing the efficiency of the process and avoiding losses to the environment.

The proposed processing schemes for various wastes containing thorium highlight the possibility of removal, recovery and valorization of thorium, suggesting possible urban mining of this element.

Author Contributions: Conceptualization, G.T.M., A.C.N., A.R.G., V.-A.G. and G.N.; methodology P.C.A., S.-K.T. and M.-R.I.; writing—original draft preparation, G.T.M., P.C.A., S.-K.T, M.-R.I., A.C.N., A.R.G., V.-A.G. and G.N.; resources, G.N.; writing—review and editing, V.-A.G. and G.N.; supervision, A.R.G., A.C.N. and G.N. All authors have read and agreed to the published version of the manuscript.

Funding: This research received no external funding.

Institutional Review Board Statement: Not applicable.

Data Availability Statement: Data are contained within the article.

Conflicts of Interest: The authors declare no conflict of interest.

Abbreviations

AAS	Atomic Absorption Spectrometry
BLM (MLV)	Bulk Liquid Membrane
CME	Chemically Modified Electrode
CV	Cyclic Voltammetry
D	Dialysis
DM	Membrane Distillation
E	Extraction
ED	Electrodialysis
EDI	Electro-deionization
ELM	Emulsion Liquid Membrane
F	Filtration
G	Grinding
HFLM	Hollow Fiber Liquid Membrane
HLM	Hybrid Liquid Membrane
ICP-AES	Inductively Coupled Plasma–Atomic Emission Spectrometry
ICP-MS	Inductively Coupled Plasma–Mass Spectrometry
ICP-OES	Inductively Coupled Plasma–Optical Emission Spectrometry
IE	Ion Exchange
M	Milling
MF	Microfiltration
MOF	Metal-Organic Framework
MUF	Micellar Ultra-Filtration system
N	Neutralization
NF	Nanofiltration
P	Precipitation
PV	Pervaporation
RE	Re-Extraction
REE	Rare Earth Element
RO	Reverse Osmosis
S	Striping
SG	Gas Separation
TBP	Tri-Butyl Phosphate
UF	Ultrafiltration

References

1. Jyothi, R.K.; De Melo, L.G.T.C.; Santos, R.M.; Yoon, H.S. An overview of thorium as a prospective natural resource for future energy. *Front. Energy Res.* **2023**, *11*, 1132611. [CrossRef]
2. Takaki, N.; Mardiansah, D. Core Design and Deployment Strategy of Heavy Water Cooled Sustainable Thorium Reactor. *Sustainability* **2012**, *4*, 1933–1945. [CrossRef]
3. Takaki, N.; Sidik, P.; Sekimoto, H. *Feasibility of Water Cooled Thorium Breeder Reactor Based on LWR Technology*; American Nuclear Society: La Grange Park, IL, USA, 2007; pp. 1733–1738.
4. Sidik, P.; Takaki, N.; Sekimoto, H. Feasible region of design parameters for water cooled thorium breeder reactor. *J. Nucl. Sci. Technol.* **2007**, *44*, 946–957.
5. Sidik, P.; Takaki, N.; Sekimoto, H. Breeding capability and void reactivity analysis of heavy-water-cooled thorium reactor. *J. Nucl. Sci. Technol.* **2008**, *45*, 589–600.
6. Ault, T.; Krahn, S.; Croff, A. Comparing the environmental impacts of uranium- and thorium-based fuel cycles with different recycle options. *Prog. Nucl. Energy* **2017**, *100*, 114–134. [CrossRef]
7. Perron, R.; Gendron, D.; Causey, P.W. Construction of a thorium/actinium generator at the Canadian Nuclear Laboratories. *Appl. Radiat. Isot.* **2020**, *164*, 109262. [CrossRef]
8. Von Gunten, H.R.; Roessler, E.; Lowson, R.T.; Reid, P.D.; Short, S.A. Distribution of uranium- and thorium series radionuclides in mineral phases of a weathered lateritic transect of a uranium ore body. *Chem. Geol.* **1999**, *160*, 225–240. [CrossRef]
9. Available online: https://scholar.google.com/scholar?q=Thorium+reserve&hl=en&as_sdt=0,5 (accessed on 24 June 2023).
10. Available online: https://scholar.google.com/scholar?hl=en&as_sdt=0%2C5&q=%22Thorium%22&btnG= (accessed on 24 June 2023).

11. Corb Aron, R.A.; Abid, A.; Vesa, C.M.; Nechifor, A.C.; Behl, T.; Ghitea, T.C.; Munteanu, M.A.; Fratila, O.; Andronie-Cioara, F.L.; Toma, M.M.; et al. Recognizing the Benefits of Pre-/Probiotics in Metabolic Syndrome and Type 2 Diabetes Mellitus Considering the Influence of *Akkermansia muciniphila* as a Key Gut Bacterium. *Microorganisms* **2021**, *9*, 618. [CrossRef] [PubMed]
12. Phillip, E.; Choo, T.F.; Khairuddin, N.W.A.; Abdel Rahman, R.O. On the Sustainable Utilization of Geopolymers for Safe Management of Radioactive Waste: A Review. *Sustainability* **2023**, *15*, 1117. [CrossRef]
13. Kusumkar, V.V.; Galamboš, M.; Viglašová, E.; Daňo, M.; Šmelková, J. Ion-Imprinted Polymers: Synthesis, Characterization, and Adsorption of Radionuclides. *Materials* **2021**, *14*, 1083. [CrossRef]
14. Santiago-Aliste, A.; Sánchez-Hernández, E.; Langa-Lomba, N.; González-García, V.; Casanova-Gascón, J.; Martín-Gil, J.; Martín-Ramos, P. Multifunctional Nanocarriers Based on Chitosan Oligomers and Graphitic Carbon Nitride Assembly. *Materials* **2022**, *15*, 8981. [CrossRef]
15. Chen, Y.; Chen, Y.; Lu, D.; Qiu, Y. Synthesis of a Novel Water-Soluble Polymer Complexant Phosphorylated Chitosan for Rare Earth Complexation. *Polymers* **2022**, *14*, 419. [CrossRef] [PubMed]
16. Inman, G.; Nlebedim, I.C.; Prodius, D. Application of Ionic Liquids for the Recycling and Recovery of Technologically Critical and Valuable Metals. *Energies* **2022**, *15*, 628. [CrossRef]
17. Brewer, A.; Florek, J.; Kleitz, F. A perspective on developing solid-phase extraction technologies for industrial-scale critical materials recovery. *Green Chem.* **2022**, *24*, 2752–2765. [CrossRef]
18. Yudaev, P.; Chistyakov, E. Chelating Extractants for Metals. *Metals* **2022**, *12*, 1275. [CrossRef]
19. Varbanov, S.; Tashev, E.; Vassilev, N.; Atanassova, M.; Lachkova, V.; Tosheva, T.; Shenkov, S.; Dukov, I. Synthesis, characterization and implementation as a synergistic agent in the solvent extraction of lanthanoids. *Polyhedron* **2017**, *134*, 135. [CrossRef]
20. Xiong, X.H.; Tao, Y.; Yu, Z.W.; Yang, L.X.; Sun, L.J.; Fan, Y.L.; Luo, F. Selective extraction of thorium from uranium and rare earth elements using sulfonated covalent organic framework and its membrane derivate. *Chem. Eng. J.* **2020**, *15*, 123240. [CrossRef]
21. Cazalaà, J.B. Radium and thorium applications for the general public: Unexpected consequences of the discovery from Pierre and Marie Curie. *J. Am. Soc. Anesthesiol.* **2012**, *117*, 1202. [CrossRef]
22. Pohjalainen, I.; Moore, I.D.; Geldhof, S.; Rosecker, V.; Sterba, J.; Schumm, T. Gas cell studies of thorium using filament dispensers at IGISOL. *Nucl. Instrum. Methods Phys. Res. Sect. B Beam Interact. Mater. At.* **2020**, *484*, 59–70. [CrossRef]
23. Lavi, N.; Alfassi, Z.B. Development of Marinelli beaker standards containing thorium oxide and application for measurements of radioactive environmental samples. *Radiat. Meas.* **2005**, *39*, 15–19. [CrossRef]
24. Li, Z.J.; Guo, X.; Qiu, J.; Lu, H.; Wang, J.Q.; Lin, J. Recent advances in the applications of thorium-based metal–organic frameworks and molecular clusters. *Dalton Trans.* **2022**, *51*, 7376–7389. [CrossRef]
25. Hassan, H.J.; Hashim, S.; Sanusi, M.S.M.; Bradley, D.A.; Alsubaie, A.; Tenorio, R.G.; Bakri, N.F.; Tahar, R.M. The Radioactivity of Thorium Incandescent Gas Lantern Mantles. *Appl. Sci.* **2021**, *11*, 1311. [CrossRef]
26. Das, S.; Kaity, S.; Kumar, R.; Banerjee, J.; Roy, S.B.; Chaudhari, G.P.; Daniel, B.S.S. Determination of Room Temperature Thermal Conductivity of Thorium—Uranium Alloys. In *Thorium—Energy for the Future: Select Papers from ThEC15*; Springer: Singapore, 2019; pp. 277–286.
27. Dabare, P.R.; Mahakumara, P.D.; Mahawatte, P. Method Validation of In-Situ Gamma Spectroscopy for Quantification of Naturally Occurring Radioactive Materials (NORM) K-40, TH-232 and U-238 in Soil. In Proceedings of the International Conference on Management of Naturally Occurring Radioactive Material (NORM) in Industry, Vienna, Austria, 18–30 October 2020.
28. Vasiliev, A.N.; Severin, A.; Lapshina, E.; Chernykh, E.; Ermolaev, S.; Kalmykov, S. Hydroxyapatite particles as carriers for 223 Ra. *J. Radioanal. Nucl. Chem.* **2017**, *311*, 1503–1509. [CrossRef]
29. Peterson, S.; Adams, R.E.; Douglas, D.A., Jr. Properties of Thorium, Its Alloys, and Its Compounds 1965, (No. ORNL-TM-1144), Oak Ridge National Lab., Tenn. Available online: https://www.osti.gov/servlets/purl/4622065 (accessed on 24 June 2023).
30. Zhang, Y.; Shao, X.; Yin, L.; Ji, Y. Estimation of Inhaled Effective Doses of Uranium and Thorium for Workers in Bayan Obo Ore and the Surrounding Public, Inner Mongolia, China. *Int. J. Environ. Res. Public Health* **2021**, *18*, 987. [CrossRef] [PubMed]
31. Stojsavljević, A.; Borković-Mitić, S.; Vujotić, L.; Grujičić, D.; Gavrović-Jankulović, M.; Manojlović, D. The human biomonitoring study in Serbia: Background levels for arsenic, cadmium, lead, thorium and uranium in the whole blood of adult Serbian population. *Ecotoxicol. Environ. Saf.* **2019**, *169*, 402–409. [CrossRef] [PubMed]
32. International Commission on Radiological Protection (ICRP). ICRP Publication 142: Radiological protection from naturally occurring radioactive material (NORM) in industrial processes. *Ann. ICRP* **2019**, *48*, 5–67. [CrossRef] [PubMed]
33. Guo, P.; Duan, T.; Song, X.; Xu, J.; Chen, H. Effects of soil pH and organic matter on distribution of thorium fractions in soil contaminated by rare-earth industries. *Talanta* **2008**, *77*, 624–627. [CrossRef]
34. Hayes, C.T.; Fitzsimmons, J.N.; Boyle, E.A.; McGee, D.; Anderson, R.F.; Weisend, R.; Morton, P.L. Thorium isotopes tracing the iron cycle at the Hawaii Ocean Time-series Station ALOHA. *Geochim. Cosmochim. Acta* **2015**, *169*, 1–16. [CrossRef]
35. Suarez-Navarro, J.A.; Pujol, L.; Suarez-Navarro, M.J. Determination of specific alpha-emitting radionuclides (uranium, plutonium, thorium and polonium) in water using [Ba+Fe]-coprecipitation method. *Appl. Radiat. Isot.* **2017**, *130*, 162–171. [CrossRef]
36. Chaudhury, D.; Sen, U.; Sahoo, B.K.; Bhat, N.N.; Kumara, S.; Karunakara, N.; Biswas, S.; Shenoy, S.; Bose, B. Thorium promotes lung, liver and kidney damage in BALB/c mouse via alterations in antioxidant systems. *Chem. Biol. Interact.* **2022**, *363*, 109977. [CrossRef]

37. Wickstroem, K.; Hagemann, U.B.; Kristian, A.; Ellingsen, C.; Sommer, A.; Ellinger-Ziegelbauer, H.; Wirnitzer, U.; Hagelin, E.-M.; Larsen, A.; Smeets, R.; et al. Preclinical Combination Studies of an FGFR2 Targeted Thorium-227 Conjugate and the ATR Inhibitor BAY 1895344. *Int. J. Radiat. Oncol. Biol. Phys.* **2019**, *105*, 410–422. [CrossRef] [PubMed]
38. Yantasee, W.; Sangvanich, T.; Creim, J.A.; Pattamakomsan, K.; Wiacek, R.J.; Fryxell, G.E.; Addleman, R.S.; Timchalk, C. Functional sorbents for selective capture of plutonium, americium, uranium, and thorium in blood. *Health Phys.* **2010**, *99*, 413–419. [CrossRef]
39. Kumar, A.; Mishra, P.; Ghosh, S.; Sharma, P.; Ali, M.; Pandey, B.N.; Mishra, K.P. Thorium-induced oxidative stress mediated toxicity in mice and its abrogation by diethylenetriamine pentaacetate. *Int. J. Radiat. Biol.* **2008**, *84*, 337–349. [CrossRef]
40. Chen, X.A.; Cheng, Y.E.; Rong, Z. Recent results from a study of thorium lung burdens and health effects among miners in China. *J. Radiol. Prot.* **2005**, *25*, 451. [CrossRef] [PubMed]
41. Zapadinskaia, E.E.; Gasteva, G.N.; Titiova, I.N. Analysis of health state in individuals exposed to thorium and chemical hazards in occupational environment. *Med. Tr. Promyshlennaia Ekol.* **2005**, *11*, 14–19.
42. Polednak, A.P.; Stehney, A.F.; Lucas, H.F. Mortality among male workers at a thorium-processing plant. *Health Phys.* **1983**, *44*, 239–251. [CrossRef] [PubMed]
43. Yin, L.L.; Tian, Q.; Shao, X.Z.; Shen, B.M.; Su, X.; Ji, Y.Q. ICP-MS measurement of uranium and thorium contents in minerals in China. *Nucl. Sci. Tech.* **2016**, *27*, 10. [CrossRef]
44. Shao, X.Z.; Xu, Y.; Zhang, Y.; Yin, L.L.; Kong, X.Y.; Ji, Y.Q. Monitoring of ultra-trace uranium and thorium in six-grade particles. *Chemosphere* **2019**, *233*, 76–80. [CrossRef]
45. Sharmin, F.; Ovi, M.H.; Shelley, A. Radiotoxicity analysis of the spent nuclear fuel of VVER-1200 reactor. *Prog. Nucl. Energy* **2023**, *156*, 104538. [CrossRef]
46. Gao, L.; Kano, N.; Sato, Y.; Li, C.; Zhang, S.; Imaizumi, H. Behavior and distribution of heavy metals including rare earth elements, thorium, and uranium in sludge from industry water treatment plant and recovery method of metals by biosurfactants application. *Bioinorg. Chem. Appl.* **2012**, *2012*, 173819. [CrossRef]
47. Lapidus, G.T.; Doyle, F.M. Selective thorium and uranium extraction from monazite: I. Single-stage oxalate leaching. *Hydrometallurgy* **2015**, *154*, 102–110. [CrossRef]
48. Lapidus, G.T.; Doyle, F.M. Selective thorium and uranium extraction from monazite: II. Approaches to enhance the removal of radioactive contaminants. *Hydrometallurgy* **2015**, *155*, 161–167. [CrossRef]
49. El-Nadi, Y.A.; Daoud, J.A.; Aly, H.F. Modified leaching and extraction of uranium from hydrous oxide cake of Egyptian monazite. *Int. J. Miner. Process.* **2005**, *76*, 101–110. [CrossRef]
50. Amer, T.E.; Abdella, W.M.; Wahab, G.M.A.; El-Sheikh, E.M. A suggested alternative procedure for processing of monazite mineral concentrate. *Int. J. Miner. Process.* **2013**, *125*, 106–111. [CrossRef]
51. Eyal, Y.; Olander, D.R. Leaching of uranium and thorium from monazite: I. Initial leaching. *Geochim. Cosmochim. Acta* **1990**, *54*, 1867–1877. [CrossRef]
52. Borai, E.H.; Abd El-Ghany, M.S.; Ahmed, I.M.; Hamed, M.M.; Shahr El-Din, A.M.; Aly, H.F. Modified acidic leaching for selective separation of thorium, phosphate and rare earth concentrates from Egyptian crude monazite. *Int. J. Miner. Process.* **2016**, *149*, 34–41. [CrossRef]
53. Parzentny, H.R.; Róg, L. The Role of Mineral Matter in Concentrating Uranium and Thorium in Coal and Combustion Residues from Power Plants in Poland. *Minerals* **2019**, *9*, 312. [CrossRef]
54. Selvig, L.K.; Inn, K.G.W.; Outola, I.M.J.; Kurosaki, H.; Lee, K.A. Dissolution of resistate minerals containing uranium and thorium: Environmental implications. *J. Radioanal. Nucl. Chem.* **2005**, *263*, 341–348. [CrossRef]
55. Hamed, M.G.; El-khalafawy, A.; Youssef, M.A.; Borai, E.H. Competitive sorption behavior of mono, di, and trivalent ions in highly acidic waste by polymeric resin based on crotonic acid prepared by gamma radiation. *Radiat. Phys. Chem.* **2023**, *212*, 111159. [CrossRef]
56. Shaeri, M.; Torab-Mostaedi, M.; Rahbar Kelishami, A. Solvent extraction of thorium from nitrate medium by TBP, Cyanex272 and their mixture. *J. Radioanal. Nucl. Chem.* **2015**, *303*, 2093–2099. [CrossRef]
57. Nasab, M.E. Solvent extraction separation of uranium (VI) and thorium (IV) with neutral organophosphorus and amine ligands. *Fuel* **2015**, *116*, 595–600. [CrossRef]
58. Alex, P.; Hubli, R.C.; Suri, A.K. Processing of rare earth concentrates. *Rare Met.* **2005**, *24*, 210–215.
59. Panda, R.; Kumari, A.; Jha, M.K.; Hait, J.; Kumar, V.; Kumar, J.R.; Lee, J.Y. Leaching of rare earth metals (REMs) from Korean monazite concentrate. *J. Ind. Eng. Chem.* **2014**, *20*, 2035–2042. [CrossRef]
60. Kul, M.; Topkaya, Y.; Karakaya, I. Rare earth double sulfates from pre-concentrated bastnasite. *Hydrometallurgy* **2008**, *93*, 129–135. [CrossRef]
61. Fourest, B.; Lagarde, G.; Perrone, J.; Brandel, V.; Dacheux, N.; Genet, M. Solubility of thorium phosphate-diphosphate. *New J. Chem.* **1999**, *23*, 645–649. [CrossRef]
62. Chi, R.; Xu, Z. A solution chemistry approach to the study of rare earth element precipitation by oxalic acid. *Met. Mater. Trans. B* **1999**, *30*, 189–195. [CrossRef]
63. Tomazic, B.; Branica, M. Separation of uranium(VI) from rare earths(III) by hydrolytic precipitation. *Inorg. Nucl. Chem. Lett.* **1968**, *4*, 377–380. [CrossRef]
64. Kang, M.J.; Han, B.E.; Hahn, P.S. Precipitation and adsorption of uranium (VI) under various aqueous conditions. *Environ. Eng. Res.* **2002**, *7*, 149–157.

65. Bearse, A.E.; Calkins, G.D.; Clegg, J.W.; Filbert, J.R.B. Thorium and rare earths from monazite. *Chem. Eng. Prog.* **1954**, *50*, 235–239.
66. Crouse, D.J.; Brown, K.B. The Amex Process for Extracting Thorium Ores with Alkyl Amines. *Ind. Eng. Chem.* **1959**, *51*, 1461–1464. [CrossRef]
67. Wang, X.; Huang, W.; Gong, Y.; Jiang, F.; Zheng, H.; Zhu, T.; Long, D.; Li, Q. Electrochemical behavior of Th (IV) and its electrodeposition from ThF4-LiCl-KCl melt. *Electrochim. Acta* **2016**, *196*, 286–293. [CrossRef]
68. Kumar, R.; Gupta, S.; Wajhal, S.; Satpati, S.K.; Sahu, M.L. Effect of process parameters on the recovery of thorium tetrafluoride prepared by hydrofluorination of thorium oxide, and their optimization. *Nucl. Eng. Technol.* **2022**, *54*, 1560–1569. [CrossRef]
69. Udayakumar, S.; Baharun, N.; Rezan, S.A.; Ismail, A.F.; Takip, K.M. Economic evaluation of thorium oxide production from monazite using alkaline fusion method. *Nucl. Eng. Technol.* **2021**, *53*, 2418–2425. [CrossRef]
70. Bahri, C.N.A.C.Z.; Al-Areqi, W.M.; Majid, A.A.; Ruf, M.I.F.M. Production of rare earth elements from Malaysian monazite by selective precipitation. *Malays. J. Anal. Sci.* **2016**, *20*, 44–50. [CrossRef]
71. Megawati, H.P.; Prasetiawan, H.; Triwibowo, B.; Hisyam, A. Simulation Study of Thorium Separation from Monazite Mineral. In Proceedings of the 7th Engineering International Conference on Education, Concept and Application on Green Technology (EIC 2018), Semarang, Indonesia, 18 October 2018; pp. 455–459, ISBN 978-989-758-411-4. [CrossRef]
72. Cabral, A.R.; Zeh, A. Karst-bauxite formation during the Great Oxidation Event indicated by dating of authigenic rutile and its thorium content. *Sci. Rep.* **2023**, *13*, 8368. [CrossRef]
73. Hak, C.R.C.; Januri, Z.; Mustapha, I.; Mahat, S. Extraction of Thorium Oxide (ThO$_2$) from Malaysian Monazite through Alkali Digestion: Physical and Chemical Characterization Using X-Ray Analysis. *Key Eng. Mater.* **2022**, *908*, 509–514. [CrossRef]
74. Alotaibi, A.M.; Ismail, A.F.; Aziman, E.S. Ultra-effective modified clinoptilolite adsorbent for selective thorium removal from radioactive residue. *Sci. Rep.* **2023**, *13*, 9316. [CrossRef]
75. Manchanda, V.K. Thorium as an abundant source of nuclear energy and challenges in separation science. *Radiochim. Acta* **2023**, *111*, 243–263. [CrossRef]
76. Rump, A.; Hermann, C.; Lamkowski, A.; Popp, T.; Port, M. A comparison of the chemo-and radiotoxicity of thorium and uranium at different enrichment grades. *Arch. Toxicol.* **2023**, *97*, 1577–1598. [CrossRef]
77. Schmidt, M.; Lee, S.S.; Wilson, R.E.; Soderholm, L.; Fenter, P. Sorption of tetravalent thorium on muscovite. *Geochim. Cosmochim. Acta* **2012**, *88*, 66–76. [CrossRef]
78. Chadirji-Martinez, K.; Grosvenor, A.P.; Crawford, A.; Chernikov, R.; Heredia, E.; Feng, R.; Pan, Y. Thorium speciation in synthetic anhydrite: Implications for remediation and recovery of thorium from rare-earth mine tailings. *Hydrometallurgy* **2022**, *214*, 105965. [CrossRef]
79. Khan, A.A.; Zada, Z.; Reshak, A.H.; Ishaq, M.; Zada, S.; Saqib, M.; Ismail, M.; Fazal-ur-Rehman, M.; Murtaza, G.; Zada, S.; et al. GGA and GGA + U Study of ThMn2Si2 and ThMn2Ge2 Compounds in a Body-Centered Tetragonal Ferromagnetic Phase. *Molecules* **2022**, *27*, 7070. [CrossRef] [PubMed]
80. Nisbet, H.; Migdisov, A.; Xu, H.; Guo, X.; van Hinsberg, V.; Williams-Jones, A.E.; Boukhalfa, H.; Roback, R. An experimental study of the solubility and speciation of thorium in chloride-bearing aqueous solutions at temperatures up to 250 C. *Geochim. Cosmochim. Acta* **2018**, *239*, 363–373. [CrossRef]
81. Oliveira, M.S.; Duarte, I.M.; Paiva, A.V.; Yunes, S.N.; Almeida, C.E. The role of chemical interactions between thorium, cerium, and lanthanum in lymphocyte toxicity. *Arch. Environ. Occup. Health* **2014**, *69*, 40–45. [CrossRef]
82. Ma, Y.; Wang, J.; Peng, C.; Ding, Y.; He, X.; Zhang, P.; Li, N.; Lan, T.; Wang, D.; Zhang, Z. Toxicity of cerium and thorium on daphnia magna. *Ecotoxicol. Environ. Saf.* **2016**, *134*, 226–232. [CrossRef]
83. Peng, C.; Ma, Y.; Ding, Y.; He, X.; Zhang, P.; Lan, T.; Wang, D.; Zhang, Z.; Zhang, Z. Influence of Speciation of Thorium on Toxic Effects to Green Algae *Chlorella pyrenoidosa*. *Int. J. Mol. Sci.* **2017**, *18*, 795. [CrossRef] [PubMed]
84. Hirose, K. Chemical speciation of trace metals in seawater: A review. *Anal. Sci.* **2006**, *22*, 1055–1063. [CrossRef] [PubMed]
85. Huang, W.; Hongwei, L.; Xin, X.; Guocheng, W. Development status and research progress in rare earth hydrometallurgy in China. *J. Chin. Rare Earth Soc. Chin.* **2006**, *24*, 129.
86. Santschi, P.H.; Murray, J.W.; Baskaran, M.; Benitez-Nelson, C.R.; Guo, L.D.; Hung, C.C.; Lamborg, C.; Moran, S.B.; Passow, U.; Roy-Barman, M. Thorium speciation in seawater. *Mar. Chem.* **2006**, *100*, 250–268. [CrossRef]
87. Vengosh, A.; Coyte, R.M.; Podgorski, J.; Johnson, T.M. A critical review on the occurrence and distribution of the uranium-and thorium-decay nuclides and their effect on the quality of groundwater. *Sci. Total Environ.* **2022**, *808*, 151914. [CrossRef]
88. Rao, T.P.; Metilda, P.; Gladis, J.M. Overview of analytical methodologies for sea water analysis: Part I—Metals. *Crit. Rev. in Anal. Chem.* **2005**, *35*, 247–288. [CrossRef]
89. Niazi, A.; Ghasemi, N.; Goodarzi, M.; Ebadi, A. Simultaneous spectrophotometric determination of uranium and thorium using arsenazo III by H-point standard addition method and partial least squares regression. *J. Chin. Chem. Soc.* **2007**, *54*, 411–418. [CrossRef]
90. Sai, Y.; Zhang, J. Spectrophotometric Determination of Microamounts of Thorium with 3,5-Dibromosalicylfluorone by Ion-exchange Separation. *Spectrosc. Spectr. Anal.* **2001**, *21*, 843–845.
91. Xia, C.B. Spectrophotometric Determination of Trace Thorium in Coal Gangue. *Chin. J. Rare Met.* **2003**, *27*, 416–417. [CrossRef]
92. Pavón, J.L.P.; Cordero, B.M. Micellar systems in flow injection: Determination of gadolinium with 1-(2-pyridylazo)-2-naphthol in the presence of Triton X-100. *Analyst* **1992**, *117*, 215–217. [CrossRef]

93. Khan, M.H.; Hafeez, M.; Bukhari, S.M.H.; Ali, A. Spectrophotometric determination of microamounts of thorium with thorin in the presence of cetylpyridinium chloride as surfactant in perchloric acid. *J. Radioanal. Nucl. Chem.* **2014**, *301*, 703–709. [CrossRef]
94. Agnihotri, N.K.; Singh, V.K.; Singh, H.B. Simultaneous derivative spectrophotometric determination of thorium and uranium in a micellar medium. *Talanta* **1993**, *40*, 1851–1859. [CrossRef]
95. Abu-Bakr, M.S.; Sedaira, H.; Hashem, E.Y. Complexation equilibria and spectrophotometric determination of iron (III) with 1-amino-4-hydroxyanthraquinone. *Talanta* **1994**, *41*, 1669–1674. [CrossRef]
96. Ramesh, A.; Krishnamacharyulu, J.; Ravindranath, L.K.; Brahmaji Rao, S. Simultaneous determination of uranium and thorium using 4-(2′-thiazolylazo) resacetophenone oxime as analytical reagent. A derivative spectrophotometric method. *J. Radioanal. Nucl. Chem.* **1993**, *170*, 181–187. [CrossRef]
97. Kuroda, R.; Kurosaki, M.; Hayashibe, Y.; Ishimaru, S. Simultaneous determination of uranium and thorium with Arsenazo III by second-derivative spectrophotometry. *Talanta* **1990**, *37*, 619–624. [CrossRef]
98. Sharma, C.; Eshwar, M. Rapid spectrophotometric determination of thorium (IV) with 1-(2′-thiazolylazo)-2-naphthol. *J. Radioanal. Nucl. Chem.* **1985**, *91*, 323–328. [CrossRef]
99. Mori, I.; Fujita, Y.; Fujita, K.; Kitano, S.; Ogawa, I.; Kawabe, H.; Koshiyama, Y.; Tanaka, T. Color reaction among pyrogallol red, thorium (IV) and samarium (III), and its application to the determination of these metals. *Bull. Chem. Soc. Jpn.* **1986**, *59*, 955–957. [CrossRef]
100. Jarosz, M. Study of ternary thorium complexes with some triphenylmethane reagents and cationic surfactants. *Analyst* **1986**, *111*, 681–683. [CrossRef]
101. Mori, I.; Fujita, Y.; Sakaguchi, K. Solvent extraction and spectrophotometry of Th (IV) with 3,4,5,6-tetrachlorogallein in the presence of La (III) and cetyltrimethylammonium chloride. *Bunseki Kagaku* **1982**, *31*, 99–102. [CrossRef]
102. Zaki, M.T.M.; El-Sayed, A.Y. Determination of thorium using phenylfluorone and quercetin in the presence of surfactants and protective colloids. *Anal. Lett.* **1995**, *28*, 1525–1539. [CrossRef]
103. Dacheux, N.; Aupiais, J. Determination of uranium, thorium, plutonium, americium, and curium ultratraces by photon electron rejecting α liquid scintillation. *Anal. Chem.* **1997**, *69*, 2275–2282. [CrossRef] [PubMed]
104. Amrane, M.; Oufni, L. Determination for levels of uranium and thorium in water along Oum Er-Rabia River using alpha track detectors. *J. Radiat. Res. Appl. Sci.* **2017**, *10*, 246–251. [CrossRef]
105. Pantelica, A.I.; Georgesecu, I.; Murariu-magureanu, D.M.; Margaritescu, I.; Cincu, E. Thorium determination in intercomparison samples and in some Romanian building materials by gamma ray spectrometry. *Radiat. Protect. Dosim.* **2001**, *97*, 187–191. [CrossRef]
106. Chapter 58. Thorium. *Compr. Anal. Chem.* **1996**, *30*, 721–728. [CrossRef]
107. Rao, T.P.; Metilda, P.; Gladis, J.M. Preconcentration techniques for uranium(VI) and thorium(IV) prior to analytical determination—An overview. *Talanta* **2006**, *68*, 1047–1064. [CrossRef]
108. De Carvalho, M.S.; Domingues, M.D.L.F.; Mantovano, J.L.; Da Cunha, J.W.S.D. Preconcentration method for the determination of thorium in natural water by wavelength dispersive X-ray fluorescence spectrometry. *J. Radioanal. Nucl. Chem.* **2002**, *253*, 253–256. [CrossRef]
109. Misra, N.L.; Dhara, S.; Adya, V.C.; Godbole, S.V.; Mudher, K.S.; Aggarwal, S.K. Trace element determination in thorium oxide using total reflection X-ray fluorescence spectrometry. *Spectrochim. Acta Part B At. Spectrosc.* **2008**, *63*, 81–85. [CrossRef]
110. Kasar, S.; Murugan, R.; Arae, H.; Aono, T.; Sahoo, S.K. A Microwave Digestion Technique for the Analysis of Rare Earth Elements, Thorium and Uranium in Geochemical Certified Reference Materials and Soils by Inductively Coupled Plasma Mass Spectrometry. *Molecules* **2020**, *25*, 5178. [CrossRef]
111. Mas, J.L.; Villa, M.; Hurtado, S.; García-Tenorio, R. Determination of trace element concentrations and stable lead, uranium and thorium isotope ratios by quadrupole-ICP-MS in NORM and NORM-polluted sample leachates. *J. Hazard. Mater.* **2012**, *205–206*, 198–207. [CrossRef]
112. Kulkarni, M.J.; Argekar, A.A.; Mathur, J.N.; Page, A.G. Chemical separation and inductively coupled plasma–atomic emission spectrometric determination of seventeen trace metals in thorium oxide matrix using a novel extractant–Cyanex-923. *Anal. Chim. Acta* **1998**, *370*, 163–171. [CrossRef]
113. Luo, Y.; Cong, H.X.; Zhao, Z.Q.; Hu, W.Q.; Zhou, W.; He, S.H. ICP-AES with MSF for Determination of Th and U. *J. Nucl. Radiochem.* **2015**, *37*, 37–40. [CrossRef]
114. Hou, L.Q.; Luo, S.H.; Wang, S.A.; Sheng, H.W.; Xi, Y.F. Determination of trace thorium in uranium dioxide by inductively coupled plasma-atomic emission spectrometry. *Metall. Anal.* **2006**, *26*, 50–52.
115. Ayranov, M.; Cobos, J.; Popa, K.; Rondinella, V.V. Determination of REE, U, Th, Ba, and Zr in simulated hydrogeological leachates by ICP-AES after matrix solvent extraction. *J. Rare Earths* **2009**, *27*, 123–127. [CrossRef]
116. Casartelli, E.A.; Miekeley, N. Determination of thorium and light rare-earth elements in soil water and its high molecular mass organic fractions by inductively coupled plasma mass spectrometry and on-line-coupled size-exclusion chromatography. *Anal. Bioanal. Chem.* **2003**, *377*, 58–64. [CrossRef]
117. She, Z.; Li, M.; Feng, Z.; Xu, Y.; Wang, M.; Pan, X.; Yang, Z. Determination of Trace Thorium and Uranium Impurities in Scandium with High Matrix by ICP-OES. *Materials* **2023**, *16*, 3023. [CrossRef]

118. Aydin, F.A.; Soylak, M. Separation, preconcentration and inductively coupled plasma-mass spectrometric (ICP-MS) determination of thorium(IV), titanium(IV), iron(III), lead(II) and chromium(III) on 2-nitroso-1-naphthol impregnated MCI GEL CHP20P resin. *J. Hazard. Mater.* **2010**, *173*, 669–674. [CrossRef]
119. Purohit, P.J.; Goyal, N.; Thulasidas, S.K.; Page, A.G.; Sastry, M.D. Electrothermal vaporization—Inductively coupled plasma-atomic emission spectrometry for trace metal determination in uranium and thorium compounds without prior matrix separation. *Spectrochim. Acta Part B At. Spectrosc.* **2000**, *55*, 1257–1270. [CrossRef]
120. Castro, M.A.; García-Olalla, C.; Robles, L.C.; Aller, A.J. Behavior of thorium, zirconium, and vanadium as chemical modifiers in the determination of arsenic by electrothermal atomization atomic absorption spectrometry. *Spectrochim. Acta Part B At. Spectrosc.* **2002**, *57*, 1–14. [CrossRef]
121. Casadio, S.; Lorenzinl, L. Cyclic voltammetric behavior of uranyl ion in sulfuric acid solutions. Application to some nuclear materials characterization. *Anal. Lett.* **1973**, *6*, 809–820. [CrossRef]
122. Jin, L.T.; Shan, Y.; Tong, W.; Fang, Y.Z. Determination of thorium by adsorptive type chemically modified electrode with a polycomplex system. *Microchim. Acta* **1989**, *97*, 97–104. [CrossRef]
123. Wen, J.; Dong, L.; Hu, S.; Li, W.; Li, S.; Wang, X. Fluorogenic thorium sensors based on 2,6-pyridinedicarboxylic acid-substituted tetraphenylethenes with aggregation-induced emission characteristics. *Chem. Asian J.* **2016**, *11*, 49–53. [CrossRef]
124. Safavi, A.; Sadeghi, M. Design and evaluation of a thorium (IV) selective optode. *Anal. Chim. Acta* **2006**, *567*, 184–188. [CrossRef]
125. Mirza, M.A.; Khuhawar, M.Y.; Arain, R.; Ch, M.A. Micellar electrokinetic chromatographic analysis of thorium, uranium, copper, nickel, cobalt and iron in ore and fish samples. *Arab. J. Chem.* **2018**, *11*, 305–312. [CrossRef]
126. Sarkar, A.; Alamelu, D.; Aggarwal, S.K. Determination of trace constituents in thoria by laser induced breakdown spectrometry. *J. Nucl. Mater.* **2009**, *384*, 158–162. [CrossRef]
127. Sarkar, A.; Alamelu, D.; Aggarwal, S.K. Determination of thorium and uranium in solution by laser-induced breakdown spectrometry. *Appl. Opt.* **2008**, *47*, G58–G64. [CrossRef]
128. Kadish, K.M.; Liu, Y.H.; Anderson, J.E.; Dormond, A.; Belkalem, M.; Guilard, R. Electrochemical and spectroelectrochemical studies of bis(diketonate) thorium(IV) and uranium(IV) porphyrins. *Inorg. Chim. Acta* **1989**, *163*, 201–205. [CrossRef]
129. De Diego Almeida, R.H.; Monroy-Guzmán, F.; Juárez, C.R.A.; Rocha, J.M.; Bustos, E.B. Electrochemical detector based on a modified graphite electrode with phthalocyanine for the elemental analysis of actinides. *Chemosphere* **2021**, *276*, 130114. [CrossRef] [PubMed]
130. Moghaddam, Z.S.; Kaykhaii, M.; Khajeh, M.; Oveisi, A.R. Synthesis of UiO-66-OH zirconium metal-organic framework and its application for selective extraction and trace determination of thorium in water samples by spectrophotometry. *Spectrochim. Acta Part A Mol. Biomol. Spectrosc.* **2018**, *194*, 76–82. [CrossRef]
131. Pakhui, G.; Ghosh, S.; Reddy, B.P. Th4+ | Th couple in LiCl-KCl eutectic: Anodic polarization of thorium and electrochemical impedance spectroscopy study at tungsten, cadmium and thorium electrodes. *Electrochim. Acta* **2019**, *295*, 354–366. [CrossRef]
132. Akhila Maheswari, M.; Prabhakaran, D.; Subramanian, M.S.; Sivaraman, N.; Srinivasan, T.G.; Vasudeva Rao, P.R. High performance liquid chromatographic studies on lanthanides, uranium and thorium on amide modified reversed phase supports. *Talanta* **2007**, *72*, 730–740. [CrossRef]
133. Shimada-Fujiwara, A.; Hoshi, A.; Kameo, Y.; Nakashima, M. Influence of hydrofluoric acid on extraction of thorium using a commercially available extraction chromatographic resin. *J. Chromatogr. A* **2009**, *1216*, 4125–4127. [CrossRef]
134. Radchenko, V.; Engle, J.W.; Wilson, J.J.; Maassen, J.R.; Nortier, F.M.; Taylor, W.A.; Birnbaum, E.R.; Hudston, L.A.; John, K.D.; Fassbender, M.E. Application of ion exchange and extraction chromatography to the separation of actinium from proton-irradiated thorium metal for analytical purposes. *J. Chromatogr. A* **2015**, *1380*, 55–63. [CrossRef]
135. Singh, D.; Basu, S.; Mishra, B.; Prusty, S.; Kundu, T.; Rao, R. Textural and Chemical Characters of Lean Grade Placer Monazite of Bramhagiri Coast, Odisha, India. *Minerals* **2023**, *13*, 742. [CrossRef]
136. Lazar, M.M.; Ghiorghita, C.-A.; Dragan, E.S.; Humelnicu, D.; Dinu, M.V. Ion-Imprinted Polymeric Materials for Selective Adsorption of Heavy Metal Ions from Aqueous Solution. *Molecules* **2023**, *28*, 2798. [CrossRef]
137. Lee, J.; Yi, S.C. Accurate measurement of uranium and thorium in naturally occurring radioactive materials to overcome complex matrix interference. *Appl. Radiat. Isot.* **2023**, *193*, 110649. [CrossRef]
138. Dumpala, R.M.R.; Sahu, M.; Nagar, B.K.; Raut, V.V.; Raje, N.H.; Rawat, N.; Subbiah, J.; Saxena, M.K.; Tomar, B.S. Accountancy for intrinsic colloids on thorium solubility: The fractionation of soluble species and the characterization of solubility limiting phase. *Chemosphere* **2021**, *269*, 129327. [CrossRef]
139. Zhijun, G.; Wangsuo, W.; Dadong, S.; Minyu, T. Liquid-liquid extraction of uranium (VI) and thorium (IV) by two open-chain crown ethers with terminal quinolyl groups in chloroform. *J. Radioanal. Nucl. Chem.* **2003**, *258*, 199–203. [CrossRef]
140. Metaxas, M.; Kasselouri-Rigopoulou, V.; Galiatsatou, P.; Konstantopoulou, C.; Oikonomou, D. Thorium Removal by Different Adsorbents. *J. Hazard. Mater.* **2003**, *97*, 71–82. [CrossRef]
141. Alotaibi, A.M.; Ismail, A.F. Modification of Clinoptilolite as a Robust Adsorbent for Highly-Efficient Removal of Thorium (IV) from Aqueous Solutions. *Int. J. Environ. Res. Public Health* **2022**, *19*, 13774. [CrossRef] [PubMed]
142. Felix, C.S.A.; Chagas, A.V.B.; de Jesus, R.F.; Barbosa, W.T.; Barbosa, J.D.V.; Ferreira, S.L.C.; Cerdà, V. Synthesis and Application of a New Polymer with Imprinted Ions for the Preconcentration of Uranium in Natural Water Samples and Determination by Digital Imaging. *Molecules* **2023**, *28*, 4065. [CrossRef] [PubMed]

143. Liu, H.; Qi, C.; Feng, Z.; Lei, L.; Deng, S. Adsorption of Trace Thorium (IV) from Aqueous Solution by Mono-Modified β-Cyclodextrin Polyrotaxane Using Response Surface Methodology (RSM). *J. Radioanal. Nucl. Chem.* **2017**, *314*, 1607–1618. [CrossRef]
144. Ansari, S.A.; Mohapatra, P.K.; Manchanda, V.K. A Novel Malonamide Grafted Polystyrene-Divinyl Benzene Resin for Extraction, Pre-Concentration and Separation of Actinides. *J. Hazard. Mater.* **2009**, *161*, 1323–1329. [CrossRef]
145. Karmakar, R.; Singh, P.; Sen, K. Selectivity of Th (IV) Adsorption as Compared to U (VI), La (III), Ce (III), Sm (III) and Gd (III) Using Mesoporous Al_2O_3. *Sep. Sci. Technol.* **2021**, *56*, 2369–2384. [CrossRef]
146. Kukkonen, E.; Virtanen, E.J.; Moilanen, J.O. α-Aminophosphonates, -Phosphinates, and -Phosphine Oxides as Extraction and Precipitation Agents for Rare Earth Metals, Thorium, and Uranium: A Review. *Molecules* **2022**, *27*, 3465. [CrossRef] [PubMed]
147. Georgiou, E.; Pashalidis, I.; Raptopoulos, G.; Paraskevopoulou, P. Efficient Removal of Polyvalent Metal Ions (Eu(III) and Th(IV)) from Aqueous Solutions by Polyurea-Crosslinked Alginate Aerogels. *Gels* **2022**, *8*, 478. [CrossRef]
148. Liao, W.; Zhang, Z.; Li, Y.; Wu, G.; Lu, Y. CN105734286A—Method for Separating Cerium-Fluoride and Thorium. China Patent CN105734286A, 11 December 2014.
149. Lu, Y.; Zhang, Z.; Li, Y.; Liao, W. Extraction and recovery of cerium(IV) and thorium(IV) from sulphate medium by an α-aminophosphonate extractant. *J. Rare Earths* **2017**, *35*, 34–40. [CrossRef]
150. Wei, H.; Li, Y.; Zhang, Z.; Xue, T.; Kuang, S.; Liao, W. Selective Extraction and Separation of Ce (IV) and Th (IV) from RE(III) in Sulfate Medium Using Di(2-Ethylhexyl)-N-Heptylaminomethylphosphonate. *Solvent Extr. Ion Exch.* **2017**, *35*, 117–129. [CrossRef]
151. Kuang, S.; Zhang, Z.; Li, Y.; Wu, G.; Wei, H.; Liao, W. Selective Extraction and Separation of Ce(IV) from Thorium and Trivalent Rare Earths in Sulfate Medium by an α-Aminophosphonate Extractant. *Hydrometallurgy* **2017**, *167*, 107–114. [CrossRef]
152. Kuang, S.; Zhang, Z.; Li, Y.; Wei, H.; Liao, W. Extraction and Separation of Heavy Rare Earths from Chloride Medium by α-Aminophosphonic Acid HEHAPP. *J. Rare Earths* **2018**, *36*, 304–310. [CrossRef]
153. Zhao, Q.; Zhang, Z.; Li, Y.; Bian, X.; Liao, W. Solvent Extraction and Separation of Rare Earths from Chloride Media Using α-Aminophosphonic Acid Extractant HEHAMP. *Solvent Extr. Ion Exch.* **2018**, *36*, 136–149. [CrossRef]
154. Kaygun, A.K.; Akyil, S. Study of the Behaviour of Thorium Adsorption on PAN/Zeolite Composite Adsorbent. *J. Hazard. Mater.* **2007**, *147*, 357–362. [CrossRef]
155. Al-shaybe, M.; Khalili, F. Adsorption of Thorium (IV) and Uranium (VI) by Tulul Al- Shabba Zeolitic Tuff, Jordan. *Jordan J. Earth Environ. Sci.* **2009**, *2*, 108–119.
156. Khazaei, Y.; Faghihian, H.; Kamali, M. Removal of Thorium from Aqueous Solutions by Sodium Clinoptilolite. *J. Radioanal. Nucl. Chem.* **2011**, *289*, 529–536. [CrossRef]
157. Nurliati, G.; Krisnandi, Y.K.; Sihombing, R.; Salimin, Z. Studies of Modification of Zeolite by Tandem Acid-Base Treatments and Its Adsorptions Performance towards Thorium. *At. Indones.* **2015**, *41*, 87. [CrossRef]
158. Akl, Z.F.; Hegazy, M.A. Selective cloud point extraction of thorium (IV) using tetraazonium based ionic liquid. *J. Environ. Chem. Eng.* **2020**, *8*, 104185. [CrossRef]
159. Varala, S.; Kumari, A.; Dharanija, B.; Bhargava, S.K.; Parthasarathy, R.; Satyavathi, B. Removal of thorium (IV) from aqueous solutions by deoiled karanja seed cake: Optimization using Taguchi method, equilibrium, kinetic and thermodynamic studies. *J. Environ. Chem. Eng.* **2016**, *4*, 405–417. [CrossRef]
160. Humelnicu, D.; Bulgariu, L.; Macoveanu, M. On the retention of uranyl and thorium ions from radioactive solution on peat moss. *J. Hazard. Mater.* **2010**, *174*, 782–787. [CrossRef]
161. Scandura, G.; Eid, S.; Alnajjar, A.A.; Paul, T.; Karanikolos, G.N.; Shetty, D.; Omer, K.; Alqerem, R.; Juma, A.; Wang, H.; et al. Photo-responsive metal–organic frameworks–design strategies and emerging applications in photocatalysis and adsorption. *Mater. Adv.* **2023**, *4*, 1258–1285. [CrossRef]
162. Wang, X.; Zheng, H.; Xu, Q.; Zhu, T.; Jiang, F.; She, C.; Wang, C.; Cong, H.; Gong, Y.; Huang, W.; et al. Electrochemical behaviors and electrolytic separation of Th (IV) and Ce (III) in ThF4-CeF3-LiCl-KCl quaternary melt. *Sep. Purif. Technol.* **2019**, *210*, 236–241. [CrossRef]
163. Gomaa, H.; Emran, M.Y.; Elsenety, M.M.; Abdel-Rahim, R.D.; Deng, Q.; Gadallah, M.I.; Saad, M.; ALMohiy, H.; Ali, H.R.H.; Faraghally, F.A.; et al. Selective removal of thorium ions from aqueous solutions using a hybrid mesoporous adsorbent as benzenesulfonamide-derivative@ZrO_2. *J. Water Process Eng.* **2023**, *51*, 103436. [CrossRef]
164. Allam, E.M.; Lashen, T.A.; Abou El-Enein, S.A.; Hassanin, M.A.; Sakr, A.K.; Cheira, M.F.; Almuqrin, A.; Hanfi, M.Y.; Sayyed, M.I. Rare Earth Group Separation after Extraction Using Sodium Diethyldithiocarbamate/Polyvinyl Chloride from Lamprophyre Dykes Leachate. *Materials* **2022**, *15*, 1211. [CrossRef]
165. Zhao, D.; Yu, S.; Jiang, W.-J.; Cai, Z.-H.; Li, D.-L.; Liu, Y.-L.; Chen, Z.-Z. Recent Progress in Metal-Organic Framework Based Fluorescent Sensors for Hazardous Materials Detection. *Molecules* **2022**, *27*, 2226. [CrossRef]
166. Talan, D.; Huang, Q. Separation of Radionuclides from a Rare Earth-Containing Solution by Zeolite Adsorption. *Minerals* **2021**, *11*, 20. [CrossRef]
167. Atanassova, M. Assessment of the Equilibrium Constants of Mixed Complexes of Rare Earth Elements with Acidic (Chelating) and Organophosphorus Ligands. *Separations* **2022**, *9*, 371. [CrossRef]
168. Atanassova, M. Thenoyltrifluoroacetone: Preferable Molecule for Solvent Extraction of Metals—Ancient Twists to New Approaches. *Separations* **2022**, *9*, 154. [CrossRef]

169. Salah, B.A.; Gaber, M.S.; Kandil, A.h.T. The Removal of Uranium and Thorium from Their Aqueous Solutions by 8-Hydroxyquinoline Immobilized Bentonite. *Minerals* **2019**, *9*, 626. [CrossRef]
170. Nechifor, A.C.; Cotorcea, S.; Bungău, C.; Albu, P.C.; Pașcu, D.; Oprea, O.; Grosu, A.R.; Pîrțac, A.; Nechifor, G. Removing of the Sulfur Compounds by Impregnated Polypropylene Fibers with Silver Nanoparticles-Cellulose Derivatives for Air Odor Correction. *Membranes* **2021**, *11*, 256. [CrossRef] [PubMed]
171. Ibrar, I.; Yadav, S.; Naji, O.; Alanezi, A.A.; Ghaffour, N.; Déon, S.; Subbiah, S.; Altaee, A. Development in forward Osmosis-Membrane distillation hybrid system for wastewater treatment. *Sep. Purif. Technol.* **2022**, *286*, 120498. [CrossRef]
172. Imtiaz, A.; Othman, M.H.D.; Jilani, A.; Khan, I.U.; Kamaludin, R.; Iqbal, J.; Al-Sehemi, A.G. Challenges, Opportunities and Future Directions of Membrane Technology for Natural Gas Purification: A Critical Review. *Membranes* **2022**, *12*, 646. [CrossRef]
173. Cimbru, A.M.; Rikabi, A.A.K.K.; Oprea, O.; Grosu, A.R.; Tanczos, S.-K.; Simonescu, M.C.; Pașcu, D.; Grosu, V.-A.; Dumitru, F.; Nechifor, G. pH and pCl Operational Parameters in Some Metallic Ions Separation with Composite Chitosan/Sulfonated Polyether Ether Ketone/Polypropylene Hollow Fibers Membranes. *Membranes* **2022**, *12*, 833. [CrossRef]
174. Paun, G.; Neagu, E.; Parvulescu, V.; Anastasescu, M.; Petrescu, S.; Albu, C.; Nechifor, G.; Radu, G.L. New Hybrid Nanofiltration Membranes with Enhanced Flux and Separation Performances Based on Polyphenylene Ether-Ether-Sulfone/Polyacrylonitrile/SBA-15. *Membranes* **2022**, *12*, 689. [CrossRef]
175. Baker, W. *Membrane Technology and Applications*, 3rd ed.; John Wiley & Sons Ltd.: Chichester, UK, 2012; pp. 148–149. ISBN 9780470743720.
176. Mulder, M. The Use of Membrane Processes in Environmental Problems. An Introduction. In *Membrane Processes in Separation and Purification*; NATO ASI Series (Series E: Applied Sciences); Crespo, J.G., Böddeker, K.W., Eds.; Springer: Dordrecht, The Netherlands, 1994; Volume 272. [CrossRef]
177. Strathmann, H.; Giorno, L.; Drioli, E. *Introduction to Membrane Science and Technology*; Institute on Membrane Technology, CNR-ITMat University of Calabria: Rende, Italy, 2011; pp. 27–58.
178. Van Der Bruggen, B.; Vandecasteele, C.; Van Gestel, T.; Doyen, W.; Leysen, R. A review of pressure-driven membrane processes in wastewater treatment and drinking water production. *Environ. Prog.* **2003**, *22*, 46–56. [CrossRef]
179. Drioli, E.; Stankiewicz, A.I.; Macedonio, F. Membrane engineering in process intensification—An overview. *J. Membr. Sci.* **2011**, *380*, 1–8. [CrossRef]
180. Iulianelli, A.; Drioli, E. Membrane engineering: Latest advancements in gas separation and pre-treatment processes, petrochemical industry and refinery, and future perspectives in emerging applications. *Fuel Process. Technol.* **2020**, *206*, 106464. [CrossRef]
181. Li, N.N.; Fane, A.G.; Ho, W.W.; Matsuura, T. (Eds.) *Advanced Membrane Technology and Applications*; John Wiley & Sons: Hoboken, NJ, USA, 2011; Available online: https://toc.library.ethz.ch/objects/pdf/e16_978-0-471-73167-2_01.pdf (accessed on 24 June 2023).
182. Drioli, E.; Criscuoli, A.; Curcio, E. *Membrane Contactors: Fundamentals, Applications and Potentialities*; Elsevier: Amsterdam, The Netherlands, 2011.
183. Bernardoa, P.; Drioli, E. Membrane Gas Separation Progresses for Process Intensification Strategy in the Petrochemical Industry. *Pet. Chem.* **2010**, *50*, 271–282. [CrossRef]
184. Van der Bruggen, B.; Lejon, L.; Vandecasteele, C. Reuse, treatment, and discharge of the concentrate of pressure-driven membrane processes. *Environ. Sci. Technol.* **2003**, *37*, 3733–3738. [CrossRef]
185. Van der Bruggen, B.; Everaert, K.; Wilms, D.; Vandecasteele, C. Application of nanofiltration for the removal of pesticides, nitrate and hardness from ground water: Retention properties and economic evaluation. *J. Membr. Sci.* **2001**, *193*, 239–248. [CrossRef]
186. Peng, N.; Widjojo, N.; Sukitpaneenit, P.; Teoh, M.M.; Lipscomb, G.G.; Chung, T.S.; Lai, J.Y. Evolution of polymeric hollow fibers as sustainable technologies: Past, present, and future. *Prog. Polym. Sci.* **2012**, *37*, 1401–1424. [CrossRef]
187. Kamolov, A.; Turakulov, Z.; Rejabov, S.; Díaz-Sainz, G.; Gómez-Coma, L.; Norkobilov, A.; Fallanza, M.; Irabien, A. Decarbonization of Power and Industrial Sectors: The Role of Membrane Processes. *Membranes* **2023**, *13*, 130. [CrossRef] [PubMed]
188. Zanco, S.E.; Pérez-Calvo, J.-F.; Gasós, A.; Cordiano, B.; Becattini, V.; Mazzotti, M. Postcombustion CO_2 Capture: A Comparative Techno-Economic Assessment of Three Technologies Using a Solvent, an Adsorbent, and a Membrane. *ACS Eng. Au* **2021**, *1*, 50–72. [CrossRef]
189. Rahman, T.U.; Roy, H.; Islam, M.R.; Tahmid, M.; Fariha, A.; Mazumder, A.; Tasnim, N.; Pervez, M.N.; Cai, Y.; Naddeo, V.; et al. The Advancement in Membrane Bioreactor (MBR) Technology toward Sustainable Industrial Wastewater Management. *Membranes* **2023**, *13*, 181. [CrossRef]
190. Dunn, R.O., Jr.; Scamehorn, J.F.; Christian, S.D. Use of micellar-enhanced ultrafiltration to remove dissolved organics from aqueous streams. *Sep. Sci. Technol.* **1985**, *20*, 257–284. [CrossRef]
191. Christian, S.D.; Bhat, S.N.; Tucker, E.E.; Scamehorn, J.F.; El-Sayed, D.A. Micellar-enhanced ultrafiltration of chromate anion from aqueous streams. *AIChE J.* **1988**, *34*, 189–194. [CrossRef]
192. Dunn, R.O., Jr.; Scamehorn, J.F.; Christian, S.D. Concentration polarization effects in the use of micellar-enhanced ultrafiltration to remove dissolved organic pollutants from wastewater. *Sep. Sci. Technol.* **1987**, *22*, 763–789. [CrossRef]
193. Chen, M.; Jafvert, C.T.; Wu, Y.; Cao, X.; Hankins, N.P. Inorganic anion removal using micellar enhanced ultrafiltration (MEUF), modeling anion distribution and suggested improvements of MEUF: A review. *Chem. Eng. J.* **2020**, *398*, 125413. [CrossRef]

194. Nechifor, G.; Păncescu, F.M.; Albu, P.C.; Grosu, A.R.; Oprea, O.; Tanczos, S.-K.; Bungău, C.; Grosu, V.-A.; Ioan, M.-R.; Nechifor, A.C. Transport and Separation of the Silver Ion with n–decanol Liquid Membranes Based on 10–undecylenic Acid, 10–undecen–1–ol and Magnetic Nanoparticles. *Membranes* **2021**, *11*, 936. [CrossRef]
195. Burts, K.S.; Plisko, T.V.; Sjölin, M.; Rodrigues, G.; Bildyukevich, A.V.; Lipnizki, F.; Ulbricht, M. Development of Antifouling Polysulfone Membranes by Synergistic Modification with Two Different Additives in Casting Solution and Coagulation Bath: Synperonic F108 and Polyacrylic Acid. *Materials* **2022**, *15*, 359. [CrossRef]
196. Nagarale, R.K.; Gohil, G.S.; Shahi, V.K. Recent developments on ion-exchange membranes and electro-membrane processes. *Adv. Colloid Interface Sci.* **2006**, *119*, 97–130. [CrossRef]
197. Doornbusch, G.; van der Wal, M.; Tedesco, M.; Post, J.; Nijmeijer, K.; Borneman, Z. Multistage electrodialysis for desalination of natural seawater. *Desalination* **2021**, *505*, 114973. [CrossRef]
198. Tekinalp, Ö.; Zimmermann, P.; Holdcroft, S.; Burheim, O.S.; Deng, L. Cation Exchange Membranes and Process Optimizations in Electrodialysis for Selective Metal Separation: A Review. *Membranes* **2023**, *13*, 566. [CrossRef]
199. Cournoyer, A.; Bazinet, L. Electrodialysis Processes an Answer to Industrial Sustainability: Toward the Concept of Eco-Circular Economy? —A Review. *Membranes* **2023**, *13*, 205. [CrossRef]
200. Dammak, L.; Fouilloux, J.; Bdiri, M.; Larchet, C.; Renard, E.; Baklouti, L.; Sarapulova, V.; Kozmai, A.; Pismenskaya, N. A Review on Ion-Exchange Membrane Fouling during the Electrodialysis Process in the Food Industry, Part 1: Types, Effects, Characterization Methods, Fouling Mechanisms and Interactions. *Membranes* **2021**, *11*, 789. [CrossRef]
201. Solonchenko, K.; Kirichenko, A.; Kirichenko, K. Stability of Ion Exchange Membranes in Electrodialysis. *Membranes* **2023**, *13*, 52. [CrossRef]
202. Veerman, J.; Gómez-Coma, L.; Ortiz, A.; Ortiz, I. Resistance of Ion Exchange Membranes in Aqueous Mixtures of Monovalent and Divalent Ions and the Effect on Reverse Electrodialysis. *Membranes* **2023**, *13*, 322. [CrossRef]
203. Gao, W.; Fang, Q.; Yan, H.; Wei, X.; Wu, K. Recovery of Acid and Base from Sodium Sulfate Containing Lithium Carbonate Using Bipolar Membrane Electrodialysis. *Membranes* **2021**, *11*, 152. [CrossRef]
204. Zhou, H.; Ju, P.; Hu, S.; Shi, L.; Yuan, W.; Chen, D.; Wang, Y.; Shi, S. Separation of Hydrochloric Acid and Oxalic Acid from Rare Earth Oxalic Acid Precipitation Mother Liquor by Electrodialysis. *Membranes* **2023**, *13*, 162. [CrossRef]
205. Gurreri, L.; Tamburini, A.; Cipollina, A.; Micale, G. Electrodialysis Applications in Wastewater Treatment for Environmental Protection and Resources Recovery: A Systematic Review on Progress and Perspectives. *Membranes* **2020**, *10*, 146. [CrossRef]
206. Drioli, E.; Romano, M. Progress and new perspectives on integrated membrane operations for sustainable industrial growth. *Ind. Eng. Chem. Res.* **2001**, *40*, 1277–1300. [CrossRef]
207. Christensen, J.J.; Lamb, J.D.; Brown, P.R.; Oscarson, J.L.; Izatt, R.M. Liquid Membrane Separations of Metal Cations Using Macrocyclic Carriers. *Sep. Sci. Technol.* **1981**, *16*, 1193–1215. [CrossRef]
208. Brown, P.R.; Izatt, R.M.; Christensenand, J.J.; Lamb, J.D. Transport of Eu^{2+} in a H2O-CHCl3-H2O liquid membrane system containing the macrocyclic polyether 18-crown-6. *J. Membr.Sci.* **1983**, *13*, 85–88. [CrossRef]
209. Burgard, M.; Elisoamiadana, P.; Leroy, M.J.F. Liquid membrane studies: Transport against the concentration gradient of AuCl. In Proceedings of the International Solvent Extraction Conference (ISEC'83), Denver, CO, USA, 26 August–2 September 1983; Volume II, pp. 399–400.
210. Kislik, V.S.; Eyal, A.M. Hybrid liquid membrane (HLM) system in separation technologies. *J. Membr. Sci.* **1996**, *111*, 259–272. [CrossRef]
211. Majumdar, S.; Sirkar, K.K.; Sengupta, A. Hollow-Fiber Contained Liquid Membrane. In *Membrane Handbook*; Ho, W.S.W., Sirkar, K.K., Eds.; Springer: Boston, MA, USA, 1992. [CrossRef]
212. Schlosser, S.; Sabol, E. Three-phase contactor with distributed U-shaped bundles of hollow-fibers for pertraction. *J. Membr. Sci.* **2002**, *210*, 331–347. [CrossRef]
213. Wodzki, R.; Nowaczyk, J. Propionic and acetic acid pertraction through a multimembrane hybrid system containing TOPO or TBP. *Sep. Purif. Technol.* **2002**, *26*, 207–220. [CrossRef]
214. Nechifor, A.C.; Goran, A.; Grosu, V.-A.; Bungău, C.; Albu, P.C.; Grosu, A.R.; Oprea, O.; Păncescu, F.M.; Nechifor, G. Improving the Performance of Composite Hollow Fiber Membranes with Magnetic Field Generated Convection Application on pH Correction. *Membranes* **2021**, *11*, 445. [CrossRef]
215. Kubisova, L.; Sabolova, E.; Schlosser, S.; Martak, J.; Kertesz, R. Mass-transfer in membrane based solvent extraction and stripping of 5-methyl-2-pyrazinecarboxylic acid and co-transport of sulphuric acid in HF contactors. *Desalination* **2004**, *163*, 27–38. [CrossRef]
216. Nechifor, A.C.; Goran, A.; Tanczos, S.-K.; Păncescu, F.M.; Oprea, O.-C.; Grosu, A.R.; Matei, C.; Grosu, V.-A.; Vasile, B.Ș.; Albu, P.C. Obtaining and Characterizing the Osmium Nanoparticles/n–Decanol Bulk Membrane Used for the p–Nitrophenol Reduction and Separation System. *Membranes* **2022**, *12*, 1024. [CrossRef]
217. Ferencz, A.; Grosu, A.R.; Al-Ani, H.N.A.; Nechifor, A.C.; Tanczos, S.-K.; Albu, P.C.; Crăciun, M.E.; Ioan, M.-R.; Grosu, V.-A.; Nechifor, G. Operational Limits of the Bulk Hybrid Liquid Membranes Based on Dispersion Systems. *Membranes* **2022**, *12*, 190. [CrossRef] [PubMed]
218. Al-Ani, F.H.; Alsalhy, Q.F.; Al-Dahhan, M.H. Enhancing Emulsion Liquid Membrane System (ELM) Stability and Performance for the Extraction of Phenol from Wastewater using Various Nanoparticles. *Desalin. Water Treat.* **2021**, *210*, 180–191. [CrossRef]

219. Pavón, S.; Blaesing, L.; Jahn, A.; Aubel, I.; Bertau, M. Liquid Membranes for Efficient Recovery of Phenolic Compounds Such as Vanillin and Catechol. *Membranes* **2021**, *11*, 20. [CrossRef] [PubMed]
220. Wang, B.-Y.; Zhang, N.; Li, Z.-Y.; Lang, Q.-L.; Yan, B.-H.; Liu, Y.; Zhang, Y. Selective Separation of Acetic and Hexanoic Acids across Polymer Inclusion Membrane with Ionic Liquids as Carrier. *Int. J. Mol. Sci.* **2019**, *20*, 3915. [CrossRef] [PubMed]
221. Yang, B.; Bai, L.; Li, T.; Deng, L.; Liu, L.; Zeng, S.; Han, J.; Zhang, X. Super selective ammonia separation through multiple-site interaction with ionic liquid-based hybrid membranes. *J. Membr. Sci.* **2021**, *628*, 119264. [CrossRef]
222. Jean, E.; Villemin, D.; Hlaibi, M.; Lebrun, L. Heavy metal ions extraction using new supported liquid membranes containing ionic liquid as carrier. *Sep. Pur. Technol.* **2018**, *201*, 1–9. [CrossRef]
223. Craveiro, R.; Neves, L.A.; Duarte, A.R.C.; Paiva, A. Supported liquid membranes based on deep eutectic solvents for gas separation processes. *Sep. Pur. Technol.* **2021**, *254*, 117593. [CrossRef]
224. Wang, Z.Y.; Sun, Y.; Tang, N.; Miao, C.L.; Wang, Y.T.; Tang, L.H.; Wang, S.X.; Yang, X.J. Simultaneous extraction and recovery of gold(I) from alkaline solutions using an environmentally benign polymer inclusion membrane with ionic liquid as the carrier. *Sep. Purif. Technol.* **2019**, *222*, 136–144. [CrossRef]
225. Bazhenov, S.D.; Bildyukevich, A.V.; Volkov, A.V. Gas-liquid hollow fiber membrane contactors for different applications. *Fibers* **2018**, *6*, 76. [CrossRef]
226. Diaconu, I.; Nechifor, G.; Nechifor, A.C.; Ruse, E.; Totu, E.E. Membranary techniques used at the separation of some phenolic compounds from aqueous media. *UPB Sci. Bull. Ser. B Chem. Mater. Sci.* **2009**, *71*, 39–46.
227. Diaconu, I.; Gîrdea, R.; Cristea, C.; Nechifor, G.; Ruse, E.; Totu, E.E. Removal and recovery of some phenolic pollutants using liquid membranes. *Rom. Biotechnol. Lett.* **2010**, *15*, 5702–5708.
228. Koter, S.; Szczepański, P.; Mateescu, M.; Nechifor, G.; Badalau, L.; Koter, I. Modeling of the cadmium transport through a bulk liquid membrane. *Sep. Purif. Technol.* **2013**, *107*, 135–143. [CrossRef]
229. Szczepański, P.; Tanczos, S.K.; Ghindeanu, D.L.; Wódzki, R. Transport of p-nitrophenol in an agitated bulk liquid membrane system—Experimental and theoretical study by network analysis. *Sep. Pur. Technol.* **2014**, *132*, 616–626. [CrossRef]
230. Craciun, M.E.; Mihai, M.; Nechifor, G. Characteristics of double jet imobilized membrane. *Environ. Eng. Manag. J.* **2009**, *8*, 771–776.
231. Yoshida, W.; Baba, Y.; Kubota, F.; Kolev, S.D.; Goto, M. Selective transport of scandium(III) across polymer inclusion membranes with improved stability which contain an amic acid carrier. *J. Membr. Sci.* **2019**, *572*, 291–299. [CrossRef]
232. Rahman, R.O.A.; Ibrahium, H.A.; Hung, Y.-T. Liquid Radioactive Wastes Treatment: A Review. *Water* **2011**, *3*, 551–565. [CrossRef]
233. Barambu, N.U.; Bilad, M.R.; Wibisono, Y.; Jaafar, J.; Mahlia, T.M.I.; Khan, A.L. Membrane Surface Patterning as a Fouling Mitigation Strategy in Liquid Filtration: A Review. *Polymers* **2019**, *11*, 1687. [CrossRef]
234. Karuppasamy, K.; Theerthagiri, J.; Vikraman, D.; Yim, C.-J.; Hussain, S.; Sharma, R.; Maiyalagan, T.; Qin, J.; Kim, H.-S. Ionic Liquid-Based Electrolytes for Energy Storage Devices: A Brief Review on Their Limits and Applications. *Polymers* **2020**, *12*, 918. [CrossRef]
235. Gonçalves, A.R.P.; Paredes, X.; Cristino, A.F.; Santos, F.J.V.; Queirós, C.S.G.P. Ionic Liquids—A Review of Their Toxicity to Living Organisms. *Int. J. Mol. Sci.* **2021**, *22*, 5612. [CrossRef]
236. Friess, K.; Izák, P.; Kárászová, M.; Pasichnyk, M.; Lanč, M.; Nikolaeva, D.; Luis, P.; Jansen, J.C. A Review on Ionic Liquid Gas Separation Membranes. *Membranes* **2021**, *11*, 97. [CrossRef]
237. Alashkar, A.; Al-Othman, A.; Tawalbeh, M.; Qasim, M. A Critical Review on the Use of Ionic Liquids in Proton Exchange Membrane Fuel Cells. *Membranes* **2022**, *12*, 178. [CrossRef] [PubMed]
238. Rosli, N.A.H.; Loh, K.S.; Wong, W.Y.; Yunus, R.M.; Lee, T.K.; Ahmad, A.; Chong, S.T. Review of Chitosan-Based Polymers as Proton Exchange Membranes and Roles of Chitosan-Supported Ionic Liquids. *Int. J. Mol. Sci.* **2020**, *21*, 632. [CrossRef]
239. Ribas, T.C.F.; Croft, C.F.; Almeida, M.I.G.S.; Mesquita, R.B.R.; Kolev, S.D.; Rangel, A.O.S.S. Use of a Polymer Inclusion Membrane and a Chelating Resin for the Flow-Based Sequential Determination of Copper (II) and Zinc (II) in Natural Waters and Soil Leachates. *Molecules* **2020**, *25*, 5062. [CrossRef] [PubMed]
240. Radzyminska-Lenarcik, E.; Ulewicz, M. Polymer Inclusion Membranes (PIMs) Doped with Alkylimidazole and their Application in the Separation of Non-Ferrous Metal Ions. *Polymers* **2019**, *11*, 1780. [CrossRef] [PubMed]
241. Bahrami, S.; Dolatyari, L.; Shayani-Jam, H.; Yaftian, M.R.; Kolev, S.D. On the Potential of a Poly(vinylidenefluoride-co-hexafluoropropylene) Polymer Inclusion Membrane Containing Aliquat® 336 and Dibutyl Phthalate for V(V) Extraction from Sulfate Solutions. *Membranes* **2022**, *12*, 90. [CrossRef]
242. Eyal, A.; Kislik, V. Aqueous hybrid liquid membrane a novel system for separation of solutes using water-soluble polymers as carriers. *J. Membr. Sci.* **1999**, *161*, 207–221. [CrossRef]
243. Alamdar Milani, S.; Zahakifar, F.; Charkhi, A. Continuous bulk liquid membrane technique for thorium transport: Modeling and experimental validation. *J. Iran. Chem. Soc.* **2018**, *16*, 455–464. [CrossRef]
244. Dong, X.; Lu, D.; Harris, T.A.L.; Escobar, I.C. Polymers and Solvents Used in Membrane Fabrication: A Review Focusing on Sustainable Membrane Development. *Membranes* **2021**, *11*, 309. [CrossRef]
245. Nicoletta, F.P.; Cupelli, D.; Formoso, P.; De Filpo, G.; Colella, V.; Gugliuzza, A. Light Responsive Polymer Membranes: A Review. *Membranes* **2012**, *2*, 134–197. [CrossRef]
246. Rybak, A.; Rybak, A.; Kolev, S.D. A Modern Computer Application to Model Rare Earth Element Ion Behavior in Adsorptive Membranes and Materials. *Membranes* **2023**, *13*, 175. [CrossRef] [PubMed]

247. Sabbatovskii, K.G. The effect of the adsorption of multicharge cations on the selectivity of a nanofiltration membrane. *Colloid J.* **2003**, *65*, 237–243. [CrossRef]
248. Kaptakov, V.O.; Milyutin, V.V.; Nekrasova, N.A.; Zelenin, P.G.; Kozlitin, E.A. Nanofiltration Extraction of Uranium and Thorium from Aqueous Solutions. *Radiochemistry* **2021**, *63*, 169–172. [CrossRef]
249. Xu, L.; Gao, X.; Li, Z.; Gao, C. Removal of fluoride by nature diatomite from high-fluorine water: An appropriate pretreatment for nanofiltration process. *Desalination* **2015**, *369*, 97–104. [CrossRef]
250. Khedr, M.G. Radioactive contamination of groundwater, special aspects and advantages of removal by reverse osmosis and nanofiltration. *Desalination* **2013**, *321*, 47–54. [CrossRef]
251. Favre-Reguillon, A.; Lebuzit, G.; Foos, J.; Guy, A.; Draye, M.; Lemaire, M. Selective concentration of uranium from seawater by nanofiltration. *Ind. Eng. Chem. Res.* **2003**, *42*, 5900–5904. [CrossRef]
252. Sharma, M.; Sharma, P.; Yadav, L.; Janu, V.C.; Gupta, R. Sequestration and recovery of thorium ions using a recyclable, low-cost, glutathione-based magnetic nanocomposite: Experimental study and statistical modeling. *Sep. Pur. Technol.* **2023**, *322*, 124264. [CrossRef]
253. Zahakifar, F.; Keshtkar, A.R.; Talebi, M. Performance evaluation of sodium alginate/polyvinyl alcohol/polyethylene oxide/ZSM5 zeolite hybrid adsorbent for ion uptake from aqueous solutions: A case study of thorium (IV). *J. Radioanal. Nucl. Chem.* **2021**, *327*, 65–72. [CrossRef]
254. Hamza, M.F.; Guibal, E.; Wei, Y.; Ning, S. Synthesis, characterization, and evaluation of thiocarbazide-functionalized maleic-based polymer for thorium (IV) removal from aqueous solutions. *Chem. Eng. J.* **2023**, *464*, 142638. [CrossRef]
255. Zahakifar, F.; Keshtkar, A.; Souderjani, E.Z.; Moosavian, M. Use of response surface methodology for optimization of thorium (IV) removal from aqueous solutions by electrodeionization (EDI). *Prog. Nucl. Energy* **2020**, *124*, 103335. [CrossRef]
256. Cheira, M.F.; Orabi, A.S.; Atia, B.M.; Hassan, S.M. Solvent extraction and separation of thorium (IV) from chloride media by a Schiff base. *J. Solut. Chem.* **2018**, *47*, 611–633. [CrossRef]
257. Yousef, L.; Saad, M.; Afifi, S.; Ismail, A. Leaching and precipitation of thorium ions from Cataclastic rocks. Abu Rusheid Area, Southeastern Desert. *Egypt Arab. J. Nucl. Sci. Appl.* **2019**, *51*, 10–19. [CrossRef]
258. Fitzsimmons, J.; Abraham, A.; Catalano, D.; Younes, A.; Cutler, C.S.; Medvedev, D. Evaluation of inorganic ion exchange materials for purification of 225Ac from thorium and radium radioisotopes. *J. Med. Imaging Radiat. Sci.* **2019**, *50*, S11. [CrossRef]
259. Xiu, T.; Liu, Z.; Wang, Y.; Wu, P.; Du, Y.; Cai, Z. Thorium adsorption on graphene oxide nanoribbons/manganese dioxide composite material. *J. Radioanal. Nucl. Chem.* **2019**, *319*, 1059–1067. [CrossRef]
260. Liatsou, I.; Christodoulou, E.; Pashalidis, I. Thorium adsorption by oxidized biochar fibres derived from Luffa cylindrica sponges. *J. Radioanal. Nucl. Chem.* **2018**, *317*, 1065–1070. [CrossRef]
261. Yin, Z.; Pan, D.; Liu, P.; Wu, H.; Li, Z.; Wu, W. Sorption behavior of thorium (IV) onto activated bentonite. of thorium (IV) onto activated bentonite. *J. Radioanal. Nucl. Chem.* **2018**, *316*, 301–312. [CrossRef]
262. Kaynar, U.H.; Şabikoğlu, İ. Adsorption of thorium (IV) by amorphous silica; response surface modelling and optimization. *J. Radioanal. Nucl. Chem.* **2018**, *318*, 823–834. [CrossRef]
263. Xiong, X.H.; Yuan, Y.H.; Huang, B.; He, M.; Chen, H.; Luo, Y.C.; Zhu, Y.A.; Luo, T.A.; Chen, Q.S. Th (IV) adsorption onto titanium tetrachloride modified sodium bentonite. *J. Radioanal. Nucl. Chem.* **2019**, *319*, 805–815. [CrossRef]
264. Talebi, M.; Abbasizadeh, S.; Keshtkar, A.R. Evaluation of single and simultaneous thorium and uranium sorption from water systems by an electrospun PVA/SA/PEO/HZSM5 nanofiber. *Process. Saf. Environ. Prot.* **2017**, *109*, 340–356. [CrossRef]
265. Wang, Y.; Chen, X.; Hu, X.; Wu, P.; Lan, T.; Li, Y.; Tu, Y.; Liu, Y.; Yuan, D.; Wu, Z.; et al. Synthesis and characterization of poly (TRIM/VPA) functionalized graphene oxide nanoribbons aerogel for highly efficient capture of thorium (IV) from aqueous solutions. *Appl. Surf. Sci.* **2021**, *536*, 147829. [CrossRef]
266. Zhang, H.; Yang, F.; Bai, R.; Zhao, Z.; Cai, C.; Li, J.; Ma, Y. Facile preparation of Ce enhanced vinyl-functionalized silica aerogel-like monoliths for selective separation of radioactive thorium from monazite. *Mater. Des.* **2020**, *186*, 108333. [CrossRef]
267. Sharma, M.; Laddha, H.; Yadav, P.; Jain, Y.; Sachdev, K.; Janu, V.C.; Gupta, R. Selective removal of uranium from an aqueous solution of mixed radionuclides of uranium, cesium, and strontium via a viable recyclable GO@chitosan based magnetic nanocomposite. *Mater. Today Commun.* **2022**, *32*, 104020. [CrossRef]
268. Alamdarlo, F.V.; Solookinejad, G.; Zahakifar, F.; Jalal, M.R.; Jabbari, M. Study of kinetic, thermodynamic, and isotherm of Sr adsorption from aqueous solutions on graphene oxide (GO) and (aminomethyl) phosphonic acid–graphene oxide (AMPA–GO). *J. Radioanal. Nucl. Chem.* **2021**, *329*, 1033–1043. [CrossRef]
269. Zahakifar, F.; Charkhi, A.; Torab-Mostaedi, M.; Davarkhah, R. Kinetic study of uranium transport via a bulk liquid membrane containing Alamine 336 as a carrier. *J. Radioanal. Nucl. Chem.* **2018**, *316*, 247–255. [CrossRef]
270. Nechifor, G.; Grosu, A.R.; Ferencz, A.; Tanczos, S.-K.; Goran, A.; Grosu, V.-A.; Bungău, S.G.; Păncescu, F.M.; Albu, P.C.; Nechifor, A.C. Simultaneous Release of Silver Ions and 10–Undecenoic Acid from Silver Iron–Oxide Nanoparticles Impregnated Membranes. *Membranes* **2022**, *12*, 557. [CrossRef]
271. Karimi, M.; Milani, S.A.; Abolgashemi, H. Kinetic and isotherm analyses for thorium (IV) adsorptive removal from aqueous solutions by modified magnetite nanoparticle using response surface methodology (RSM). *J. Nucl. Mater.* **2016**, *479*, 174–183. [CrossRef]
272. Agrawal, Y.K.; Vora, S.B. Selective extraction and separation of thorium from monazite using N-phenylbenzo-18-crown-6-hydroxamic acid. *Microchim. Acta* **2003**, *142*, 255–261. [CrossRef]

273. Alharbi, H.F.; Haddad, M.Y.; Aijaz, M.O.; Assaifan, A.K.; Karim, M.R. Electrospun Bilayer PAN/Chitosan Nanofiber Membranes Incorporated with Metal Oxide Nanoparticles for Heavy Metal Ion Adsorption. *Coatings* **2020**, *10*, 285. [CrossRef]
274. Goh, P.S.; Samavati, Z.; Ismail, A.F.; Ng, B.C.; Abdullah, M.S.; Hilal, N. Modification of Liquid Separation Membranes Using Multidimensional Nanomaterials: Revealing the Roles of Dimension Based on Classical Titanium Dioxide. *Nanomaterials* **2023**, *13*, 448. [CrossRef]
275. Jakubski, Ł.; Dudek, G.; Turczyn, R. Applicability of Composite Magnetic Membranes in Separation Processes of Gaseous and Liquid Mixtures—A Review. *Membranes* **2023**, *13*, 384. [CrossRef]
276. Wu, W.; Shi, Y.; Liu, G.; Fan, X.; Yu, Y. Recent development of graphene oxide based forward osmosis membrane for water treatment: A critical review. *Desalination* **2020**, *491*, 114452. [CrossRef]
277. Worku, L.A.; Bachheti, A.; Bachheti, R.K.; Rodrigues Reis, C.E.; Chandel, A.K. Agricultural Residues as Raw Materials for Pulp and Paper Production: Overview and Applications on Membrane Fabrication. *Membranes* **2023**, *13*, 228. [CrossRef]
278. Charcosset, C. Classical and Recent Developments of Membrane Processes for Desalination and Natural Water Treatment. *Membranes* **2022**, *12*, 267. [CrossRef]
279. Shahid, M.K.; Mainali, B.; Rout, P.R.; Lim, J.W.; Aslam, M.; Al-Rawajfeh, A.E.; Choi, Y. A Review of Membrane-Based Desalination Systems Powered by Renewable Energy Sources. *Water* **2023**, *15*, 534. [CrossRef]
280. Qasim, M.; Badrelzaman, M.; Darwish, N.N.; Darwish, N.A.; Hilal, N. Reverse osmosis desalination: A state-of-the-art review. *Desalination* **2019**, *459*, 59–104. [CrossRef]
281. Bundschuh, J.; Kaczmarczyk, M.; Ghaffour, N.; Tomaszewska, B. State-of-the-art of renewable energy sources used in water desalination: Present and future prospects. *Desalination* **2021**, *508*, 115035. [CrossRef]
282. Awad, A.M.; Jalab, R.; Minier-Matar, J.; Adham, S.; Nasser, M.S.; Judd, S.J. The status of forward osmosis technology implementation. *Desalination* **2019**, *461*, 10–21. [CrossRef]
283. Skuse, C.; Gallego-Schmid, A.; Azapagic, A.; Gorgojo, P. Can emerging membrane-based desalination technologies replace reverse osmosis? *Desalination* **2021**, *500*, 114844. [CrossRef]
284. Zhao, S.; Liao, Z.; Fane, A.; Li, J.; Tang, C.; Zheng, C.; Lin, J.; Kong, L. Engineering antifouling reverse osmosis membranes: A review. *Desalination* **2021**, *499*, 114857. [CrossRef]
285. Santoro, S.; Timpano, P.; Avci, A.H.; Argurio, P.; Chidichimo, F.; De Biase, M.; Straface, S.; Curcio, E. An integrated membrane distillation, photocatalysis and polyelectrolyte-enhanced ultrafiltration process for arsenic remediation at point-of-use. *Desalination* **2021**, *520*, 115378. [CrossRef]
286. Moon, H.-C. Equilibrium ultrafiltration of hydrolyzed thorium (IV) solutions. *Bull. Korean Chem. Soc.* **1989**, *10*, 270–272.
287. Nilchi, A.; Dehaghan, T.S.; Garmarodi, S.R. Kinetics, isotherm and thermodynamics for uranium and thorium ions adsorption from aqueous solutions by crystalline tin oxide nanoparticles. *Desalination* **2013**, *321*, 67–71. [CrossRef]
288. Abbasizadeh, S.; Keshtkar, A.R.; Mousavian, M.A. Preparation of a novel electrospun polyvinyl alcohol/titanium oxide nanofiber adsorbent modified with mercapto groups for uranium (VI) and thorium (IV) removal from aqueous solution. *Chem. Eng. J.* **2013**, *220*, 161–171. [CrossRef]
289. Tsezos, M.; Volesky, B. Biosorption of uranium and thorium. *Biotechnol. Bioeng.* **1981**, *23*, 583–604. [CrossRef]
290. Hritcu, D.; Humelnicu, D.; Dodi, G.; Popa, M.I. Magnetic chitosan composite particles: Evaluation of thorium and uranyl ion adsorption from aqueous solutions. *Carbohydr. Polym.* **2012**, *87*, 1185–1191. [CrossRef]
291. Zolfonoun, E.; Yousefi, S.R. Sorption and preconcentration of uranium and thorium from aqueous solutions using multi-walled carbon nanotubes decorated with magnetic nanoparticles. *Radiochim. Acta* **2015**, *103*, 835–841. [CrossRef]
292. García, A.C.; Latifi, M.; Amini, A.; Chaouki, J. Separation of Radioactive Elements from Rare Earth Element-Bearing Minerals. *Metals* **2020**, *10*, 1524. [CrossRef]
293. Bejanidze, I.; Petrov, O.; Kharebava, T.; Pohrebennyk, V.; Davitadze, N.; Didmanidze, N. Study of the Healing Properties of Natural Sources of Georgia and Modeling of Their Purification Processes. *Appl. Sci.* **2020**, *10*, 6529. [CrossRef]

Disclaimer/Publisher's Note: The statements, opinions and data contained in all publications are solely those of the individual author(s) and contributor(s) and not of MDPI and/or the editor(s). MDPI and/or the editor(s) disclaim responsibility for any injury to people or property resulting from any ideas, methods, instructions or products referred to in the content.

Disclaimer/Publisher's Note: The title and front matter of this reprint are at the discretion of the Guest Editors. The publisher is not responsible for their content or any associated concerns. The statements, opinions and data contained in all individual articles are solely those of the individual Editors and contributors and not of MDPI. MDPI disclaims responsibility for any injury to people or property resulting from any ideas, methods, instructions or products referred to in the content.

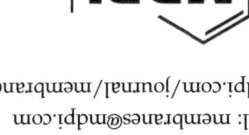

Membranes Editorial Office
E-mail: membranes@mdpi.com
www.mdpi.com/journal/membranes

MDPI AG
Grosspeteranlage 5
4052 Basel
Switzerland
Tel.: +41 61 683 77 34

www.ingramcontent.com/pod-product-compliance
Lightning Source LLC
LaVergne TN
LVHW072331090526
838202LV00019B/2394